Standard error of estimate	$s_{y \cdot x} = \sqrt{\left[\dfrac{1}{N(N-2)} \right]\left[(\text{II}) - \dfrac{(\text{III})^2}{(\text{I})} \right]}$	126
Correlation coefficient	$r = \dfrac{(\text{III})}{\sqrt{(\text{I})(\text{II})}}$	132

Probability

Probability of A	$P(A) = \dfrac{\#(A)}{\#(S)}$	167
Probability of B given A (conditional probability)	$P(B|A) = \dfrac{P(A \cap B)}{P(A)}$	168
Probability of A and B (intersection)	$P(A \cap B) = P(A)P(B|A)$	171
Probability of A or B (union)	$P(A \cup B) = P(A) + P(B) - P(A \cap B)$	175
Permutations	$_nP_r = \dfrac{n!}{(n-r)!}$	179
Combinations	$_nC_r = \dfrac{n!}{(n-r)!r!}$	182
Binomial probability	$P(r, n; p) = \dfrac{n!}{(n-r)!r!}p^r q^{n-r}$	183

Elementary Parametric Techniques

Standard score forms	$z_{\text{obs}} = \dfrac{X - \mu}{\sigma}$ or $z_{\text{obs}} = \dfrac{\bar{X} - \mu}{\sigma_{\bar{x}}} = \dfrac{\bar{X} - \mu}{\sigma_x / \sqrt{N}}$	202
	$t_{\text{obs}} = \dfrac{\bar{X} - \mu}{s_{\bar{x}}} = \dfrac{\bar{X} - \mu}{s_x / \sqrt{N}}$	221
Confidence intervals for the mean	Upper limit $= \bar{X} + t_{\alpha/2}(s_{\bar{x}})$ Lower limit $= \bar{X} - t_{\alpha/2}(s_{\bar{x}})$	227
Difference between two independent groups	$t_{\text{obs}} = \dfrac{\bar{X}_1 - \bar{X}_2}{\sqrt{\left[\dfrac{(N_1 - 1)s_1^2 + (N_2 - 1)s_2^2}{N_1 + N_2 - 2} \right] \cdot \left[\dfrac{1}{N_1} + \dfrac{1}{N_2} \right]}}$ $df = N_1 + N_2 - 2$	235

(continued on back endpaper)

THIRD EDITION

FUNDAMENTAL STATISTICS FOR PSYCHOLOGY

THIRD EDITION

FUNDAMENTAL STATISTICS FOR PSYCHOLOGY

ROBERT B. McCALL

Boys Town Center for the Study of Youth Development

Under the General Editorship of
JEROME KAGAN
Harvard University

HARCOURT BRACE JOVANOVICH, PUBLISHERS
San Diego   New York   Chicago   Washington, D.C.   Atlanta
London   Sydney   Toronto

preface

A veteran scholar of statistics once remarked that there are two ways to teach statistics—accurately or understandably. Disproving that "either-or" statement is the challenge that has guided my writing of this book.

As a textbook for the first course in applied statistics, *Fundamental Statistics for Psychology* is used primarily by students majoring in psychology, the other social sciences, and education. In writing for this audience, my earliest and most basic decision was to emphasize the purpose, rationale, and application of important statistical concepts over rote memorization and the mechanical application of formulas. I believe that students at the introductory level, regardless of whether they plan to take advanced courses in statistics, are served better by a book that fosters an understanding of statistical logic than by one that stresses mechanics.

When understanding is emphasized, elementary statistics is neither dull nor mathematically difficult. *Fundamental Statistics for Psychology* does not require much background in mathematics. The student need only be familiar with the thinking patterns learned in high school algebra and geometry; all relevant terms and operations are reviewed in Appendix I. To be sure, the book contains many computations and problems to solve, but most statistical formulas rely heavily on simple arithmetic—addition, subtraction, multiplication, division, and the taking of square roots—and can be worked out quickly with the aid of a hand calculator. In addition, I have kept the data for computational problems simple so that the emphasis remains on the rationale and outcome of techniques instead of on calculation for its own sake.

Understanding a statistical concept or strategy can be a formidable mental exercise, but once it is done details that boggle the minds of mechanical learners become obvious deductions. The goal of understanding concepts is not hard to reach if students also understand as much of the mathematical reasoning as is within their grasp. Whenever possible, I have explained in mathematically simple terms the logic that undergirds the basic concepts and techniques, although a few items require advanced mathematics and must therefore be taken on faith at this level. Beginning in Chapter 3, optional tables that show full algebraic derivations and proofs supplement the text explanations. (These tables can be omitted without loss of continuity.)

For the sake of students, and contrary to the traditional practice of mathematical writers, I have included and explained every step in proofs, however "obvious." I have also avoided excessive use of symbols, since symbols require an extra mental translation and thus often confuse students. Deviation scores ($X_i - \overline{X}$), for instance, are *not* abbreviated by x. Further, each new symbol is carefully introduced and is accompanied by its verbal equivalent.

Anyone who has analyzed his or her own data knows the anticipation that accompanies the final calculation of the r, t, or F hidden in a mass of numbers that took months to gather. Students, too—even though it is not their own data they are analyzing—can experience the excitement of seeing meaning emerge as they manipulate an apparently patternless collection of numbers. Yet they sometimes fail to see the fascination of statistical analysis because it is presented more as a numbers game played in a vacuum than as a crucial part of the scientific investigation of real phenomena.

The Prologue, a significant new feature in the third edition, places statistics in its proper real-life context. The Prologue introduces the basic principles and techniques of research design and stresses the complementary relationship between design and statistical analysis, thus integrating these two phases of the scientific endeavor.

Also, many end-of-chapter exercises and examples in the text are drawn from actual studies (modified for numerical simplicity). For example, the distinction between a correlation and a difference between means is demonstrated through findings that relate the IQs of adopted children to those of the biological and adoptive mothers, and Freedman's work on the feeding behavior of dogs reared under indulged and disciplined conditions is used to illustrate interaction effects in the two-factor analysis of variance. Although most of these studies were performed by psychologists, many of them concern developmental and educational issues of interest to future teachers and school administrators.

As another means of showing that numbers can have real applications, I have tried to give students a feel for the behavior of a statistic by providing several data sets that display obvious contrasts. Before calculating an analysis of variance in Chapter 11, for example, I present a set of random numbers, introduce a main effect, and then add an interaction treatment effect. Many exercises ask students to alter a given set of scores in some way and to observe how the change affects the value of a statistic.

The third edition includes two major improvements in format. First, the formulas most commonly used in computations are printed on the endpapers inside the front and back covers of the book. These formulas are provided mainly as a reference and study aid for students, but instructors might also find them useful as, say, the reference material for a closed-book test. Second, the first page of each chapter contains an outline of its contents for students to use as a preview and study device. In addition, the format of the tables has been made more consistent, and, as in earlier editions, definitions are set off in special type. Numerous graphics throughout the book will help students who learn geometrically more easily than algebraically.

Not so apparent, but more extensive than the changes in format, are improvements in the written presentation. The text itself has been thoroughly reworked for clarity and precision. New headings and subheadings will give students a clearer picture of each chapter's progression, and the relationships between topics covered in different chapters have been clarified (for example, a new table in Chapter 13 summarizes the uses of the parametric and nonparametric tests covered in the last four chapters of the book).

In the *Study Guide,* which was introduced with the second edition, the format and phrasing have similarly been reworked and refined. The main section of each guide chapter is a semi-programed unit that reviews the basic terms, concepts, and computational routines described in the text. The presentation is stripped of excess detail, and its tone is very concrete and applied. A feature students found particularly valuable has been retained: guided computational examples—step-by-step outlines that show students how to organize the operations required in complicated calculations and that clarify the logical and computational details.

Individual professors emphasize different aspects of elementary statistics, and many must select a subset of chapters that can be covered in the available time. The organization of the third edition is straightforward. Part 1 (Chapters 1–6) presents descriptive statistics, including central tendency, variability, relative position, regression, and correlation. Part 2 (Chapters 7–13) deals with inferential statistics, including probability, sampling distributions, the logic of hypothesis testing, elementary parametric tests, one- and two-factor analysis of variance, and a selection of nonparametric techniques. For most courses the core chapters are 1–4 and 8–10; to these chapters instructors may add at their option material on experimental design (Prologue), regression and correlation (Chapters 5 and 6), probability theory (Chapter 7), the analysis of variance (Chapters 11 and 12), and nonparametric tests (Chapter 13).

If the third edition reads and teaches better, much of the credit belongs to its manuscript editor, Catherine Caldwell Brown, who not only improved the clarity of the prose but also insisted on good pedagogy and a logical sequence of presentation. Her accurate perception of the best way to present and illustrate a concept and her gentle insistence that no point was too small to warrant attention led to innumerable improvements in the text.

In addition to the colleagues and friends who contributed their advice to earlier editions, I want to thank Dr. Mark Appelbaum of the Psychometric Laboratory of the Department of Psychology at the University of North Carolina, Chapel Hill. He was my chief technical consultant, helping to blend my personal style with the demands of formal theory. Dr. Arnold Well of the Department of Psychology at the University of Massachusetts provided a very useful critique of the entire manuscript. I am indebted to the many whose commitment to the course and to the book led them to communicate directly with me over the years; their comments, criticisms, and suggestions stimulated many of the improvements in this new edition.

I am grateful to the Literary Executor of the late Sir Ronald A. Fisher, F.R.S., to Dr. Frank Yates, F.R.S. and to Longman Group Ltd., London, for permission to reprint Tables III, IV, VII, and part of XXXIII from their book *Statistical Tables for Biological, Agricultural and Medical Research* (sixth edition, 1974).

Producing a statistics text is a complex job. It was made much easier by the technical assistance of Cindy Bellows Kennedy, the typing and manuscript preparation of Pat Mordeson and Bess Melvin, the proofreading of Carol Barth, and the diverse skills of Judith Greissman, Nola Healy Lynch, Jeremiah Lighter, Nina Ackerman Indig, and Richard Lewis at Harcourt Brace Jovanovich, who translated this work into book form. Moreover, I am indebted to the Center for the Study of Youth Development, Boys Town, Nebraska, for supporting me in this endeavor.

Special thanks go to my wife, Rozanne, for her encouragement and for the absence of complaints while she was temporarily widowed for this cause.

ROBERT B. McCALL

contents

PART 2
INFERENTIAL STATISTICS

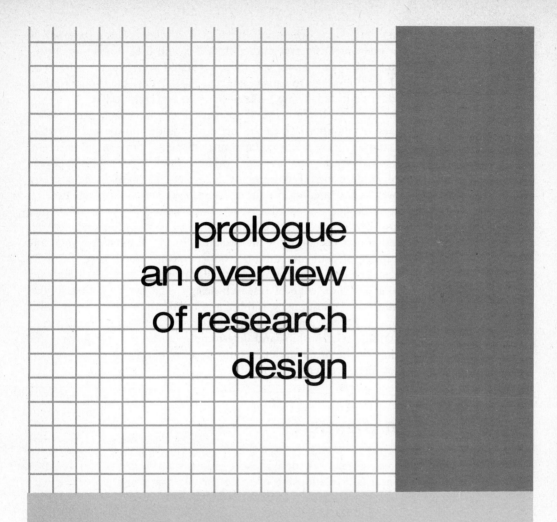

prologue
an overview
of research
design

Variables and Relationships

Defining and Measuring Variables
Reliability and Validity
Causal and Noncausal
   Relationships

Research Strategies

Observational Versus
   Experimental Research
Sampling

   Sample and Population
   Attributes of Samples

Random Assignment of Subjects
   to Experimental Conditions

   Extraneous Variables
   Confounding

Interpreting Research Results

   The Uncertainty of Inference
   The Role of Statistics

SCIENTIFIC RESEARCH, whether in physics or psychology, is an attempt to answer questions in a systematic, objective, and precise manner. In the first phase of this endeavor, a question is stated precisely, measurements planned, a logical strategy for answering the question selected, observations made, and data collected. These tasks are part of **research design**.

But the job is not finished. The measurements, which are often a jumbled mass of numbers, must be organized and summarized. Perhaps graphs and charts will be made, calculations will reveal the general nature of the measurements (e.g., the average values), and relationships between measures will be described. Finally, some decisions are usually made concerning the answer to the research question. These tasks are part of **statistics**.

Research design and statistics are both necessary to answer most questions in social science. But statistical analysis comes after and depends upon the design and execution of the research. Even the most thorough and brilliant statistical treatment cannot produce meaningful answers from a poorly designed or conducted study. For these reasons, students of statistics should have some knowledge of research design. The overview that follows describes the sorts of questions, variables, and relationships that are typical in the behavioral sciences, summarizes the distinctions between different strategies for describing relationships and determining the causes of behavior, and discusses some of the problems that confront researchers as they collect and interpret their measurements and results.

VARIABLES AND RELATIONSHIPS

The first step in research design is to ask a question. Often a research question concerns a relationship of some sort: Is something associated with something else? Are school behavior problems in a young child associated with the divorce of the child's parents? Is the level of achievement motivation in young women associated with the amount of encouragement they received as children from their fathers? Is the prognosis for people suffering from certain neuroses associated with the kind of therapy they receive? Is aggressive behavior in young children associated with watching violence on television?

Defining and measuring variables

A relationship is an association between two things. In research, these things are called variables. A **variable** is a particular set of objects, characteristics, or events; the members of the set may be assigned different numerical values.

When a researcher sets out to define a variable that is to be investigated, the first requirement is the specification of the general type of objects, characteristics, or events—what kind of school behavior problems, achievement motivation, parental encouragement, neuroses, therapies, aggressive behavior, or violence on television? The second requirement is a definition that will permit the assignment of numbers to particular instances of the object, characteristic, or event. Systematic and precise observation and measurement is the heart of science, and a question is not suitable for scientific inquiry unless the variables of interest can be measured. There may be many ways to measure a given variable. For example, the variable *social class* might be measured by years of education, annual income, social status of occupation, or size of home. These measures are similar to one another but not identical, yet they are all possible measures of the variable social class.

The researcher's specification of the type of object, characteristic, or event to be studied is called the **conceptual variable**. The procedure selected for quantifying or measuring particular instances of the conceptual variable is called the variable's **operational definition**, because the measurement *operations* concretely *define* the variable. For example, social class is a conceptual variable that may be operationally defined to be years of education.

Often it is not easy to determine a good operational definition for a conceptual variable. The conceptual variable of achievement motivation in young women could be operationally defined by grade average, the social status of the woman's occupation, how long she will persist in trying to solve a problem, or a score on a paper-and-pencil test of how many different activities she would give up to have a successful career. But none of these operational definitions is a perfect measure of the conceptual variable of achievement motivation. For example, some women get good grades but are not strongly interested in a career. Some hold high-status jobs for reasons other than being strongly motivated toward achievement. Some may have "retired" to raise a child but wish to return to work a few years later; these women may be no less achievement oriented than those with uninterrupted careers. Because of difficulties like these, there may not be total agreement on the most appropriate operational definition of a given conceptual variable, and the results of the research will depend to a considerable extent on the particular operational definition adopted.

Reliability and validity

A good operational definition of a conceptual variable will produce measurements that are reliable and valid. **Reliability** refers to whether the measurement procedures assign the same value to a characteristic each time that it is measured under essentially the same circumstances. For example, when a person is given an IQ test on two successive days, he or she may or may not have precisely the same score on the two occasions. The conceptual variable —the individual's intelligence—has not changed from one day to the next, but the value assigned by its operational definition depends not only on the person's intelligence but on other factors as well. Perhaps the person is less alert one day than the other; one form of the test may be easier for the person than another; and so forth.

Some operational definitions produce perfectly reliable measurements. Given the availability of birth certificates, the variable *age* can be determined with perfect reliability. But many psychological measurements are not perfectly reliable and can probably never be made so. In these cases, scientists attempt to evaluate just how reliable their measurement procedures are. Perhaps two observers watch the same behavior and make their measurements independently. If the measurements are the same, they are said to be reliable. If they are not, they are unreliable. Measurement in psychology is most often neither perfectly reliable nor totally unreliable. Statistical procedures exist to evaluate the degree of reliability. Then the scientist must judge whether the reliability is sufficiently high for the measurements to be meaningful.

Validity refers to the extent to which the measurement procedures accurately reflect the conceptual variable being measured. Sometimes a measurement is obviously valid. Years since birth is a perfectly valid measure of age. The number of hours a child watches television is a valid measure of amount of television watching. The validity of other measurements is less certain. Is one's score on an IQ test a valid measure of one's intelligence? Is the number of times a child hits other children in nursery school a valid index of how aggressive that child is? Is the length of time an individual persists in trying to solve a difficult problem a valid measure of achievement motivation?

A given measurement may be valid for some purposes but not for others. A score on an IQ test may be a valid predictor of a child's future achievement in school but not a valid index of the much broader concept, intelligence. The length of time a nine-month-old infant cries when a parent leaves the room may be a valid measure of the conceptual variable separation anxiety but not of the strength of the infant's attachment to the parent. A given operational definition is not valid by itself; it can be valid only as a measure of some specific conceptual variable.

Causal and noncausal relationships

Two major tasks of research are, first, to identify a relationship between two (or more) variables and, second, to determine whether the observed relationship is causal or not.

The two variables of a research question are called the dependent variable and the independent variable. The **dependent variable** is the variable whose character or value *depends* on the character or value of the **independent variable**. Presumably, hours of violence watched on television—the independent variable—influences how much social aggressiveness—the dependent variable—the child shows. A father's encouragement presumably influences his daughter's achievement motivation. Thus, a research question typically has the form: Is X related to Y, where Y is the dependent variable whose character or value may or may not depend on the character or value of X?

Relationships may be causal or noncausal. In a **causal relationship** the independent variable in some way actually influences or determines the value of the dependent variable. For example, watching violent behavior on television makes some children more aggressive under some circumstances. It produces, in part, the aggressiveness. Sometimes an independent variable completely determines the value of the dependent variable. When the temperature of water at sea level is brought below 0° Celsius, the water freezes—essentially every time and for each sample of pure water. But such perfect causality rarely occurs in behavioral science. Research indicates that watching violent behavior on television probably makes *some* children (especially those who are already somewhat aggressive) a little more aggressive in some situations (but not necessarily all), and that watching television is not the only cause of aggressiveness. The relationship is causal, but imperfectly so.

A **noncausal relationship** is one in which the independent variable does not actually produce, cause, or influence the dependent variable, although the values of the independent and dependent variables are in some way related. For example, some mothers give their six-month-old infants a great deal of intellectual stimulation (they provide mobiles for the infants, talk to them, play with them, and so on). The children grow up to be slightly more intelligent than those of mothers who do not provide such stimulation. However, it is unlikely that the stimulation provided at six months of age *makes* the children brighter. Rather, the mothers who stimulate their infants at six months also provide a good language model, require their children to think and ask questions, and offer them books and diverse experiences when the children are older. It is the stimulation in early childhood—not at six months of age—that likely produces somewhat higher intelligence. But even this relationship may not be perfectly causal, because mothers who provide such stimulation for their young children also tend to have better genes for intelligence, and the intellectual advantage of their children may be caused by genetic inheritance as much as or more than by environmental circumstances. Research indicates that both factors are involved.

A similar issue has clouded interpretations of research on smoking and lung cancer. Early studies indicated that people who smoked had higher rates of lung cancer, and the more they smoked the more likely they were to get the disease. The tobacco industry countered that such an association did not necessarily prove that smoking causes cancer. Perhaps, it was argued, anxiety and nervousness are the culprits. People who are tense tend to smoke more than people who are relaxed, and tension might produce cancer through some biochemical change or genetic disposition. In this case, little evidence exists that tension causes lung cancer (though it may play a role in heart disease, which also occurs at a higher rate in smokers).

RESEARCH STRATEGIES

Observational versus experimental research

The purpose of **observational research** is to find out whether there is a relationship—causal or noncausal—between the variables of interest. This was the first step in research on television and aggressiveness. Scientists simply observed that preschool boys who spent more hours watching violence on television tended to display a larger number of aggressive acts in nursery school. Observational research cannot by itself determine whether an association is causal or not; its purpose is to determine whether a relationship exists at all and if so, how strong it is. If no association had been found between watching violent behavior on television and displaying aggressive behavior in nursery schools, researchers might have abandoned the whole issue.

But these associations did occur, thus provoking a second question: Is the association causal? The observed association by itself cannot answer this question. For example, the television networks claimed that youngsters who have aggressive personalities tend to watch more violent behavior on television and are more aggressive in nursery schools. (There is some truth to this statement.) They argued that watching violence on television does not cause aggressiveness in school; rather, both these behaviors are caused by an aggressive personality.

To determine if an association is causal, it is usually necessary to conduct an **experiment**. Whereas a researcher using the observational method controls neither kind of variable, in an experiment the scientist controls and deliberately manipulates the independent variable and notes the effects of the manipulation on the dependent variable. For example, in an observational study of violent behavior on television and aggressive social behavior, parents might be asked to keep a record of the shows their children watch over a two-week period. During the same period the number of aggressive acts by these children in nursery school would be recorded by observers, so that the scientist could determine whether aggressive behavior increased with the amount of violent behavior watched. In such a case, the researcher does not control whether the children see many or few violent programs. In contrast, an experiment might be conducted in which the researcher showed some children a certain number of hours of programs with some violent behavior (such as *Batman*), while other children did not see these shows. In this case, who sees violence on television, what kind of violence, and how much violence are determined by the researcher. He or she is manipulating the independent variable; that manipulation is one of the defining characteristics of experimental research.

In an experiment in which some children are exposed to television programs that depict violent behavior while other children are not, the youngsters exposed to violent behavior on television are called the **experimental group**, and those receiving no such exposure are the **control group**. If there were no control group, one would not be able to tell whether the aggressive behavior displayed by the experimental group was caused by viewing television or whether such behavior was merely normal nursery school antics that would be expected even if the group did not have the special television experience. The control group *controls* for this possibility. But even if the children in the experimental group commit many more aggressive acts than those in the control group, the researcher cannot yet conclude that the cause of these acts was watching violent acts on television. Perhaps the act of watching television itself, regardless of the particular program, produces aggressive behavior. To control for this possibility a third group might be added to the experiment. This group would be exposed to nonviolent programs on television (such as *Mister Rogers' Neighborhood*). Sometimes, many control groups are necessary before the researcher can conclude that a particular causal relationship exists.

Although experimental research is the best method for determining causality, it is not always possible to investigate questions experimentally. Consider the question of achievement motivation in females. From a scientific standpoint, it would be ideal to put some young girls into new families with fathers who encourage them to achieve, while other girls would be assigned to fathers who are not so encouraging. Obviously, such an experiment is not possible, either practically or ethically. That is often the case.

Although it may seem that experimental research is always to be preferred over observational research because of its potential for determining causality, in fact the appropriate kind of research depends on the question being asked. If one wants to know whether the viewing of violent television programs can cause increased aggressiveness among nursery-school children, then an experiment is probably required. But suppose one is interested in knowing whether the everyday aggressiveness typically displayed in nursery schools is caused by the everyday television viewing habits of children. In this case, the question itself stipulates that nothing be changed from the circumstances that actually exist now in American homes, in the television industry, and in nursery schools, and it therefore eliminates most forms of experimental manipulation. A naturalistic observational study is required. If no relationship between television viewing habits and aggressiveness were found, then despite experiments showing that watching violent behavior on television *can cause* aggressiveness, one would have to conclude that in current circumstances it *does not cause* the aggressiveness typically observed in nursery schools. If, on the other hand, the observational study did reveal a relationship, one still would not be perfectly certain of the causal combination. Watching violent behavior on television *could* produce aggressiveness, but perhaps everyday nursery-school fighting is caused by aggressive children who happen to watch more violence on television than do other youngsters. To assess that possibility, an experiment could be performed in which some aggressive children would be given more violent television programs to watch; they might turn out to be even more disruptive than matched aggressive children who view action programs without violence.

In the long run, a combination of experimental and observational research like the one just described, consisting of studies conducted under progressively more natural circumstances, has led scientists to believe that there probably is a causal connection: that some aggressiveness in society is produced by violent programs on television.

Sampling

Once the question has been formulated, the variables defined, and the type of research specified, cases for observation must be selected. In psychological studies, these cases are often individual persons called **subjects**, but the cases can also be individual nerves in an animal under study, parent-child pairs, families, or entire classrooms of students.

Sample and population   Most often, a research question is initially stated in very general terms. For example, does encouragement by fathers increase their daughters' later achievement? The researcher would really like to know if this is true for a very large group of fathers and daughters, perhaps all American father-daughter pairs. This maximum group of people to which the researcher hopes to apply the results of the study is called a **population**. But entire populations usually cannot be studied, either because they are unavailable or because using so many subjects would be too expensive. Therefore, a **sample** of father-daughter pairs is selected, and the results of observations on this smaller group of subjects are applied to the larger population.

Attributes of samples   It is obvious that if information about a sample is to be used to draw conclusions about a population, the sample that is selected must be **representative** of the population. If only upper-middle-class father-daughter pairs are studied, then the results can be applied only to a larger group of upper-middle-class pairs. The relationship between paternal encouragement and achievement motivation in women may be quite different for individuals from poverty circumstances, for mothers and daughters, and for fathers or mothers and sons. To make sure that the results of a study can be applied to a certain population, special efforts are sometimes made to ensure that various groups in a population are proportionately represented in the sample, especially groups suspected of differing from one another with respect to the variables under study. Such efforts are often made by political pollsters, who try to make sure that their samples include representative numbers of men and women, Democrats and Republicans, blue- and white-collar workers, various races and religions, and so forth.

Ideally, subjects should be **randomly** and **independently selected** from the population. This means that each member of the population should have an equal chance of being selected. For example, a random sample could be selected by assigning each person in the population a number, consulting a random-number table, such as Appendix II, Table J of this book, and choosing subjects corresponding to the first numbers encountered in the table. Such a procedure would yield a truly random sample but is rarely feasible. For practical reasons, less-than-perfect sampling methods are often used. For example, much psychological research has been performed using college sophomores and the white rat—and even so, the sophomores and rats actually used in research are not, on the whole, randomly selected from their respective groups. In short, practical considerations often limit the scientist's ability to obtain a random or representative sample. When that is the case, the researcher must be very careful to specify precisely how the sample was selected so that the results can be limited to the correct population.

Sampling that is random is also likely to be **representative**, at least in large samples, because each person from each special subgroup has an equal chance of being selected. Notice that the definition of randomness pertains to how the subjects are selected, not to how representative the final sample turns out to be. Four coins flipped simultaneously may all come up heads, but, despite the apparent rarity of the outcome, the sample of coin flips is still random. Similarly, a randomly selected sample of voters may turn out to overrepresent one political interest group or another. Therefore, to ensure that a sample will be representative, the researcher may randomly select an appropriate number of subjects from each interest group in the population.

Random assignment of subjects to experimental conditions

Once subjects for an experiment have been randomly selected from a popula-tion, they must be **randomly assigned to experimental conditions**. In the study of the effects of viewing violence on television, nursery-school children would be randomly assigned to one of the three groups—the children who would be shown violent television programs, those who would be shown nonviolent television programs, and those who would be shown no television. In observational research, random assignment is usually not possible. A scientist who studies sex differences, for example, cannot assign subjects to one sex or the other. In this case, nature, and not the scientist, controls the assignment to groups, and nature often introduces differences between males and females other than their biological sex. For this reason, observational studies are more susceptible than are experiments to the effects of extraneous variables and confounding.

Extraneous variables   As we have discussed, children who view much and those who view little violence on television may be different in many ways besides their television viewing habits; these other factors, not the watching of violence itself, may cause the difference in their aggressive behavior. They may have personalities that are prone to aggressiveness; they may have parents who approve of aggressive behavior and condone violent portrayals of it; and so on. These other factors are called **extraneous variables**.

When the subjects in an experiment are randomly assigned to groups, all factors associated with aggressive behavior have an equal chance of being represented in all the groups. If the groups are large, it can be assumed that they are nearly equal with respect to these other factors. That is, it can be assumed that all groups will include aggressive and nonaggressive children, children whose parents encourage and discourage violence, and children subject to all the other extraneous variables associated with aggressive behavior. Large groups assembled through random assignment will tend to be equal with respect to these extraneous variables.

Confounding   When extraneous variables have influenced the relationship between the variables under study, scientists say that **confounding** has occurred. In studies that related smoking and lung cancer, tension was a possible extraneous variable whose effects could not be equalized across groups because smoking and lung cancer must be studied observationally rather than experimentally. Therefore, the relationship between smoking and lung cancer is confounded or confused.

Confounding can occur for several reasons. The type discussed in the preceding paragraph is sometimes known as **subject bias**, since extraneous variables associated with the subjects may have confounded the relationship between smoking and lung cancer. In the observational study of violent television programs and aggression, too, the aggressive personalities of the children who watch violent television programs would be classified as subject bias.

Subject bias can occur even in an experiment in which subjects are randomly assigned to the experimental and control groups. For example, if subjects know they are receiving a drug that is supposed to make them tense or nervous, they may actually become tense or nervous because they *expect* they should feel that way, not because the drug makes them so. Some people believe that the effects of marijuana are enhanced by such expectations, especially when a group of people smoke together. To prevent this kind of subject bias in research on drugs, subjects who are not to be given the drug under study may be administered a **placebo**, such as a cigarette that tastes and smells like marijuana but has no corresponding physiological effect. The subjects given the placebo are led to believe that they are taking the drug so that their expectations will match those of the subjects who actually receive the drug.

Another possible source of confounding is **observer bias**, which occurs when the behavior of the researcher—instead of (or in addition to) the independent variable—influences the results of the study. Observers who know the purpose of an experiment, or who know which groups are supposed to do "better" than other groups, may treat the subjects differently and may actually produce behavioral differences between the groups. Researchers may handle more gently rats who are expected to learn very quickly; they may be more courteous and encouraging to people who are expected to perform better; they may even perceive and score the performance of some subjects in accordance with their expectations. Suppose you are observing the aggressive behavior of children in a nursery school: You may be more prone to decide that a slight nudge between pupils is an aggressive act if you know that the nudger saw violent television programs. To guard against observer bias, many researchers attempt to use observers and research assistants who do not know the purpose of the study and who do not know which experimental or control group the subjects belong to (e.g., they do not know what television group the children are in). Such observers are called *blind* because they are blind to which subjects are in which group.

When researchers use subjects and observers who are both blind to the research conditions, they are using a **double blind** research design. Sometimes it is not possible to have both subjects and observers blind. If either subjects or observers are blind, but not both, the experiment is called a **single blind** design.

Finally, confounding can occur because the independent variable was not defined specifically enough. For example, *Batman* is not only a more violent television show than *Mister Rogers' Neighborhood* but a louder one; there is more physical activity; Batman and Mister Rogers are very different kinds of people; and so forth. Watching *Batman* may cause an increase in nursery-school aggression relative to watching *Mister Rogers*, but the increase may occur, not because of the violence portrayed on *Batman*, but because of these extraneous variables that are confounded with violence. To minimize this possibility, a researcher might use aggressive and nonaggressive programs that are in some way corrected for these extraneous factors. For example, two programs that use the same characters and action-packed episodes could be specially designed so that only one of the programs would portray violent behavior.

INTERPRETING RESEARCH RESULTS

The goal of psychological research is to design observations and experiments that will permit the scientist to conclude as specifically as possible that a systematic association between variables exists for a particular population. Obviously, the scientist would like to conclude that the research results hold true for the population of interest, not just for those sample cases that were actually studied. This process of generalizing from sample to population is called **inference**, because the scientist *infers* that something is true about the population on the basis of results obtained on a sample.

The uncertainty of inference

But studying 50 cases and saying the results apply to 5000 or 5 million cases is a chancy business. Suppose you are asked to determine if a coin is fair or biased. If it is fair, then in the population of all possible flips of the coin, half should be heads and half tails. But suppose you try 4 flips that result in 3 heads and a tail. Are you willing to infer from your sample of 4 flips that the coin is biased? Probably not. Perhaps after 12 flips, 8 are heads. The bias seems less than before, but you have more cases. Are you ready to announce your decision? Suppose after 100 flips, 61 are heads. Is the coin really biased? Scientific inference is the researcher's tool for deciding whether something is true for a population as a whole on the basis of a sample of observations; that sample may be flips of a single coin, numerous father-daughter pairs, and so on. Whether a particular inference is justified depends on the number of cases in the sample, how they were selected, how representative they are of the relevant population, and how the research was designed and carried out.

Even when a study has been well conceived and properly conducted, it is desirable to repeat it on new samples from the same population to see if the results will occur again. Such a double-check procedure is called **replication**. In addition, a researcher's conclusions become more persuasive if many studies are conducted on different samples, in different contexts, and under different circumstances in order to determine if the association occurs under a variety of conditions. If the relationship is not found in each case, under which particular circumstances is it found and under which is it absent? When a relationship is observed under many different sampling and research conditions, scientists call the agreement **converging themes** because the research results seem to point toward a common interpretation.

The role of statistics

A good research design creates a set of circumstances that permits the researcher to say that one variable of interest is associated with another and, to the extent possible, not with extraneous factors. But often the measurements collected during the study are merely a mass of numbers until they are treated statistically. Research requires not only design but statistical analysis.

Part 1 of this book introduces the techniques of **descriptive statistics**, which organize, summarize, and describe a set of measurements. For example, in the television–aggressiveness study, observers presumably counted the number of aggressive acts displayed in nursery school by children in the three groups. Various descriptive techniques could be applied to the resulting collection of numbers. Scores for each group might be arranged in order, and the number of children in each group who exhibited no, some, or large numbers of aggressive acts might be tallied and portrayed graphically (Chapter 2). The scientist might calculate the central tendency, or the average number of aggressive acts for each group, and also the extent to which the scores differed from one another (Chapter 3). The scores of individual subjects could be described in terms of a comparison with individuals who displayed less aggressive behavior (Chapter 4). The strength of the relationship between the viewing of violent television programs and aggressive behavior could be indexed numerically and predictions made for individual subjects (Chapters 5 and 6).

Often the research question can be answered only by using **inferential statistics**, techniques that help the researcher infer something about the population on the basis of a sample. Typically, such techniques involve quantifying the ever-present uncertainty associated with making decisions about the influence of the independent variable, a process that relies on probability (Chapters 7–9). A variety of specific techniques for drawing inferences can be used, depending on the research question and the research design (Chapters 10–13).

Research is inherently an uncertain enterprise. Statistics cannot make up for deficiencies in the questions asked; in the precision, reliability, or validity of measurements; or in other aspects of the research design; but statistics can clarify and quantify some aspects of the inherent uncertainty so that the data can be more easily summarized and conclusions drawn. Research design and statistics are two sides of the scientific coin, and students need to be acquainted with both.

EXERCISES

1. Distinguish between the tasks of research design and those of statistics, and discuss the necessary role each aspect of research plays in the conduct of social science.

2. Distinguish between a conceptual variable and its operational definition. Discuss, using examples, the problem of obtaining good measurement of a variable.

3. Define and contrast reliability and validity.
4. Define independent and dependent variable, and give several illustrations.
5. Discuss the difference between observational and experimental research, and consider the uses and advantages of each, especially with respect to the practical tasks of investigating causal and noncausal relationships.
6. As a rule, subjects are randomly and independently selected and assigned to experimental groups or conditions. Why?
7. Speculate on sources of potential confounding in the conclusions drawn from the following experiments.
   a. In order to determine whether marijuana actually produces unusual feelings and experiences, a researcher gives 20 young adults marijuana and 20 other young adults regular cigarettes to smoke, and then he or she asks the subjects to describe their sensations and perceptions.
   b. A scientist has a theory that music, brightly colored offices, and the availability of magazines and other entertainment devices for coffee breaks will raise the morale of workers and increase productivity. Therefore, two comparable sections of a large electronics factory are selected for study. In one, the new amenities are installed; nothing is done to the other section. Productivity increases in the improved section, and the scientist concludes that music, bright colors, and more entertaining coffee breaks increase productivity.
   c. Women who live together are known to synchronize their menstrual cycles over a period of several months, but the reasons for this phenomenon are not understood. An experiment is performed[1] in which one group of women rub a special "perfume" under their noses each day. The perfume is distilled each day from the sweat of women who live elsewhere in town; then it is mixed with alcohol. Another group of women receive only the alcohol. The women in the first group slowly change their cycles to match the cycles of their respective perfume donors, while the control women do not change. The researcher concludes that the sweat scent is stronger during certain portions of the menstrual cycle and that synchrony depends on smelling this odor.

[1]Based on, but not identical to, research conducted by Michael Russell and reported in *Science News*, 2 July 1977, p. 5.

DESCRIPTIVE STATISTICS

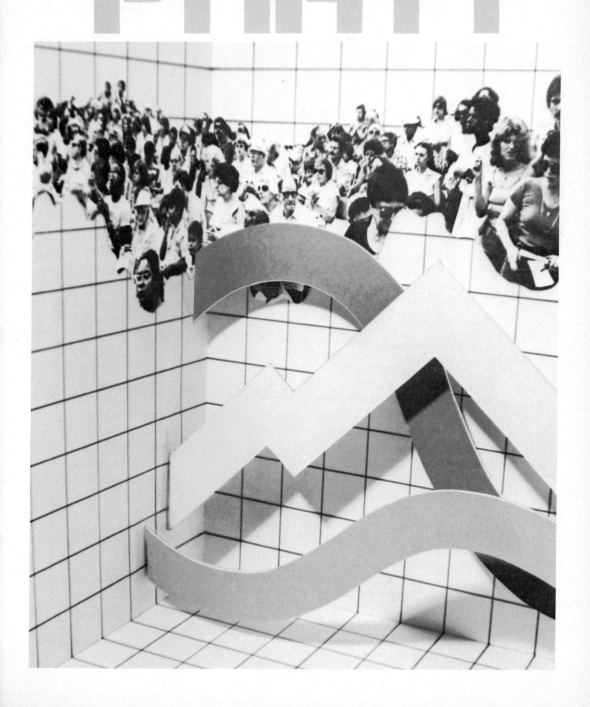

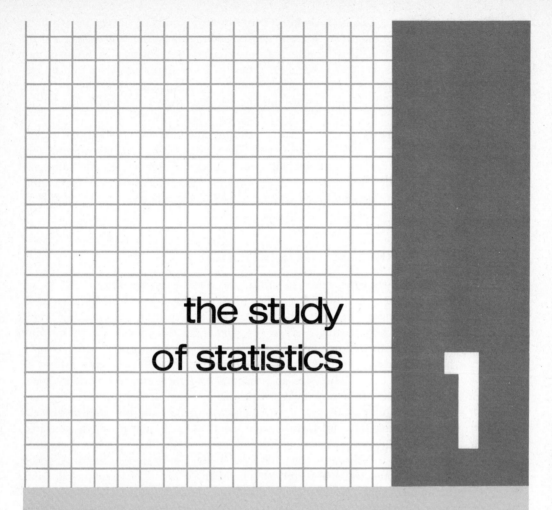

the study
of statistics

1

N ORDER TO BECOME an object for scientific study, an event or attribute must be measurable. **Measurement**, which is the systematic assignment of numbers to objects or events, forms the very basis of science. Some things (such as height, weight, and the time it takes a rat to run a maze) are relatively easy to measure; others (such as aggressiveness, "femininity," and amount of ill feeling toward another ethnic group) are much harder to quantify.

The measurements collected during scientific observation are called **data**. **Statistics** is the study of methods for describing and interpreting this quantitative information, including techniques for organizing and summarizing data and techniques for making generalizations and inferences from data. The first of these two broad classes of methods is called descriptive statistics, and the second is called inferential statistics.

DESCRIPTIVE AND INFERENTIAL STATISTICS

Descriptive statistics refers to procedures for organizing, summarizing, and describing quantitative information or data. Most people are familiar with some statistics of this sort. The baseball fan is accustomed to checking over a favorite player's batting average, the sales manager relies on charts showing the sales distribution and cost-efficiency of an enterprise, the head of a household may consult tables showing the average domestic expenditures of families of comparable size and income, and the actuary possesses charts outlining the life expectancies of people in various professions. These are relatively simple statistical tools for characterizing data, but additional techniques are available to describe such things as the extent to which measured values deviate from one another and the relationship between individual differences in performance of two different kinds. For example, one can describe statistically the degree of relationship between the scores of a group of students on a college entrance examination and their later performance in college.

The second major class of statistics, **inferential statistics**, includes methods for making inferences about a larger group of individuals on the basis of data concerning a small group. For example, people who develop diabetes in adulthood are often treated with the drug insulin. The insulin is thought not only to regulate blood sugar level, preventing coma and possibly death, but to retard the progressive damage to blood vessels that is the major cause of blindness, heart attacks, disability, and death among diabetics. Suppose a test had been made to determine whether people who were treated with insulin suffered fewer of these major consequences than patients who were treated with a special diet thought to be beneficial, but not with insulin.[1]

[1]Inspired by, but not identical to, the study reported by Genell L. Knatterud, Christian R. Klimt, Marvin E. Levin, Maynard E. Jacobson, and Martin G. Goldner, "Effects of Hypoglycemic Agents on Vascular Complications in Patients with Adult-Onset Diabetes," *Journal of the American Medical Association* 240 (1978): 37–42.

Two groups, each consisting of 200 diabetic volunteers judged medically able to participate in the study, followed a special diet and received "medication." For one group the medication was insulin, but for the other group it was a substance that was known to have no effect (a placebo). Both groups of patients were observed for 9 to 11 years. Suppose that during those years 56 of the insulin group and 63 of the non-insulin group suffered at least one of the known consequences of diabetes. The question is: Did the insulin treatment and diet together do more to help prevent undesirable complications than the special diet alone?

Note that whether the 200 people who received insulin had fewer problems than the 200 who took the placebo is *not* the question. The results of the experiment itself have answered that—56 versus 63 patients developed later problems. What we would really like to know is whether insulin is likely to have this effect generally, for all such people, even though we must make that decision on the basis of results determined with only 400 people. We must make a decision about a larger group based upon observations of a much smaller group. That is, we want to draw a statistical inference.

However, we feel some uncertainty about deciding whether insulin is generally helpful or not. While 56 cases is certainly less than 63, perhaps such a difference is the result of chance. Imagine for a moment that the insulin in fact had no effect whatsoever. Even in this case, we would not expect *exactly* the same number of complications in each group, because one group of 200 people and the circumstances of their disease are not identical in every way to another group of 200. One group might have 58 cases of complications and the other 61, or perhaps 56 and 63. So the question we wish to answer through the use of inferential statistics is: How much of a difference between these two groups must there be before it is safe to decide that insulin helps?

The uncertainty that we feel can be quantified, and this quantification relies on **probability**. When we say that the likelihood of a fair coin coming up heads is .50, we have made a probability statement that quantifies our uncertainty about the outcome of flipping the coin. The probability value really says that in an uncountable number of flips, .50 (or half) will turn up heads. It is in this sense that probability quantifies our uncertainty about what will happen in a large number of cases.

Probability is used in statistical inference in the same way as in daily life. For instance, you might consider the probability of rain in deciding whether or not to go on a picnic. Suppose you decide that if the probability of rain is greater than .40 (i.e., if there is more than a 40 % chance of rain), you will not go. The forecaster announces that the chance for rain is .60, so you decide to postpone your picnic.

In studying the effectiveness of insulin, the scientist decides in advance that he or she will conclude that insulin has no effect unless the probability of no effect is less than .05. Notice that the probability applies to the negative case—that insulin has no effect. In other words, the evidence that insulin is effective must be reasonably certain, which in this case is defined quantitatively by the requirement that the probability that insulin has *no* effect be very small—specifically, less than .05 (or one chance in 20). Then the experiment is conducted, and the probability value is determined on the basis of the research data by procedures described in detail in Part 2. If the observed probability that insulin has no effect turns out to be .15, the probability is not small enough and the scientist will conclude that no evidence was found that insulin helps. If it is .03, then the researcher will conclude that insulin probably does help prevent complications.

To recapitulate, statistics is the study of methods of handling data. Descriptive statistics organizes, summarizes, and describes data; making inferences from small to larger groups of subjects or events is the province of inferential statistics.

VARIABILITY

Social scientists must use statistical techniques primarily because almost all data in the social sciences contain variability. **Variability** refers to the fact that the scores or measurement values obtained in a study differ from one another, even when all the subjects in the study are assessed under the same circumstances. If your teacher were to give a test to all the students in your statistics class, they would not all obtain the same score. Similarly, all rats tested in a multiple-T maze do not make the same number of errors.

There is variability in social science data for at least three reasons. First, the units (people, usually) that social scientists study are rarely identical to one another. A chemist or a physicist assumes that each of the units under study (molecules, atoms, electrons, etc.) is identical in its composition and behavior to every other unit of its kind. But the social scientist cannot assume that all people will respond identically in a given situation. In fact, the social scientist can count on the fact that they will not respond in the same way. The first major source of variability, then, is **individual differences** in the behavior of different subjects.

Second, as noted in the Prologue, social scientists cannot always measure as accurately as they would like the attribute or behavior they wish to study. Scores on a classroom examination are supposed to be a measure of how much students have learned, but even a good test has flaws that make it an imperfect measure of learning. Perhaps some questions are ambiguous, or perhaps one part of the material is overemphasized. All factors other than knowledge of the material that influence test performance constitute **measurement error** and contribute to variability. Almost all behavioral measurements contain some measurement error.

Third, even a single unit (person, animal) usually will not respond exactly the same way on two different occasions. If you have a person step on a precise scale 10 times in succession, the scale will probably record the same weight each time. But if you give a person several opportunities to rate the attractiveness of advertising displays or the degree of aggressiveness in filmed episodes of each of 15 nursery-school children, he or she will not assign the same scores on each occasion. This source of variability is called **unreliability**.

Variability means that less faith can be placed in a single score than one would like. Therefore, it may be necessary either to measure a single subject several times or to assess many different subjects or events in order to obtain data that faithfully reflect the characteristics being studied.

Individual differences, measurement error, and unreliability are the major reasons that it is often difficult to draw accurate conclusions from the results of controlled observations. It is the task of statistics to quantify the variability in a set of measurements; to describe the data for a group of subjects, despite the variability inherent in the measurements; and to derive precisely stated and consistent decisions about the results by quantifying the uncertainty produced by variability.

WHY STUDY STATISTICS?

There are good reasons for studying statistics. Obviously, students who plan to design and carry out experiments in their future studies or professions will also have to describe the results and make inferences from data; these operations require a knowledge of statistical techniques. Therefore, statistics is often a required course for undergraduate and graduate students in the social sciences.

A knowledge of statistics is virtually essential, too, for those who will not conduct research themselves but who want to be able to read and evaluate social science literature intelligently. The members of many professions (e.g., clinical psychology, special education teaching) are expected to read, comprehend, evaluate, and apply detailed research findings, many of which are reported in statistical terms. For students who expect to attend graduate school in any of a number of disciplines, this course will be only an introduction to a much more extensive study of statistics.

But what about the student who does not expect to attend graduate school and who probably will not read professional scientific literature? Of what benefit can the study of statistics be? Most important, rudimentary statistics finds its way into the communications media. For example, one might read in the newspaper that the **average** (mean) family income in the United States is higher than the **median** income; both are used as measures of typical income. Why are they different, and in what sense does each measure reflect typical income?

Consider another set of facts once reported in newspapers and magazines[2] across the country and not atypical of presentations that may prompt

[2] e.g., *Time*, 3 May 1968, p. 41.

unreasoned conclusions. At conception each person is ordinarily given 23 pairs of chromosomes, which in turn determine much of the individual's development. One such pair of chromosomes determines the sex of the individual. If this pair is XX a female results, whereas if the pair is XY a male results. Occasionally, an extra Y chromosome is contributed to a sex cell, yielding an XYY. This person is male, tends to be approximately six inches taller than average, and has other distinguishing characteristics. Although the incidence and implications of the XYY configuration are still being debated, it has been estimated that among male prisoners the XYY condition is some 60 times more common than in the general population. What does such a statistic imply? For example, does it argue for giving each male a chromosomal analysis at birth and then keeping track of the XYY people, perhaps even restricting their freedom? Does it mean that the XYY grouping causes hyperaggressiveness or criminal behavior? A knowledge of statistics can help one answer such questions.

A course in statistics, like a course in logic, generally breeds a healthy skepticism in its students. To continue the preceding example on genetic combinations and criminals, suppose that an XYY combination occurs once in every 2000 male births. For convenience, assume there are 100 million males in the United States and 120,000 men in prisons. Therefore, in the general population there are approximately 50,000 men in the country with an XYY combination, only 3600 (7.2%) of whom are in prison. A person with a background in statistics might well be skeptical of the suggestion that we set up a program to observe or even restrict the liberty of 50,000 men, 46,400 of whom will not be likely to cause any trouble.

MEASUREMENT

A major function of statistics is to describe efficiently the nature of (1) experimental results, (2) observations on large groups of subjects, and (3) relationships between two different types of measurement. The next chapter presents a basic method of describing a group of measurements—the frequency distribution. Since the description depends in part on how the measurements were made, it is necessary first to discuss some terminology and characteristics of the measurement process.

Scales of measurement

Properties of scales A major concern about a measurement technique is that it faithfully reflect the attribute being measured. For example, if one wants to measure the heights of people in a class, it is necessary to have a number scale that indeed reflects "tallness." Although this proposition appears trivial at first, it happens that in many sciences the concepts of interest to the scientist can be measured only in relatively rudimentary ways. For example, suppose a researcher wanted to assign to each child in a certain nursery school a number that would represent that child's degree of aggressiveness. One method might be to have a clinical psychologist interview each child and then rate him or her from 1 to 10 according to the extent of aggressiveness. The important question for the present discussion is: What are the general mathematical properties of the scale of aggressiveness created by rating the children from 1 to 10? Similarly, what are the properties of a scale of height measured in inches (or centimeters)?

What properties can a measurement scale possibly have? For this discussion there are three: magnitude, equal intervals, and an absolute zero point.

> When a scale has **magnitude**, one instance of the attribute being measured can be judged greater than, less than, or equal to another instance of the attribute.

If the clinical psychologist in the preceding example assigns a score of 8 to Jane and a score of 5 to Harry, this scale of measurement reflects a difference in the magnitude of aggressiveness of the two children—Jane is more aggressive than Harry.

> When a scale has **equal intervals**, the magnitude of the attribute represented by a unit of measurement on the scale is the same regardless of where on the scale the unit falls.

Take the measurement of height in inches as an example. One is confident that the difference in height between someone measuring 61 inches and someone measuring 60 inches is the same magnitude as the difference in height that exists between someone measuring 75 inches and someone measuring 74 inches. That is, an inch reflects a certain amount of height regardless of where that inch falls on the scale. The same is true if height is measured in centimeters. However, consider the year-end baseball standings in the National League. This simple ordering of teams is a crude scale of the baseball prowess of these teams. But if Cincinnati edges out Los Angeles by half a game for first place while the third-place San Francisco team is eight games behind the leader, we feel that the difference between Cincinnati and Los Angeles is less than the difference between Los Angeles and San Francisco, despite the fact that the standings place them equally apart—1, 2, 3. Thus, as a scale of team quality, baseball standings do not possess the property of equal intervals. The difference in quality between one pair of teams adjacent in the standings is not necessarily equal to the difference in quality between another pair of teams that are also adjacent in the standings.

An **absolute zero point** is a value which indicates that nothing at all of the attribute being measured exists.

Thus, 0 inches or 0 centimeters of height is a scale value that indicates no height whatsoever—it is absolute zero. However, in the case of rating aggressiveness, the lowest rating that the psychologist can assign does not indicate absolute zero, no aggressive tendencies whatsoever. A child who receives a rating of 1 may nonetheless hit other children or verbally abuse them on occasion. The child displays a low level of aggression only in a relative sense, and even if the value 0 were a part of the scale, it would not necessarily imply the complete absence of aggressive behavior. Thus, the rating scale for aggressiveness does not possess an absolute zero point.

Types of scales   Many of the measurement scales one uses in everyday life possess all three of these properties. As explained above, a scale used to measure height in inches or centimeters has magnitude, equal intervals, and an absolute zero point. So does a scale used to measure weight in ounces or grams, or a scale used to measure time in minutes and seconds.

Any scale of measurement possessing magnitude, equal intervals, and an absolute zero point is called a **ratio scale**.

The scale is termed ratio because the collection of properties that it possesses allows ratio statements to be made about the attribute being measured. If an adult is 70 inches tall and a child 35 inches, it is correct to infer that the adult is twice as tall as the child. Such ratio statements can be made only if the scale possesses all three properties.

Not all scales used in research in psychology, education, and sociology are ratio scales. Many attributes cannot be measured with scales that reflect all three of the properties under discussion.

An **interval scale** possesses the properties of magnitude and equal intervals but not an absolute zero point.

The most common example of an interval scale is the scale for measurement of temperature in degrees Fahrenheit (or Celsius). Although from the standpoint of physics or chemistry the absolute zero point is reached when all molecular movement ceases, for all practical purposes there is no point at which one can say that there is no temperature whatsoever.[3] Note that neither 0° Fahrenheit nor 0° Celsius denotes a point at which there is no temperature at all (i.e., absolute zero). Further, if the temperature today is 30° and

[3]The Kelvin scale of temperature does have an absolute zero and is therefore a type of ratio scale.

yesterday it was 15°, it is not correct to proclaim that it is twice as hot today as yesterday. Ratio statements cannot be made without an absolute zero point. The temperature scale does possess the properties of magnitude and equal intervals, however. For example, 25° is a warmer temperature than 19°, and the difference between 40° and 50° represents the same difference in temperature as the difference between 90° and 100°. Since the temperature scale has the properties of magnitude and equal intervals, but not of absolute zero, it is an interval scale.

Some scales have only one of the three properties discussed.

An **ordinal scale** reflects only magnitude and does not possess the properties of equal intervals or an absolute zero point.

For example, take the people in the class and line them up according to height, and then rather than measuring them with a tape measure merely rank them according to their height, the shortest receiving a rank of 1, the next tallest 2, etc. The result is an ordinal scale of height.

This scale clearly has the property of magnitude, but does it possess equal intervals? In order to assess this, consider the graphic example below:

1   2   3

Although the ranking places these people at equally spaced locations on the scale (i.e., 1, 2, 3), the difference in height between persons x and y is not the same as between y and z. Thus, the ranking scale of height does not possess equal intervals. Recall that the same was true for baseball standings. Further, the scale does not possess an absolute zero point: There is no ranking that expresses no height at all. Hence, the ranking of height (in contrast to a scale that measures height in inches or centimeters) produces an ordinal scale because it possesses magnitude but not equal intervals or an absolute zero.

It is possible to have a scale that possesses none of the three properties just discussed, although the term *scale* is usually reserved for measurements that imply differences, at least in magnitude.

> A **nominal scale** results from the classification of items into discrete (i.e., mutually exclusive) groups that do not bear any magnitude relationships to one another.

For example, if a person were to stand on a busy street corner and name cars, the classifications might be Ford, Chevrolet, Plymouth, etc. The dimension of classification is *make of car*. However, one would not say that Ford is more or less of a make of car than Chevrolet. It may be more or less expensive or appealing to the eye, but not more or less of a make of car. Hence, grouping cars according to brand name represents a nominal scale, which does not possess the properties of magnitude, equal intervals, or an absolute zero point. Examples of nominal scales used in scientific research include the classification of plants and animals by order, genus, and species in biology and the classification of human beings by personality type in psychology.

Variables

Variables versus constants   As discussed in the Prologue, most scientific data consist of values of a variable.

> A **variable** is a particular set of objects, characteristics, or events, the members of which may take on different numerical values.

If the quantity is the height in centimeters of members of a class, then height is a variable because the heights of members of the class may differ from one to another.

> A **constant** is a quantity that does not change its value within a given context.

The mathematical symbol π is a constant because it always equals approximately 3.1416; its value does not change. Also, if one measures the heights of people in a class in inches but wishes to report the measurements in centimeters, height is a variable, but the conversion factor of 2.54 centimeters to the inch is a constant: It is the same value in each case. The average height of people in the class would be another constant, but only with respect to that class. Across several classes within a school, average height would be a variable.

Discrete versus continuous variables   Variables may be either discrete or continuous.

> A **discrete variable** is one that can assume only a finite number of values between any two points.

For example, a family can have only 1, 2, 3, or more children. One does not think of a family as having $1\frac{1}{2}$ children. Thus between the values 1 and 3, there is only one other possible value, 2.

> A **continuous variable** is one that theoretically can assume an infinite number of values between any two points.

For example, consider weight. Even between 100 and 101 pounds there is an unlimited number of values possible: 100.1, 100.247, etc. Time is another example of a continuous variable. In contrast, the number of points scored in a basketball game is a discrete variable because a team can score 90 or 91 points, but not 90.5 points.

The discrete–continuous distinction actually applies to the variable being assessed (called the **conceptual variable** referred to in the Prologue) and not to the scale used to measure the variable. In practice, continuous as well as discrete variables are measured in discrete scores. For example, we measure weight in ounces, pounds, or tenths of an ounce, even though there are, theoretically, an infinite number of values between any two weights.

Continuous variables and some discrete variables may be measured with ratio, interval, or ordinal scales. For example, intelligence is a continuous variable because, given a sufficiently fine instrument, theoretically there are an infinite number of values between any two points. But the current measurement of intelligence is relatively crude, and IQ test scores probably constitute an ordinal scale if the extremely low and high ends of the scale are included.

Real limits

Since a continuous variable is one for which an infinite number of values exist between any two points on the scale, the actual measurements are rounded off and are therefore approximate. For example, the continuous variable of time may be measured in years, months, days, hours, minutes, seconds, milliseconds, etc. If one measures to the nearest second, it is clear that more refined approximations could be made with more sophisticated timers. Consequently, if a child is asked to solve a given mathematics problem and does so in "33 seconds," the value 33 probably does not mean *exactly* 33 seconds but *approximately* 33 seconds. More precisely, it means between 32.5 and 33.5 seconds. These values are called the **real limits** of 33 seconds. The **lower real limit** is 32.5 because any number lower than this (e.g., 32.4) would be rounded to a whole number other than 33 seconds (32), and the **upper real limit** is 33.5 because any number greater than this also would be rounded to a whole number other than 33 (34).

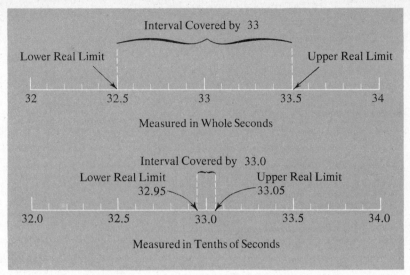

Fig. 1–1.   Real limits of 33 when the measurement unit is whole seconds
and when it is tenths of seconds

The **real limits** of a number are those points falling one-half a
measurement unit above and one-half a measurement unit below
that number.

To illustrate, if measurement is being made in whole seconds, the unit of
measurement is one second. The real limits of 33 seconds are 32.5 seconds
and 33.5 seconds because these values are one-half unit (.5 second) below and
one-half unit above 33 seconds, respectively.[4]

However, note that the definition states that the limits fall one-half *unit*
above and below the number. Therefore, the real limits of 33 seconds are
different for measurements made in units other than whole seconds. Suppose
that a stop watch is available and that the length of time to solve a problem is
recorded in tenths of a second. In this case, the unit of measurement is .1
second and one-half of this unit is .05 second. Consequently, when measuring
in tenths of a second, the real limits of 33.0 are 32.95 and 33.05. Similarly, if
measurement is made in hundredths of a second, the real limits of 33.00 are
32.995 and 33.005. These points are presented graphically in Figure 1–1.

Rounding

If measurements are taken in tenths of a second but reported in whole
seconds, they are said to be **rounded** to the nearest whole second.

---

[4]Technically, the upper real limit of 33 is not 33.5 but 33.49999 . . . . That degree of accuracy
will not be necessary here.

Numbers are **rounded** according to the following conventions (customs):

    1. If the remaining decimal fraction is less than .5 unit, drop the remaining decimal fraction.

    2. If the remaining decimal fraction is greater than .5 unit, increase the preceding digit by one.

    3. If the remaining decimal fraction is exactly .5 unit, add 1 to the preceding digit if that digit is an odd number but drop the remaining decimal fraction if the preceding digit is an even number.

For example, if the unit of measurement is a whole second and the value 33.4 is obtained, round this to 33 seconds in accordance with convention 1 above. If the obtained value is 32.6, round to 33 in accordance with convention 2. If the obtained value is 34.5, convention 3 dictates rounding to 34 because the preceding digit (the 4) is even; but if 37.5 is to be rounded, the result is 38 because the preceding digit (the 7) is odd. Note, however, that 34.51 is rounded to 35 because the remaining decimal fraction is more than .5 unit (i.e., .51 is more than .50).

    The reason that some numbers with remaining decimal fractions of exactly .5 are rounded up and some are rounded down is so that, over many instances of rounding such numbers, approximately half will be rounded up and half rounded down. If a simple rule of always rounding .5 up were invoked, more numbers would be rounded up than down. By rounding some of these cases down according to convention 3, this bias is minimized.

    Rounding numbers may seem a trivial issue, but it is not. Whether and how one rounds numbers can have a significant effect on the outcome of certain calculations. For example, suppose one were required to determine the value of G with the equation

$$G = (H)(J)$$

If H is 8247 and J is calculated from available data to be .244893, then $G = 2019.63$. However, if J is rounded to .24 and then multiplied by H, the answer is 1979.28. In this case, rounding makes a difference of 40.35.

    Some years ago, the custom was to round each number in a series of computations to one more decimal place than the number of places desired in the final answer. This is still good advice: Following the custom in solving the above problem, one would have used .245 for J if the answer was to be given to two decimals, yielding an answer of 2020.52, which is within .89 of the exact answer. However, when using a small calculator that carries several decimal places and permits the result of one calculation to be automatically entered into the next operation, it is laborious to round all numbers to a certain number of decimal places. Because some students have calculators and others do not, and one calculator may carry a different number of places than another, not all students will obtain the same numerical answers for the exercises in this book. Different rounding practices will yield different answers. All answers provided in the text and the accompanying *Study Guide* have been performed with as many places as small computing aids would carry, and final answers have been rounded to two decimal places (except when more are required). Occasionally a range of answers is given when rounding practices would make a great difference. Students who are calculating by hand should try to carry three or four decimals in intermediate calculations (the exercises have been kept as numerically simple as possible), but all students will at times obtain answers slightly different from those given here. Intermediate values as well as the final answers are offered to help determine the source of a possible error, rounding or otherwise.

SUMMATION SIGN

Statistics are measures computed on a group of scores. When formulas are given for statistics it is convenient to have some symbolic terminology to represent groups of scores and operations on groups rather than on single scores. Consider the group of five scores below:

Subject	Score Symbol	Score Value
1	X_1	2
2	X_2	3
3	X_3	8
4	X_4	4
5	X_5	7

Suppose we denote by the capital letter X the variable reflected by these scores. In actuality, it could be any variable—response time, number of correct answers, eye movements per minute, etc. But since formulas can apply to any variable, it is convenient to symbolize the variable by a capital letter (X, Y, V, etc.). In contrast, when the value of a constant is unknown it is symbolized by a lower case letter, usually c or k. There are five scores in the above group, and each score represents a specific measurement of the variable under investigation, X. In order to distinguish one specific score from another, each X symbol is given a subscript corresponding to the number of the subject who made that score on the variable. Customarily, in a group of five scores, these subscripts would be 1, 2, 3, 4, 5. Thus the score on variable X for the first subject is represented by X_1. In general, if there are N scores, the subscripts run from 1 to N. Frequently, it is useful to be able to refer to a single score, but not necessarily any particular one—just any single score in the distribution. This single score is referred to as the ith score in the distribution, and it is written X_i. In this example the distribution of all the X_i

(read "all the scores on the variable") contains five scores. Any one of them is referred to as X_i, but the score of the third subject is represented by X_3 (which has a score value of 8), the score of the fourth subject is represented by X_4 (score value of 4), and so on.

In later chapters of this book it will be necessary to consider more than one group of scores on the same variable. In such a case, in order to specify a particular score one must indicate both the group and the subject within that group. Two subscripts are used to accomplish this. For example, any particular score would be written X_{ij}, where i indicates the subject number and j specifies the group number. The score for the fourth subject in the third group would be designated as X_{43}. This topic will be considered again in the sections on the analysis of variance in Part 2.

One of the most frequently performed operations in statistics consists of summing all or some of the scores in a group. For example, in computing the average of a group of measurements, one sums all the scores and divides by the number of scores. However, when writing the formula for the average it is cumbersome to use

$$\text{Average} = \frac{\text{sum of all scores}}{\text{number of scores}}$$

Some students will be familiar with the symbol N, which is commonly used to represent the number of scores in a distribution. The operation of summing the scores, which is called for in so many formulas, also requires symbolic abbreviation. The Greek capital letter *sigma*, Σ, is employed to indicate the operation of summing; and an English letter (such as X) is used to indicate what is to be summed. Thus, another way to write the formula for the average is

$$\text{Average} = \frac{\Sigma X}{N}$$

The meaning of ΣX (sum of all the scores on variable X) can also be written out more fully, as

$$\sum_{i=1}^{N} X_i$$

The small notations under and over the Σ in this fuller expression are called the **limits of the summation**. The entire symbol is read "sum of the X_i from $i = 1$ to N." It means to add X_1 plus X_2 plus ... plus X_N. In symbols,

$$\sum_{i=1}^{N} X_i = X_1 + X_2 + X_3 + \cdots + X_N$$

Thus $\sum\limits_{i=1}^{5} X_i$ signifies the sum of the first five scores and $\sum\limits_{i=2}^{4} X_i$ means the sum of the second through the fourth scores, inclusive. In terms of the above data,

$$\sum_{i=1}^{5} X_i = 2 + 3 + 8 + 4 + 7 = 24$$

and

$$\sum_{i=2}^{4} X_i = 3 + 8 + 4 = 15$$

Often, when all the scores in a distribution are to be summed (as in the formula for the average shown above), the limits of the summation are not written and/or the subscript i is omitted:

$$\Sigma X_i \quad \text{or} \quad \Sigma X \quad \text{implies} \quad \sum_{i=1}^{N} X_i$$

There are three shortcuts for using the summation sign in algebraic operations that will help the student in working with the summation sign.

1. The sum of a constant times a variable equals the constant times the sum of the variable. If c is a constant and X_i a variable,

$$\sum_{i=1}^{N} cX_i = c \sum_{i=1}^{N} X_i$$

Suppose a father, mother, and child ran a mile in 7, 8, and 10 minutes, respectively. How many *seconds* did it take the family to run the mile? There are two ways to solve the problem. One way is to multiply each person's time in minutes (X_i) by 60 seconds (c) and then add the times together:

$$\sum_{i=1}^{N} cX_i = 60(7) + 60(8) + 60(10) = 1500 \text{ seconds}$$

The other way is to multiply the total number of minutes by 60:

$$c \sum_{i=1}^{N} X_i = 60(7 + 8 + 10) = 1500 \text{ seconds}$$

Consider what the expression $\sum\limits_{i=1}^{N} cX_i$ means algebraically:

$$\sum_{i=1}^{N} cX_i = cX_1 + cX_2 + cX_3 + \cdots + cX_N$$

But, this series of terms may be factored in the same manner as

$$ca + cb = c(a + b)$$

with the following results

$$\sum_{i=1}^{N} cX_i = c(X_1 + X_2 + X_3 + \cdots + X_N)$$

$$= c\left(\sum_{i=1}^{N} X_i\right)$$

Since the expression within the parentheses is what has been defined to be $\sum_{i=1}^{N} X_i$, the sum of a constant times a variable is the constant times the sum of the variable:

$$\sum_{i=1}^{N} cX_i = c\sum_{i=1}^{N} X_i$$

2. The sum of a constant taken N times is N times the constant. If c is a constant,

$$\sum_{i=1}^{N} c = Nc$$

In other words, multiplication is multiple addition, and it is easier to multiply 10 times 9 than to add ten 9s. That fact can be seen by writing out the expression being considered.

$$\sum_{i=1}^{N} c = \underbrace{c + c + c + \cdots + c}_{N \text{ terms}}$$

The symbol $\sum_{i=1}^{N} c$ calls for adding N c's together. However, the operation of multiplication is precisely this repetitive addition, so that adding N c's is identical to multiplying c by N. Therefore, the sum of a constant c taken N times is Nc.

3. The summation of a sum of variables is the sum of each of these variable sums. If X and Y are variables,

$$\sum_{i=1}^{N} (X_i + Y_i) = \sum_{i=1}^{N} X_i + \sum_{i=1}^{N} Y_i$$

This expression merely says that if the scores of N students on a midterm and a final exam are to be added together, they can be added in any order. For example, first you can add the two exam grades separately for each student $(X_i + Y_i)$ and then add all the students' combined totals $\left[\sum_{i=1}^{N}(X_i + Y_i)\right]$. Alternatively, you can add all the scores on the first exam to all the scores on the second exam $\left(\sum_{i=1}^{N}X_i + \sum_{i=1}^{N}Y_i\right)$.

Again, writing out the expression

$$\sum_{i=1}^{N}(X_i + Y_i) = (X_1 + Y_1) + (X_2 + Y_2) + \cdots + (X_N + Y_N)$$

removing parentheses,

$$= X_1 + Y_1 + X_2 + Y_2 + \cdots + X_N + Y_N$$

and regrouping produces

$$= X_1 + X_2 + \cdots + X_N + Y_1 + Y_2 + \cdots + Y_N$$

$$\sum_{i=1}^{N}(X_i + Y_i) = \underbrace{\sum_{i=1}^{N}X_i}_{} + \underbrace{\sum_{i=1}^{N}Y_i}_{}$$

This result may be generalized to any number of terms. For example,

$$\Sigma(X_i + Y_i + W_i) = \Sigma X_i + \Sigma Y_i + \Sigma W_i$$

Throughout this text the student must be attentive to the difference between ΣX^2 and $(\Sigma X)^2$. Suppose one had a sample of the following three subjects:

Subject	X	X^2
1	4	16
2	5	25
3	2	4
	$\Sigma X = 11$	$\Sigma X^2 = 45$
	$(\Sigma X)^2 = 121$	

The symbol ΣX^2 directs that each X score be squared first and then the squared values summed, while $(\Sigma X)^2$ tells one to sum the X scores and then to square that sum. Stated another way, ΣX^2 means the sum of the squared X scores, while $(\Sigma X)^2$ means the squared sum of the X scores.

Students should notice that $\Sigma(X_i Y_i)$ is not the same as $\Sigma(X_i + Y_i)$, and $\Sigma(X_i Y_i)$ is *not* equal to $(\Sigma X_i)(\Sigma Y_i)$. The expression $\Sigma(X_i Y_i)$ or simply ΣXY instructs one to sum a group of **cross products**. To do this, one first obtains a cross product for each subject by multiplying each person's first score (X_i) by the person's second score (Y_i). Then one adds the individual cross products of all people in the group. Notice the difference between ΣXY and $\Sigma(X + Y)$ in the following table.

Subject	X	Y	X + Y	XY
1	4	6	10	24
2	2	5	7	10
3	9	3	12	27
			$\Sigma(X + Y) = 29$	$\Sigma XY = 61$

Understanding and being able to use the summation sign is necessary in order to follow the mathematics presented in the rest of this text. The student is advised to practice the exercises for mastery in the use of the summation sign. In compensation, no new mathematics other than elementary algebra and geometry will be required to understand the material presented in this text. Moreover, most mathematical material will be presented in optional tables, which may be skipped. If the reader feels a bit shaky about basic algebra, Appendix I and the Appendix of the *Study Guide* that accompanies this text present brief reviews.

EXERCISES

1. Indicate the scale of measurement used in the studies described below, and justify your choice:
   a. A nursery-school teacher ranked 25 students according to their aggressiveness. A rank of 1 indicated little aggressiveness.
   b. A physiological psychologist studied the effects of stress on the metabolism of cholesterol. After monkeys were reared in either stressful or nonstressful circumstances, their major arteries were removed and the amount of fatty tissue was weighed in grams as a measure of fatty deposits in these arteries.
   c. Mothers who were recently divorced were interviewed as part of a study of types of post-divorce difficulty. The interviewer determined whether their greatest problem was (1) coping with all the work associated with job, household, and children; (2) financial stress; or (3) the lack of close adult relationships.

2. Discuss the difference in the scales of temperature produced by the Celsius versus the Kelvin methods. (The Kelvin method uses $-273°C$ as its zero point. This is the point at which all molecular movement ceases.)

3. The table below shows a set of scores for seven students on an examination and the rank ordering of these students on the basis of their performance. Using students 4, 5, and 6, illustrate the nature of the information lost when one resorts to ordinal scales rather than interval (or ratio) scales of measurement.

Student	Rank	Score
1	7	79
2	6	52
3	5	46
4	4	41
5	3	25
6	2	24
7	1	21

4. Which of the following variables are discrete, and which are continuous?
   a. shyness versus assertiveness of adults in a group setting, rated from 1 to 10
   b. number of live births in a hospital
   c. gross income per family
   d. a rank ordering of all members of a high school class based upon final grade average
5. What are the real limits of the following numbers? (The unit of measurement is the smallest decimal place in the number.)
   a. 8   d. 15.3   g. 2.76
   b. 8.0   e. 9.5   h. 14.01
   c. 8.00   f. 99.95   i. 200.000
6. Round the following numbers to tenths.
   a. 1.56   d. 9.45
   b. 3.22   e. 10.05
   c. 5.95   f. 6.251
7. Given the following data, determine the numerical value of each of the following expressions.

Subject (i)	X_i	Y_i
1	2	4
2	7	0
3	3	9
4	5	1
5	8	6

a. $\sum_{i=1}^{5} X_i$ e. $\sum_{i=1}^{5} (X_i + Y_i)$

b. $\sum_{i=1}^{3} Y_i$ f. $\sum_{i=1}^{5} X_i Y_i$

c. $\sum_{i=3}^{5} X_i$ g. $\sum_{i=1}^{5} X_i^2$

d. $\sum Y$ h. $\left(\sum_{i=1}^{5} X_i\right)^2$

8. Given the data from exercise 7 for X and Y and that c, a constant, equals 3, determine the numerical value of each of the following expressions.
   a. $\sum cX_i$ d. $\sum_{i=1}^{5} c$
   b. $\sum cXY$ e. $\sum_{i=1}^{3} (X + c)$
   c. $\sum c(X + Y)$
9. Simplify the following expressions (W and Z are variables, k is a constant).
   a. $\dfrac{\sum(kW + W)}{\sum W}$
   b. $\dfrac{\sum(W + k) + \sum(Z - k)}{\sum(W + Z)}$
   c. $\dfrac{\sum(k - W) + \sum(Z + W) + (\sum k)(\sum Z)}{N\left(\sum kZ + \frac{\sum k}{N}\right)}$

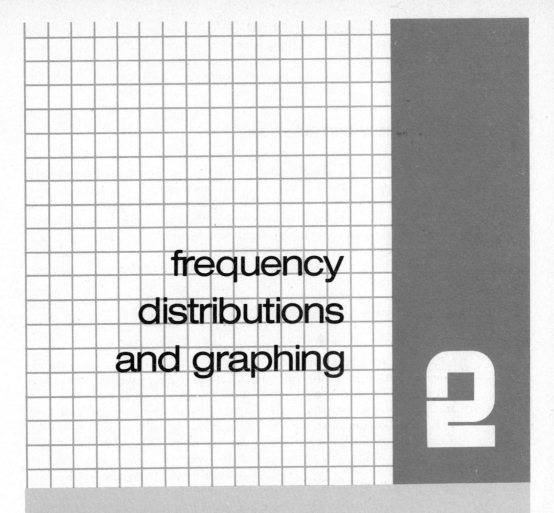

frequency distributions and graphing

2

WHEN BEHAVIORAL SCIENTISTS design a simple research study, they select a variable and a scale of measurement and then make observations on a group of subjects. Once a collection of scores has been obtained, how can one describe it efficiently?

TYPES OF FREQUENCY DISTRIBUTIONS

It is often desirable to be able to characterize an entire group of scores rather than any single value. For example, one might want to know the range of score values included in the group, the value about which most of the scores seem to cluster, or the extent to which the scores are spread over the measurement scale. It would be highly cumbersome and inefficient to enumerate all the scores each time these questions were posed; furthermore, such a listing would not provide a clear picture of the characteristics of the group.

A **frequency distribution** is a tally of the number of times each score value (or interval of score values) occurs in a group of scores.

Suppose a short history quiz is given to a class of 10 students, and the essay is graded on a 10-point scale. The scores are presented in part A of Table 2-1. The first thing to do in order to obtain a clearer picture of this small group of scores is to list them in descending order, as in part B of Table 2-1. Next, observe that there is one score of 10, two of 9, three each of 8 and 7, and one of 6. If a score value is symbolized by "X" and a tally is made next to the X for each occurrence of that value, the result is a frequency distribution. In part C of Table 2-1, f indicates the frequency for any given value of the variable X.

Consider another example of the development of a frequency distribution: An opinion researcher administers a questionnaire that asks a sample of people to what extent they approve of the way the President of the United States is performing. The researcher provides five possible responses and arbitrarily assigns a number to each level of response:

1. Disapprove greatly
2. Disapprove (generally disapprove but approve of some policies)
3. Ambivalent (disapprove of about half the President's actions and approve of half)
4. Approve (generally approve but disapprove of some policies)
5. Approve greatly

Suppose 80 people were questioned and were asked to indicate one of these five opinions. The frequency of their responses is presented in Table

2–1 Development of a Frequency Distribution
for Scores on a History Quiz

A. Scores (X)	B. Scores Arranged in Descending Order
8	10
10	9
9	9
7	8
9	8
8	8
7	7
6	7
8	7
7	6

C. Frequency Distribution of Scores

X	Tally	f
10	/	1
9	/ /	2
8	/ / /	3
7	/ / /	3
6	/	1
		$N = 10$

2–2. It is clear from this frequency distribution that people's opinions about the President are somewhat split. A sizable group generally disapproves while another sizable group approves of his actions. Few people are either ambivalent ($X = 3$) or adamantly positive or negative. Such descriptive conclusions would be difficult to reach if all 80 scores were written down without the assistance of the frequency distribution. Hence, not only does a frequency distribution save time in displaying data, it also organizes the numbers in a way that summarizes the data.

A distribution that indicates the proportion of the total number of cases which was observed at each score value (or interval of score values) is called a **relative frequency distribution.**

2–2 Frequency Distribution of Opinions on Presidential Policy

Opinion	X	f	Rel f
Approve Greatly	5	9	.11
Approve	4	30	.38
Ambivalent	3	10	.12
Disapprove	2	25	.31
Disapprove Greatly	1	6	.08
		$N = 80$	1.00

An example of a relative frequency distribution (*Rel f*) is given in the last column of Table 2–2. Because the numbers there are proportions, they are expressed as decimal fractions; moving the decimal point two places to the right will change the proportions into percentages. The advantage of a relative frequency distribution is that it expresses the pattern of scores in a manner that does not depend on the specific number of cases involved. Thus, an opinion pollster would not say that nine people greatly approved of the

2–3 Mathematics Ability Test Scores for 150 Eighth-Grade Students

79	51	67	50	78	80	77	75	55	65
62	89	83	73	80	67	74	63	32	88
88	48	60	71	79	79	47	55	70	34
89	63	55	93	71	81	72	68	75	93
41	81	46	50	61	72	86	66	54	58
59	50	90	75	61	82	73	57	87	41
75	98	53	79	80	64	67	51	36	52
70	37	42	72	74	78	91	69	95	76
67	73	79	67	85	74	70	62	76	69
91	73	77	36	77	45	39	59	63	57
53	67	85	74	77	78	73	61	47	43
76	43	42	96	83	83	84	67	81	75
70	92	59	86	53	71	49	68	42	46
32	67	67	71	71	59	80	66	39	49
82	68	30	72	57	92	50	38	73	56

President's actions, but rather that 11% did. Of course, it is always informative to know the total number of people polled as well.

The need for a frequency distribution is most obvious when one is confronted with a large number of scores and a wide range of score values. To illustrate, suppose 150 eighth-grade students were given a mathematics ability test prior to taking an algebra course. The scores for these students are presented in Table 2–3. This display emphasizes the fact that listing all the scores does not provide much immediate information. In this case, however, a frequency distribution constructed by counting the number of times each possible score value between 0 and 100 appears would be only slightly more

2–4 Ordering of 150 Mathematics Ability Test Scores

—	74, 74, 74, 74	49, 49
98	73, 73, 73, 73, 73, 73	48
—	72, 72, 72, 72	47, 47
96	71, 71, 71, 71, 71	46, 46
95	70, 70, 70, 70	45
—	69, 69	—
93, 93	68, 68, 68	43, 43
92, 92	67, 67, 67, 67, 67, 67, 67, 67, 67	42, 42, 42
91, 91	66, 66	41, 41
90	65	—
89, 89	64	39, 39
88, 88	63, 63, 63	38
87	62, 62	37
86, 86	61, 61, 61	36, 36
85, 85	60	—
84	59, 59, 59, 59	34
83, 83, 83	58	—
82, 82	57, 57, 57	32, 32
81, 81, 81	56	—
80, 80, 80, 80	55, 55, 55	30
79, 79, 79, 79, 79	54	
78, 78, 78	53, 53, 53	
77, 77, 77, 77	52	
76, 76, 76	51, 51	
75, 75, 75, 75, 75	50, 50, 50, 50	

informative than a list of all the scores. To describe the entire set of scores efficiently, the researcher must group the score values into clusters called class intervals.

A **class interval** is a segment of the measurement scale that contains several possible score values.

For example, for the scores listed above, the set of score values 30, 31, 32, 33, and 34 might be used to define the first class interval, and the frequency of scores falling within that interval in the distribution could be counted.

Table 2–4 was constructed in this way, by ordering the scores listed in Table 2–3 and grouping them into class intervals of five values each. From Table 2–4, one can see that there were only three cases of a score between 95 and 99 inclusive (namely, 95, 96, and 98), seven cases within the interval of 90 to 94 inclusive, nine between 85 and 89 inclusive, etc. A summary of this accounting is presented in the frequency distribution shown in the two leftmost columns of Table 2–5. The only difference between this frequency distribution and previous ones is that, rather than having frequencies stated for each possible score value, the score values are grouped into class intervals (e.g., 30–34, 35–39) and frequencies are determined for each interval. When scores are presented this way, they are sometimes referred to as **grouped data**.

2–5 Distributions for 150 Mathematics Ability Test Scores

Class Interval	f	Rel f	Cum f	Cum Rel f
95–99	3	.02	150	1.00
90–94	7	.05	147	.98
85–89	9	.06	140	.93
80–84	13	.09	131	.87
75–79	20	.13	118	.79
70–74	23	.15	98	.65
65–69	17	.11	75	.50
60–64	10	.07	58	.39
55–59	12	.08	48	.32
50–54	11	.07	36	.24
45–49	8	.05	25	.17
40–44	7	.05	17	.11
35–39	6	.04	10	.07
30–34	4	.03	4	.03
	$N = 150$	1.00		

Grouped data can be examined not only through a frequency distribution but through a relative frequency distribution, as shown in the third column of Table 2–5. Two other distributions are also displayed. They are the cumulative frequency and cumulative relative frequency distributions.

A **cumulative frequency distribution** is one in which the entry for any class interval is the sum of the frequencies in that interval plus the frequencies in all class intervals below (i.e., of lower score values).

A **cumulative relative frequency distribution** is one in which the entry for any class interval expresses that interval's cumulative frequency as a proportion of the total number of cases.

That is, in the cumulative versions of frequency and relative frequency distributions, the entries are progressively accumulated starting from the lowest class interval. Cumulative distributions provide a means for rapidly ascertaining the number or proportion of scores that fall within and below a given class interval.

Suppose that Johnny had a score of 64 on this mathematics test. The cumulative proportion (*Cum Rel f*) of scores within and below the interval 60–64 in Table 2–5 is .39, which says that 39% of the scores were equal to or below the score of 64. Johnny could be said to be at the 39th percentile. Percentiles will be discussed in more detail in Chapter 4.

A distribution of scores, then, may be displayed as a frequency, relative frequency, cumulative frequency, or cumulative relative frequency distribution. The advantage of such distributions is that they provide an efficient method of organizing and presenting a large group of scores so that certain characteristics of the group as a whole become apparent.

CONSTRUCTING FREQUENCY DISTRIBUTIONS WITH CLASS INTERVALS

Number of class intervals

It is somewhat easier to read and understand a frequency distribution than it is to actually organize a set of scores into that form because certain decisions must be made about the nature of the class interval to be used. In the above case, an interval such as 30–34 was selected, but would 30–32 or 30–47 have done just as well? Table 2–6 illustrates these two alternatives.

Consider the examples of Table 2–6 from the standpoint of the general goal of frequency distributions, which is to summarize data in a form that

2-6 Distributions for 150 Mathematics Ability Test Scores
Using Class Intervals of Different Sizes

Class Interval	f	Class Interval	f
96–98	2	84–101	20
93–95	3	66–83	71
90–92	5	48–65	37
87–89	5	30–47	22
84–86	5		$N = 150$
81–83	8		
78–80	12		
75–77	12		
72–74	14		
69–71	11		
66–68	14		
63–65	5		
60–62	6		
57–59	8		
54–56	5		
51–53	6		
48–50	7		
45–47	5		
42–44	5		
39–41	4		
36–38	4		
33–35	1		
30–32	3		
$N = 150$			

accurately depicts the group as a whole. In the first instance (left-hand distribution in Table 2–6), the summarization advantage of the frequency distribution is lost because there are too many class intervals. Listing 23 groups of scores is not much of an improvement over listing the scores one by one. The distribution is too spread out or diffuse to accomplish the goal of summarization.

Conversely, the right-hand distribution in Table 2–6 has too few class intervals. It is clear that most of the scores fall between 66 and 83, but considerable accuracy and detail have been lost by grouping the data into only four classes. For example, one cannot tell whether the 22 cases between

30 and 47 fall nearer to the value of 47, nearer to 30, or evenly within the interval. The lowest score in the group could be anywhere between 30 and 47, given only the information provided by this distribution.

From these examples it should be clear that too many class intervals do not summarize the group of scores adequately, whereas too few intervals reduce the accuracy of the description. One must select a number of class intervals that represents a compromise between these extremes.

It is usually suggested that between 10 and 20 class intervals be chosen; but there can be a great difference between the picture one obtains of a distribution if it is displayed with 10 intervals and the picture one obtains from 20 intervals. Hence, although this is a good rule of thumb, one has to make an intelligent decision that depends upon the nature of the data. For Table 2–5, 14 intervals were chosen. In general, if the total number of frequencies is small (e.g., 20–50) one tends to pick fewer intervals than if there are 100 or 200 scores. If the distribution had only 10 cases one would want fewer intervals than 10, perhaps 4 or 5; but if there were 2000 cases, as many as 20 or more intervals might be contemplated. Hence, the guideline of 10 to 20 intervals merely reflects a concern for an accurate summarization and description of a group of scores.

Size of the class interval

Once a tentative decision on the approximate number of class intervals has been made, the size of the interval must be determined. A good approach is to subtract the smallest score from the largest. This provides a range of values covered by the group of scores that must be broken into intervals. If this result is divided by the approximate number of intervals, an estimate of the size of the intervals is obtained. For example, if approximately 15 intervals would be appropriate for the data in Table 2–4 and the difference between the largest and smallest scores is $98 - 30 = 68$, then 68 divided by 15 yields an approximate interval size: 4.5. However, it is usually inconvenient to use intervals involving fractions like 4.5. Hence, one might use an interval of 5, thus covering a range of 68 with 14 intervals.

To determine the size of a class interval from a distribution already constructed, one must first determine the upper and lower real limits of the interval.

> The **upper real limit** of a class interval is the upper real limit of the highest score value contained in that interval; the **lower real limit** of the interval is the lower real limit of the lowest score value contained in the interval.

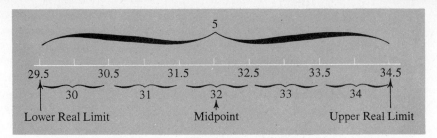

Fig. 2–1.   Demonstration that the size of the interval 30–34 is 5

In contrast:

> The **stated limits** of a class interval are the highest and lowest score values contained in the interval.

Therefore, the class interval including the scores 30, 31, 32, 33, and 34 has stated limits of 30 and 34 but real limits of 29.5 and 34.5.

> The **size of the class interval** is obtained by subtracting the lower real limit of the interval from its upper real limit.

The size of the stated interval 30–34 is 5, because

$$34.5 - 29.5 = 5$$

It may seem a bit puzzling at first that the size of the interval 30–34 is 5 and not 4, but if the scores contained within the interval are listed (30, 31, 32, 33, 34), there are clearly 5 of them, not 4. This notion is further illustrated in Figure 2–1, which shows a segment of a linear scale of measurement and the real limits for each score in the interval.

Lowest class interval

Once the number and size of the class intervals have been established, all that remains is to specify the first interval; then the scale will be completely determined. Obviously, the first interval must include the lowest score. But what should its limits be? There is a custom that the first *stated* limit be evenly divisible by the size of the interval. Thus, the size of the lowest interval, 30–34, is 5 and the lowest stated limit is 30. Since 30 divided by 5 is an even 6, this meets the requirement. If the lowest score in a distribution is 49 and an interval of size 4 is selected, the first interval would be 48–51 because 48 (but not 49) is evenly divisible by 4.

Midpoint of a class interval

It is also helpful to know the midpoint of an interval.

> The **midpoint** of a class interval is the precise center of that interval, that is, the point halfway between the interval's real limits. It can be determined by adding one-half of the size of the interval to its lower real limit.

For the interval 30–34, whose size is 5 (half of which is 2.5) and whose lower real limit is 29.5, the midpoint is

$$29.5 + 2.5 = 32$$

For an interval with the stated limits 45.6–48.3, whose size is 2.8 and whose lower real limit is 45.55, the midpoint is

$$45.55 + 1.4 = 46.95$$

The important thing to remember is to use the lower *real* limit (in this case, 45.55), not the lower *stated* limit (45.6).

2–7 Steps in Constructing Frequency Distributions with Grouped Data

1. Estimate the number of class intervals. This usually should be 10 to 20. It may be less if the total number of cases is small, and it may be more if the total is very large.

2. Estimate the size of the class interval by dividing the difference between the largest and smallest score in the distribution by the number of intervals selected in Step 1. Round this result up to the nearest whole number (or convenient fraction if a whole number is not appropriate).

3. Select the lowest class interval so that its lowest *stated* limit is evenly divisible by the size of the interval.

4. Place the lowest interval at the bottom of the table that shows the class intervals and their respective frequencies.

Table 2–7 provides a summary of the steps in constructing a frequency distribution using class intervals; Table 2–8 displays the resulting distribution for the sample data presented in Table 2–4 and also shows the real limits, size, and midpoint of each interval.

2–8 Frequency Distribution for the Data in Table 2–4 with Real Limits, Interval Size, and Midpoint

Class Interval	Real Limits	Interval Size	Midpoint	Frequency
95–99	94.5–99.5	5	97	3
90–94	89.5–94.5	5	92	7
85–89	84.5–89.5	5	87	9
80–84	79.5–84.5	5	82	13
75–79	74.5–79.5	5	77	20
70–74	69.5–74.5	5	72	23
65–69	64.5–69.5	5	67	17
60–64	59.5–64.5	5	62	10
55–59	54.5–59.5	5	57	12
50–54	49.5–54.5	5	52	11
45–49	44.5–49.5	5	47	8
40–44	39.5–44.5	5	42	7
35–39	34.5–39.5	5	37	6
30–34	29.5–34.5	5	32	4

$$N = 150$$

GRAPHS OF FREQUENCY DISTRIBUTIONS

Frequency histogram

A group of scores can be described by drawing a graph of the frequency distribution as well as by writing out the distribution as a table. One type of graph is a **frequency histogram**. Figure 2–2 presents as a frequency histogram the distribution of math ability scores shown in Tables 2–5 and 2–8.

To construct this histogram a horizontal scale was drawn corresponding to the scale of math ability scores. The horizontal dimension of such a plot is called the **abscissa**. Notice that the numbers along the abscissa are the midpoints of the intervals, as shown in Table 2–8. The vertical dimension is called the **ordinate**, and in the case of frequency distributions it will correspond to f, frequency. The abscissa and ordinate are called **axes**. Note that the axes are clearly marked with the numbers of their respective scales and then labeled. The width of each bar of the histogram covers the entire range of its class interval, from lower to upper *real* limits. Therefore, each bar exactly straddles the midpoint of its interval as designated along the abscissa. The height of a bar corresponds to the frequency of scores for that interval as

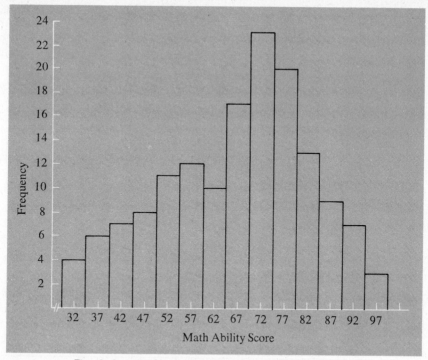

Fig. 2–2. Frequency histogram for data in Table 2–5

indicated on the ordinate. A summary of the steps in constructing a frequency histogram as well as the graphs described below is presented in Table 2–9.

Frequency polygon

An example of a **frequency polygon** is presented in Figure 2–3. The graph is constructed by placing a point above the midpoint of each class interval corresponding to the frequency within that interval. The adjacent points are connected by straight lines. Note that when drawing a frequency polygon, one adds an "empty" interval with zero frequencies to the left of the first and to the right of the last intervals that contain scores. The line that connects the points on the graph intersects the abscissa at the midpoints of these empty intervals, thus closing the polygon, from which this graph derives its name.

Relative frequency histogram and polygon

Figure 2–4 presents a relative frequency histogram and polygon for the same data. Note that these plots are graphically similar to the previously illustrated

2–9 Steps in Constructing Histograms and Polygons

Frequency Histogram

1. Mark off the abscissa with values corresponding to the midpoints of the class intervals and mark off the ordinate in frequencies. Label the axes appropriately.

2. Construct the bars of the histogram over each class interval so that their width covers the class interval from its lower to its upper real limits (not from midpoint to midpoint), and their height corresponds to the frequency of scores in the interval. There should be no space between bars.

Frequency Polygon

1. Mark off and label axes as for a frequency histogram, but add one interval below the lowest and one above the highest class interval, and assign them 0 frequencies.

2. Place points corresponding to the frequencies of each interval (including the two 0-frequency intervals) directly over the midpoints of each class interval. Connect all adjacent points (including the 0s) with straight lines.

Relative Frequency Histogram or Polygon

1. These are plotted in the same way as above, but the labels on the ordinate (and the height of the bars or points) show relative frequency, not frequency.

Cumulative Frequency Histogram or Polygon

1. Follow the steps for constructing a frequency histogram or polygon, except:

   a. Mark off and label the ordinate for cumulative frequency rather than frequency.

   b. In drawing a cumulative frequency polygon, the points are placed over the upper real limit of each class interval, including the lowest interval of 0 accumulated frequencies (note that there is no upper 0-frequency interval).

Cumulative Relative Frequency Histogram or Polygon

1. These are plotted in the same way as above, except that the labels on the ordinate (and the height of the bars or points) show cumulative relative frequency.

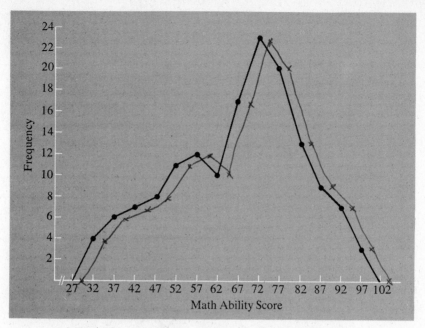

Fig. 2–3.   Frequency polygon for data in Table 2–5

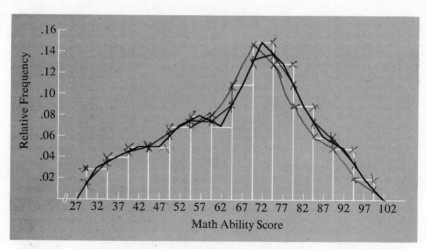

Fig. 2–4.   Relative frequency polygon and histogram for data in Table 2–5

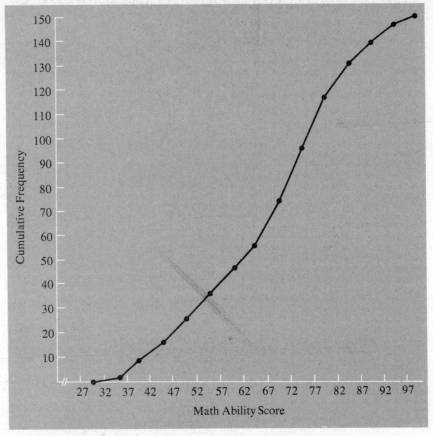

Fig. 2–5.   Cumulative frequency polygon for data in Table 2–5

plots, except that the ordinate is relative frequency and the values are taken from the column labeled *Rel f* in Table 2–5.

Cumulative frequency histogram and polygon

Just as frequency and relative frequency distributions have been plotted in the form of histograms and polygons, the two cumulative distributions in Table 2–5 can also be graphed in these ways. Steps for constructing both these graphs can be found in Table 2–9, and an example of a cumulative frequency polygon is given in Figure 2–5.

One difference between the graphs of a frequency distribution and a cumulative frequency distribution is that the ordinate changes from frequency to cumulative frequency. A second difference is that the point on a cumulative polygon is placed over the upper real limit of each interval, rather than over its midpoint as in a frequency polygon. This is because this point must indicate that up to the *end* of that interval, a certain number or proportion of cases has occurred. Since the scores that fall within a given interval may be located anywhere within that interval, the point representing the accumulation of all frequencies within and below this interval is placed at the upper real limit of the interval.

HOW DISTRIBUTIONS DIFFER

Some of the most important ways distributions differ from one another are with respect to (1) central tendency, (2) variability, (3) skewness, and (4) kurtosis. The general meaning of these concepts will be presented here. The next chapter explains how central tendency and variability can be expressed numerically.

> The **central tendency** of a distribution is a point on the scale corresponding to a typical, representative, or central score.

There are several more specific definitions of central tendency, each with its own set of characteristics and implications. Three of these will be discussed in Chapter 3: mean, median, and mode.

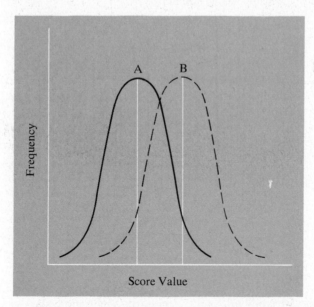

Fig. 2–6.   Distributions that differ only with respect to central tendency

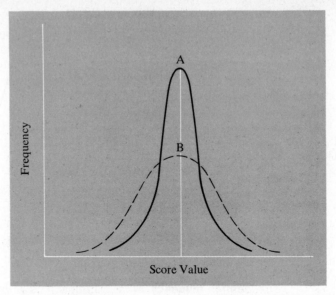

Fig. 2–7.   Distributions with the same central tendency but different
            variability

To illustrate the concept of central tendency, consider the two curves (smoothed frequency polygons) in Figure 2–6. They differ only with respect to central tendency. They have the same shape but occupy different places on the scale of measurement.

Variability is the extent to which scores in a distribution deviate from their central tendency.

Figure 2–7 shows two curves with similar central tendencies but different amounts of variability. The scores in distribution A cluster more closely about the central tendency of the distribution (which is indicated by the vertical line) than do the scores in distribution B. There is more variability in B than in A.

Skewness refers to an asymmetric distribution in which the scores are bunched on one side of the central tendency and trail out on the other.

Figure 2–8 presents two skewed distributions. Both lack symmetry, but B is more skewed than A because there is a greater tendency for the scores to bunch at one end and trail off at the other. Skewness and variability are usually related: the more the skewness, the greater the variability. But more variability does not necessarily mean more skewness (why?).

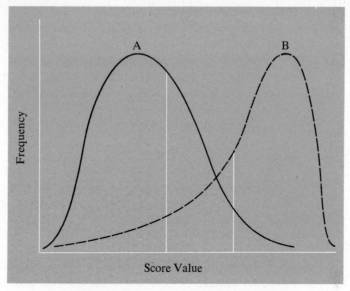

Fig. 2–8.   Distributions with different skewness

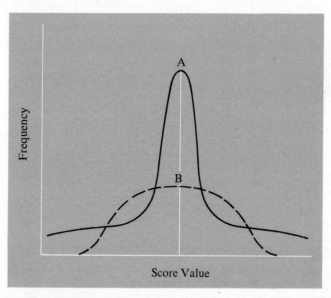

Fig. 2–9.   Distributions with the same central tendency but different kurtosis

Skewness has direction as well as magnitude. In distribution A the scores tend to trail off to the right or positive end of the scale. It is said to be **positively skewed** or **skewed to the right**. Conversely, distribution B trails off to the left or negative end of the scale (recall that the first integer to the left of 0 is −1, i.e., negative). Distribution B is said to be **negatively skewed** or **skewed to the left**.

The **kurtosis** of a distribution is the "curvedness" or "peakedness" of the graph.

Figure 2–9 depicts two curves with the same central tendencies but different kurtoses. Curve A is more peaked than B, and the changes in the height of the curve as the score value increases are more marked for A than for B. Kurtosis is frequently used in a relative sense. Curve A is more **leptokurtic** than B. The Greek *lepto* means thin, so *leptokurtic* implies a thin distribution. On the other hand, B is more **platykurtic** than A. *Platy* means flat (e.g., *platy*helminthes—flatworms, *platy*pus—a flat-billed mammal).

The terms relating to skewness and kurtosis may be used to describe the general form of a distribution. Thus one might say that a distribution is positively skewed and rather platykurtic. This verbal description gives one some idea of the shape of the curve in question, but it is not very precise. When precision is called for, as when one is comparing distributions, mathematical indices are needed. Although measures of skewness and kurtosis are available,[1] they are not often used. Therefore, only indices of central tendency and variability will be taken up in the next chapter.

[1]See, for example: G. A. Ferguson, *Statistical Analysis in Psychology and Education*, 4th ed. (New York: McGraw-Hill, 1975).

EXERCISES

1. In the right-hand column is a set of scores on a mathematics exam. Construct a frequency distribution appropriate for these data. Present the distribution in the manner of Table 2–8, with real limits, midpoint, interval size, and frequency indicated. Then compose a relative frequency distribution, cumulative frequency distribution, and cumulative relative frequency distribution for these data.

54	81	18	44	24
63	67	60	34	39
91	47	75	72	36
87	49	86	57	74
26	41	90	59	14
13	31	68	13	29
29	70	22	63	35
50	42	27	95	77
42	31	69	73	11
31	45	51	56	40

2. Construct polygons and histograms for the frequency and relative frequency distributions above. Make a polygon for the cumulative frequency distribution.

3. The procedures for constructing a frequency distribution were described for data composed of whole numbers. Obviously, many of the measurements made by scientists are in decimal form. The guidelines for constructing frequency distributions listed in Table 2–7 also apply to decimal data. Perform the tasks required in exercises 1 and 2 with the numbers below.

1.8	1.5	2.3	2.7
2.5	1.9	1.3	1.5
2.0	3.1	2.4	1.1
2.1	2.2	1.6	2.0
1.6	2.2	.4	2.0
1.3	1.4	2.2	1.2
2.8	.8	.9	2.3

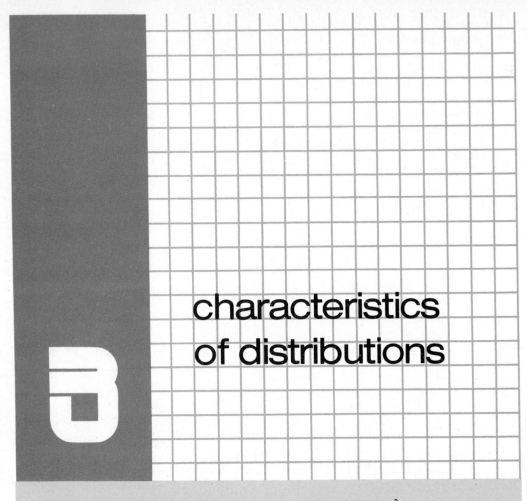

characteristics
of distributions

TWO KEY CHARACTERISTICS of a set of scores are its central tendency (the point on the measurement scale that represents a typical score) and its variability (the extent to which the scores differ from one another). When scores are arranged as a frequency distribution and graphed, the shape of the resulting curve conveys some rough information about these two characteristics. One can see, for example, that the typical score probably lies somewhere near the peak of the curve, and that a curve with gently sloping sides reflects greater variability than a curve with steep sides. However, a frequency distribution can by itself provide only a rough approximation of central tendency and variability. For precision, numerical indices are needed.

The purpose of this chapter is to introduce such numerical indices. The most commonly used measure of central tendency is the mean, or average score; others are the median and mode. Three indices are used to express variability: the range, the variance, and the standard deviation. In addition to presenting these measures as statistical concepts and explaining how to calculate their values, this chapter includes a discussion of statistical estimation that shows how values from studies with a limited number of subjects can be used to estimate characteristics of much larger groups. Estimation is a fundamental process in inferential statistics, and we will return to it in Part 2. It is introduced here because descriptive statistics, as presented in this chapter, are often used as estimators of corresponding values in larger groups of subjects.

MEASURES OF CENTRAL TENDENCY

The mean

The most common measure of the central tendency of a group of scores is the average, or mean.

> The **mean** of scores on variable X is symbolized by $\overline{X}$ (read "X bar")[1] and is computed with the formula
>
> $$\overline{X} = \frac{\sum X_i}{N}$$
>
> which instructs one to add all the scores (i.e., $\sum X_i$) and divide by N,

---

[1]Texts differ in the symbols used for different statistical concepts. Although this text uses $\overline{X}$ to denote the sample mean, some other books use M.

which is the number of scores in the distribution. For example:

$$\frac{X_i}{\begin{matrix}8\\3\\4\\10\\7\\1\end{matrix}} \qquad \bar{X} = \frac{\Sigma X_i}{N} = \frac{33}{6} = 5.5$$

$$\Sigma X_i = \overline{33}$$
$$N = 6$$

The mean is also called the arithmetic average of the scores. Notice that ΣX_i can be written without the limits $i = 1$ and N (see Chapter 1), which are understood to apply. For example, $\sum\limits_{i=1}^{N} X_i = \Sigma X_i = \Sigma X$. This abbreviated notation will be common in the remainder of the text.

Deviations about the mean   The mean possesses several properties that make it an appropriate measure of central tendency. First,

The sum of the deviations of scores about their mean is zero. In other words, if the mean is subtracted from each score in the distribution, the sum of such differences is zero. In symbols,

$$\Sigma(X_i - \bar{X}) = 0$$

Consider the following numerical example:

X_i	$\bar{X}$	$(X_i - \bar{X})$		
3	5	3 − 5	=	−2
6	5	6 − 5	=	1
5	5	5 − 5	=	0
1	5	1 − 5	=	−4
10	5	10 − 5	=	5
$\Sigma X_i = 25$		$\Sigma(X_i - \bar{X}) =$		0
$N = 5$				
$\bar{X} = 5$				

The principle can be proven true of all distributions. The algebraic proof, presented in Table 3–1 (optional), uses two of the summation rules discussed in Chapter 1 plus the fact that the mean $(\bar{X})$ of any distribution is a constant —it does not change value with respect to that distribution. The value of a

3–1 Proof that the Sum of the Deviations
about the Mean Equals Zero

OPTIONAL TABLE

Operation	Explanation
To prove $\Sigma(X_i - \bar{X}) = 0$ 1. $\Sigma(X_i - \bar{X}) = \Sigma X_i - \Sigma \bar{X}$	1. The sum of the differences between two quantities equals the difference between their sums.
2. $\phantom{\Sigma(X_i - \bar{X})} = \Sigma X_i - N\bar{X}$	2. The sum of a constant added to itself N times (i.e., $\Sigma \bar{X}$) is N times the constant (i.e., $N\bar{X}$).
3. $\phantom{\Sigma(X_i - \bar{X})} = \Sigma X_i - N\left(\dfrac{\Sigma X_i}{N}\right)$	3. Substitution of $\dfrac{\Sigma X_i}{N}$ for $\bar{X}$
4. $\Sigma(X_i - \bar{X}) = \Sigma X_i - \Sigma X_i = 0$	4. Cancellation of N's in second term.

single score (X_i) can change, depending upon which score is selected; but the mean remains constant for a particular distribution.

While the sum of the deviations of all the scores about the mean is always zero, the sum of the *squared* deviations about the mean is usually not zero. The distinction is important because formulas presented later in the chapter (e.g., for the variance) use squared deviations.

That is, whereas

$$\Sigma\left(X_i - \bar{X}\right) = 0$$

the expression

$$\Sigma\left(X_i - \bar{X}\right)^2 \text{ is } \textit{not} \text{ usually equal to } (\neq) \, 0$$

To illustrate, consider the numerical example on the previous page. If one squares the difference between each score and the mean (the numbers in the extreme right-hand column) and then sums these squared deviations, one obtains

$$(-2)^2 + (1)^2 + (0)^2 + (-4)^2 + (5)^2 = 46$$

which is obviously not zero. The reason squared deviations never add to zero (unless all the scores are the same) is that squared numbers can never be negative, and thus positive values will not be balanced by negative ones. Although this distinction between the sum of squared and unsquared deviations about the mean may seem trivial at this point, there will be many situations which will require remembering the fact that the sum of the deviations (but not of the squared deviations) of scores about their mean is zero.

Minimum variability of scores about the mean   A second property of the mean concerns the squared deviations of scores about their mean:

> The sum of the squared deviations of scores about their mean is less than the sum of the squared deviations of the same scores about any other value.

This fundamental principle will be invoked in the explanation of many subsequent concepts. It states that although the sum of the squared deviations of scores about their mean usually does not equal zero, that sum is nevertheless smaller than if the squared deviations of the same scores were taken about any value other than the mean of their distribution. For example, in the above illustration the sum of the squared deviations about the mean equaled 46. The mean of that distribution was 5.0. The sum of the squared deviations about the number 6.0 equals 51; about the number 4.0 the sum equals 51; and about the number 7.0 it equals 66. The sum of squared deviations about the mean (46) is less than any of these examples, and it can be shown (Table 3–2) that it always will be less than about any other value. It is in this sense, sometimes called the "least squares sense," that the mean is an appropriate measure of central tendency: The mean is closer (in terms of squared deviations) to the individual scores over the entire group than is any other single value.

The proof that the sum of the squared deviations about the mean is less than the sum of the squared deviations about any alternative value is presented in Table 3–2 (optional). The logic of the proof is that any other value may be expressed in terms of the mean, $\overline{X}$, plus some value, call it c. Thus, the alternative value is $(\overline{X} + c)$, where c may be a positive or a negative number. The procedure for proving that the sum of the squared deviations about the mean is a minimum involves determining the sum of squared deviations about any other value (i.e., about the value $\overline{X} + c$), and then demonstrating that this sum will always be greater than the sum of deviations about $\overline{X}$. (Again, as with all optional tables, Table 3–2 may be omitted without loss of continuity.)

OPTIONAL TABLE

3-2 Proof that the Sum of the Squared Deviations about the Mean is a Minimum

Operation	Explanation
1. $(\bar{X} + c), c \neq 0$ is a value other than $\bar{X}$.	1. Assumption.
2. (a) The sum of the squared deviations of scores about $\bar{X}$ equals $\Sigma(X_i - \bar{X})^2$ (b) The sum of the squared deviations of scores about $(\bar{X} + c)$ equals $\Sigma[X_i - (\bar{X} + c)]^2$	2. Definitions.
To prove $\Sigma(X_i - \bar{X})^2 \; < \; \Sigma[X_i - (\bar{X} + c)]^2$	
3. $\qquad\qquad < \; \Sigma[(X_i - \bar{X}) - c]^2$	3. Working with the right side of the inequality, removing parentheses and regrouping.
4. $\qquad\qquad < \; \Sigma[(X_i - \bar{X})^2 - 2c(X_i - \bar{X}) + c^2]$	4. Binomial expansion of the form: $(a - b)^2 = a^2 - 2ab + b^2$
5. $\qquad\qquad < \; \Sigma(X_i - \bar{X})^2 - 2c\underbrace{\Sigma(X_i - \bar{X})} + \underbrace{\Sigma c^2}$	5. Distributing the summation sign to all terms within the brackets, the sum (or difference) of several variables is the sum (or difference) of their sums, and the sum of a constant times a variable is the constant times the sum of the variable.
6. $\qquad\qquad < \; \Sigma(X_i - \bar{X})^2 - \quad 0 \quad + Nc^2$	6. Since $\Sigma(X_i - \bar{X}) = 0$, the second term is 0, and the sum of N c^2's equals N times c^2.
7. $\Sigma(X_i - \bar{X})^2 \; < \; \Sigma(X_i - \bar{X})^2 + Nc^2$	7. The expression is true because Nc^2 will always be greater than zero.

The median

Another measure of central tendency is the median.

> The **median**, symbolized by M_d, is the point that divides the distribu-
> tion into two parts such that an equal number of scores fall above
> and below that point.

The way the median is computed varies, depending first on whether there is
an odd or an even number of scores in the distribution and second on
whether there is a duplication of score values near the median point. The
phrase "duplication of score values" means that more than one score of the
same value exists in the distribution. The distribution (3, 4, 5, 5, 7) has a
duplication of score values (the two 5s) while the distribution (2, 3, 5, 6, 8)
does not. Duplication of score values is important only when it occurs near
the point where the median is located. Otherwise, score duplication can be
ignored.

   1. No duplication near the median; odd number of scores. When there is
an odd number of scores and no duplication of scores near the median, the
median is the middle score. For example, in the distribution (3, 5, 6, 7, 10),
the point that divides the distribution into two equal parts is 6, since two
scores fall below and two scores fall above this value.

   2. No duplication near the median; even number of scores. By convention, when there is an even number of scores in a distribution and no
duplication near the median, the average of the middle two scores is taken as
the median. Suppose the distribution is (3, 5, 6, 7, 10, 14). The point that
divides the distribution in half lies between 6 and 7. The average of these
points, 6.5, is taken as the median. Another example illustrates the convention
that is followed when the scores near the median are not adjacent values. If
the distribution is (3, 3, 4, 8, 14, 16), the median is 6 because $(4 + 8) \div 2 = 6$.
Notice that the score duplication (the two 3s) is not considered because it is
not near the median point.

   3. Duplication of scores near the median. When more than one instance
of a score value falls near the median, the median is obtained by a procedure
called *interpolation*, which proceeds in basically the same way regardless of
whether the number of scores in the distribution is odd or even. To illustrate,
suppose the distribution is (3, 4, 5, 5, 5, 6, 6, 7). Since the median is the point
dividing the distribution in such a way that an equal number of scores fall
below and above it, the median lies somewhere between the second and third
instance of the score 5: Presumably, the scores 3, 4, 5, 5 are below the median
point and 5, 6, 6, 7 are above it. A single numerical value that expresses this
situation can be determined by observing that the scores 3, 4, and two of the
three scores of 5—that is, two-thirds of the 5s—must be below the median.
The score value of 5 occupies the score interval bordered by the real limits of

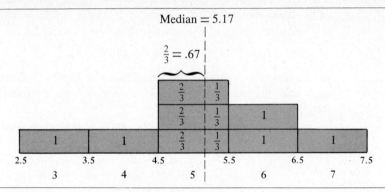

Fig. 3–1.   Computing the median when there is duplication of scores

5, namely 4.5 to 5.5. Its width is 1. If the three scores of 5 are assumed to be equally spaced within the score interval, then two of the three scores will occupy $\frac{2}{3}$ of the interval of size 1, which is $\frac{2}{3}(1) = .67$. Adding $\frac{2}{3} = .67$ to the lower real limit of this score interval, which is 4.5, gives

$$4.5 + .67 = 5.17$$

as the median. This same process is graphically illustrated in Figure 3–1, in which it can be seen that a total of four of the eight frequencies (i.e., scores) lie above and four lie below the median of 5.17. In short, the median divides the shaded area of Figure 3–1 into two equal portions.

The logic is the same if an odd number of frequencies are in the distribution. Suppose the distribution consists of (3, 4, 5, 5, 5, 6, 6, 7, 7), which case is presented in Figure 3–2. There are 9 scores, and thus the median must be the point such that $4\frac{1}{2}$ frequencies fall below and $4\frac{1}{2}$ fall above it. Counting from the low end upward, the scores 3 and 4 plus $2\frac{1}{2}$ of the 3 scores of 5 will be below the median. Therefore, $2\frac{1}{2}$ of 3 or

$$\frac{2\frac{1}{2}}{3} = \frac{\frac{5}{2}}{3} = \frac{5}{6} = .83$$

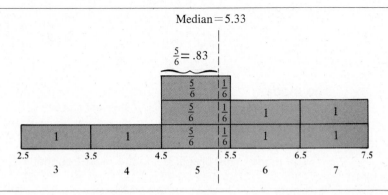

Fig. 3–2.   Computing the median when there is duplication of scores

of the score interval having a size of 1, that is, .83(1) = .83, must be added to
the lower real limit of the score interval of 5 (i.e., 4.5), which gives

$$4.5 + .83 = 5.33$$

as the median. Again, if the frequencies in Figure 3–2 are added, it can be
seen that the median point 5.33 indeed has $4\frac{1}{2}$ frequencies below and $4\frac{1}{2}$
frequencies above it.

The following formula represents these steps and can be used in
complicated situations:

$$M_d = L + \left(\frac{N/2 - n_b}{n_w} \right) i$$

in which

M_d = the median
L = the lower real limit of the score interval containing the
    median
N = the number of scores in the total distribution
n_b = the number of scores falling below the lower real limit of the
    score interval containing the median
n_w = the number of cases within the score interval containing the
    median
i = the size of the score interval ($i = 1$ if the data are in whole
    numbers[2])

In terms of the last example,

$$L = 4.5$$
$$N = 9$$
$$n_b = 2$$
$$n_w = 3$$
$$i = 1$$

$$M_d = L + \left[\frac{N/2 - n_b}{n_w} \right] i$$

$$= 4.5 + \left[\frac{\frac{9}{2} - 2}{3} \right] 1$$

$$= 4.5 + \frac{\frac{5}{2}}{3}$$

$$= 4.5 + .83$$

$$M_d = 5.33$$

The mode

A third measure of central tendency is the mode.

The **mode**, symbolized M_o, is the most frequently occurring score.

[2]If the scores are (5.1, 5.3, 5.3, 5.8), the score interval is .1.

If the distribution is (3, 4, 4, 5, 5, 5, 6, 8), the mode is 5. Sometimes a distribution will have two modes, such as the distribution (3, 4, 4, 4, 5, 6, 6, 7, 7, 7, 8). In this case, the modes are 4 and 7 and this distribution is called **bimodal**. A distribution that contains more than two modes is called **multimodal**.

Comparison of the mean, median, and mode

The essential difference between the mean and the median is that the mean reflects the value of each score in the distribution, whereas the median is based largely on where the midpoint of the distribution falls, without regard for the particular value of many of the scores. For example, consider the following illustration:

Scores	Mean	Median
1, 2, 3, 4, 5	3	3
1, 2, 3, 4, 50	12	3
1, 2, 3, 4, 100	22	3

Only the last number differs from one distribution to the other. The mean reflects these differences, but the median does not. This is because the median is the midpoint of the distribution such that an equal *number* of scores fall above and below it. The particular *value* of the extreme scores does not matter, since only the fact that those scores are above the midpoint is considered. In contrast, the mean takes into account the value of every score. Thus, changing any score value will likely change the value of the mean.

The mode reflects only the most frequently occurring score. It is not used much in the social sciences, except to describe a bimodal or highly skewed distribution.

Because the three different measures of central tendency are sensitive to different aspects of the group of scores, they are usually not the same value in a given distribution. If the distribution is symmetrical and unimodal (having one mode), then the mean, median, and mode are indeed identical. This condition is graphed in part A of Figure 3–3. If the distribution is symmetrical but has two modes as in part B, the mean and median are the same but the modes are different (the distribution is bimodal). In Chapter 2, a skewed distribution was defined as a distribution that is not symmetrical, having scores bunched at one end. Parts C and D of Figure 3–3 show two skewed distributions and the relative positions of the three measures of central tendency. The distribution in part C, which is fairly common, illustrates a condition in which most scores have moderate values but a few are very high. In this case, the mean, being sensitive to those extreme values, is somewhat

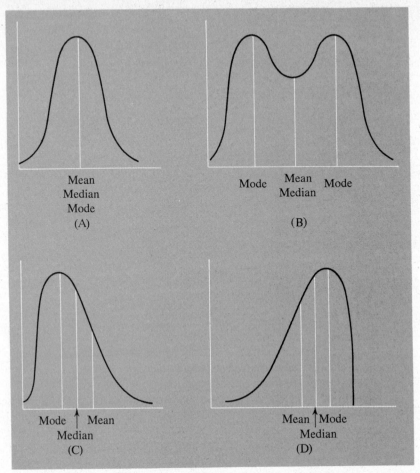

Fig. 3-3.   Mean, median, and mode in different distributions

larger than the median, which divides the area under the curve (i.e., the total number of cases) into two equal parts. A common instance of this situation occurs in the reporting of typical family income. In a given year the mean family income in the United States is usually higher than the median income because the relatively few really high incomes push the mean upward without influencing the median. For this reason, the median is closer to the incomes of most people and is thus often considered a "better" index of central tendency than mean income. Part D of Figure 3-3 illustrates the relative positions of the measures of central tendency when the skewness of the distribution is in the other direction.

Ordinarily, social science researchers select the mean as the measure of central tendency. There are several reasons for preferring the mean, but one of the major considerations is that the mean is required by so many other statistical procedures. However, sometimes the circumstances are such that the median would reflect the central tendency of the distribution more accurately than would the mean. When the distribution is very skewed, the mean may not yield a value that coincides with one's intuitive impression of the typical score. For example, the distribution (1, 2, 3, 4, 100) has a mean of 22 and a median of 3. In this case the median seems to characterize the central tendency more faithfully than does the mean, which is not close in value to any score in the distribution. Thus, in the case of a markedly skewed distribution, the median may be a better measure of central tendency than the mean would be. The mode is rarely used by itself to express central tendency. It is most often reported as a supplement to the mean or median, especially for distributions that are skewed or bimodal.

MEASURES OF VARIABILITY

In order to characterize a distribution more fully, a measure of variability is needed, in addition to an index of central tendency.

Variability refers to the extent to which the scores in a distribution differ from their central tendency.

For example, suppose two groups of scores, A and B, are defined to be

$$A = (5, 7, 9)$$

$$B = (3, 7, 11)$$

Although they both have the same mean of 7, set B has more variability because the scores differ from that mean more than do the scores in A. The purpose of this section is to discuss numerical measures of the variability of scores in a distribution.

The range

One measure of variability is the range. The range may be ascertained by taking the largest score minus the smallest score in the distribution.[3] In the distribution (3, 5, 6, 6, 8, 9), the range is 6.

---

[3]Technically, the range should probably be defined as the difference between the upper real limit of the largest score minus the lower real limit of the smallest score. Since the range is an approximate index of variability at best, it does not seem appropriate to insist upon this level of precision.

However, the range is limited in its ability to reflect the variability of a distribution. It is certainly true that as variability increases, the range of scores is also likely to increase. But the range is not sensitive to the variability of *all* the scores, only to the difference between the two most extreme values. For example,

$$C = (5, 10, 11, 12, 13, 18)$$
$$D = (5, 6, 7, 8, 16, 17, 18)$$

have the same range, but D has more variability than C. Therefore, although the range is easily computed, it is usually employed only as a crude approximation of variability.

The variance and standard deviation

Two numerical indices reflect the variability of scores in a distribution but do not suffer the limitations of the range. They are the variance and its square root, the standard deviation. Both indices increase in size as the variability between scores increases.

Though the variance and standard deviation are defined by the formulas given below, it is easier to calculate these quantities with the formulas presented in the next section of the chapter, which are mathematically equivalent to the definitions. This distinction between *definitional* and *computational* formulas will occur again in later chapters.

The variance, s^2 An index that reflects the degree of variability in a group of scores but which does not have the limitations of the range is the variance.

The **variance**, symbolized by s^2, is defined to be

$$s^2 = \frac{\Sigma\left(X_i - \overline{X}\right)^2}{N - 1}$$

In words, the variance is the sum of the squared deviations of the scores about their mean, $\Sigma(X_i - \overline{X})^2$, divided by the number of scores minus one ($N - 1$). Conceptually, it is very much like the average squared deviation of the scores about their mean, except that the squared deviations are divided by $N - 1$ rather than N.

In the example below, the variance for distribution $A = (5, 7, 9)$ is computed using the definitional formula. Notice that the mean is computed first ($\overline{X} = 7.0$). Then the mean is subtracted from each score ($X_i - \overline{X}$) and this difference is squared [$(X_i - \overline{X})^2$]. The sum of the squared deviations about the mean is divided by $N - 1$ to obtain the variance. Thus, the variance for distribution A is 4.00. For distribution $B = (3, 7, 11)$, which has greater variability, the variance is 16.

X_i	$\bar{X}$	$(X_i - \bar{X})$	$(X_i - \bar{X})^2$
5	7	-2	4
7	7	0	0
9	7	$+2$	4
$\Sigma X_i = 21$		0	$\Sigma(X_i - \bar{X})^2 = 8$
$N = 3$			
$\bar{X} = 7$			

$$s^2 = \frac{\Sigma(X_i - \bar{X})^2}{N - 1}$$

$$= \tfrac{8}{2}$$

$$s^2 = 4.00$$

The standard deviation, s   The variance measures variability in squared units. If a researcher recorded how long it took animals to find their way to a goal box at the end of a maze, the mean time would be in seconds but the variance would be in "squared seconds." This results from the fact that the formula for the mean uses the scores as they are (ΣX_i) but the formula for the variance squares the deviations $[\Sigma(X_i - \bar{X})^2]$. However, it is also useful to have a measure of variability in terms of the original units of measurement, not squared units.

The **standard deviation** (symbolized by s) is defined to be the positive square root of the variance:

$$s = \sqrt{s^2}$$

or

$$s = \sqrt{\frac{\Sigma(X_i - \bar{X})^2}{N - 1}}$$

Since the variance is in squared units, taking its square root accomplishes a return to the original units of measurement.[4]

Computational formulas for s^2 and s   Formulas like those given above are called *definitional* formulas because they define and reflect the logic behind the concepts that they express. However, definitional formulas are frequently inconvenient to use in making calculations, especially when one is dealing with large amounts of data. An expression for a statistic that is

[4]The notation SD is used in some other texts to symbolize the standard deviation.

OPTIONAL TABLE

3–3 Proof that the Definitional and Computational Formulas for the Variance Are Equivalent

Operation	Explanation
To prove $$\frac{\Sigma(X_i - \bar{X})^2}{N - 1} = \frac{N\Sigma X_i^2 - (\Sigma X_i)^2}{N(N - 1)}$$	
1. $s^2 = \dfrac{\Sigma(X_i - \bar{X})^2}{N - 1}$	1. Definition.
2. $\quad = \dfrac{\Sigma(X_i^2 - 2X_i\bar{X} + \bar{X}^2)}{N - 1}$	2. Binomial expansion of the form: $(a - b)^2 = a^2 - 2ab + b^2$
3. $\quad = \dfrac{\Sigma X_i^2 - \Sigma 2X_i\bar{X} + \Sigma\bar{X}^2}{N - 1}$	3. The sum of several terms is the sum of the separate terms: $\Sigma(X + Y + Z) = \Sigma X + \Sigma Y + \Sigma Z$
4. $\quad = \dfrac{\Sigma X_i^2 - 2\bar{X}\Sigma X_i + \Sigma\bar{X}^2}{N - 1}$	4. The sum of a constant times a variable equals the constant times the sum of the variable ($\bar{X}$ is a constant): $\Sigma kX = k\Sigma X$

5. $$= \frac{\sum X_i^2 - 2\bar{X}\sum X_i + N\bar{X}^2}{N-1}$$

6. $$= \frac{\sum X_i^2 - 2\left(\frac{\sum X_i}{N}\right)\sum X_i + N\left(\frac{\sum X_i}{N}\right)\left(\frac{\sum X_i}{N}\right)}{N-1}$$

7. $$= \frac{\sum X_i^2 - 2\left(\frac{\sum X_i}{N}\right)\sum X_i + (\sum X_i)\left(\frac{\sum X_i}{N}\right)}{N-1}$$

8. $$= \frac{\sum X_i^2 - \left(\frac{\sum X_i}{N}\right)\sum X_i}{N-1}$$

9. $$= \frac{N\sum X_i^2 - N\left(\frac{\sum X_i}{N}\right)\sum X_i}{N(N-1)}$$

10. $$s^2 = \frac{N\sum X_i^2 - (\sum X_i)^2}{N(N-1)}$$

5. The sum of a constant taken N times is N times the constant: $\sum_{i=1}^{N} k = Nk$

6. Substitution: $\bar{X} = \frac{\sum X_i}{N}$

7. Cancellation in the third term.

8. Subtraction involving the last two terms of the numerator.

9. Multiplication of numerator and denominator by N.

10. Cancellation.

mathematically equivalent to the definitional formula but is more convenient for calculating is called a *computational* formula.

The computational formula for the variance is

$$s^2 = \frac{N\Sigma X_i^2 - (\Sigma X_i)^2}{N(N-1)}$$

The computational formula for the standard deviation is

$$s = \sqrt{\frac{N\Sigma X_i^2 - (\Sigma X_i)^2}{N(N-1)}} \quad \text{or} \quad s = \sqrt{s^2}$$

It is important to realize that these computational formulas yield results that are equivalent (within rounding error) to those calculated by the definitional formulas. Table 3–3 (optional) demonstrates the algebraic equivalence of the two formulas for the variance.

The computational formula has the advantages of requiring only one division, not requiring that the mean be calculated first, and being easy to compute on a standard hand calculator. Only three quantities are needed: ΣX_i^2, ΣX_i, and N. Consider the variance and standard deviation of the following distribution:

X_i	X_i^2
3	9
4	16
7	49
8	64
8	64
9	81
10	100
$\Sigma X_i = 49$	$\Sigma X_i^2 = 383$
$N = 7$	

Variance	**Standard Deviation**
$s^2 = \dfrac{N\Sigma X_i^2 - (\Sigma X_i)^2}{N(N-1)}$	$s = \sqrt{s^2}$
$= \dfrac{7(383) - (49)^2}{7(7-1)} = \dfrac{2681 - 2401}{42}$	$= \sqrt{6.67}$
$= \dfrac{280}{42}$	$s = 2.58$
$s^2 = 6.67$	

It is very important to distinguish between two quantities used in the formula: ΣX_i^2 and $(\Sigma X_i)^2$. Recall that the first, ΣX_i^2, means the sum of all the squared scores—*first* square each score, *then* add. The second, $(\Sigma X_i)^2$, means the square of the sum of the scores—*first* add all the scores, *then* square this sum. Confusion between these two operations is often the source of computational error.

Properties of s^2 and s as measures of variability   The variance is difficult to explain because it cannot be diagramed or pointed at. Rather, the variance is an abstract numerical index that increases with the amount of variability in the group of scores.

But despite its abstractness, the variance does have a number of properties that make it (and its square root, the standard deviation) an appropriate measure of variability.

First, since the mean is the central value of the distribution, it seems natural to base a measure of variability upon the extent to which the scores deviate from their central tendency—that is, on $(X_i - \overline{X})^2$. In addition, recall from the discussion of the mean that the sum of the squared deviations about the mean is less than about any other value. This fact adds to the logic of selecting squared deviations about the mean (as opposed to some other value) as an index of variability.

Second, squared numbers are always positive because squaring a negative number results in a positive value. Therefore s^2 and s are always positive values. If the deviations were not squared, the negative deviations would cancel out the positive and their sum would be zero because

$$\Sigma\left(X_i - \overline{X}\right) = 0$$

Third, large deviations, when squared, contribute disproportionately to the total. A deviation of 4 units becomes 16 when squared, but a deviation of twice that size, that is of 8 units, contributes 64 to the total sum of squared deviations. Thus, the final index is especially sensitive to extreme departures from the mean because large deviations, when squared, become disproportionately large and make the variance greater.

Fourth, recall that variability is defined as the extent to which the scores deviate from one another. Thus, as the definition suggests, one way to measure variability would be to sum the squared deviations between each score and every other score in the distribution and divide this total by the number of such pairs $[N(N - 1)/2]$. This procedure, it is important to note, yields a value that is directly proportional to the one given by the formula for the variance. Therefore, the variance is indeed closely associated with the concept of variability which it measures.

Fifth, as the variability of the scores increases, the statistical variance also increases. This can be seen in the few examples listed below:

Scores	s^2
10, 10, 10	0
8, 10, 12	4
6, 10, 14	16
4, 10, 16	36
2, 10, 18	64

As the scores show more and more variability, the value of s^2 increases, reflecting the extent to which the scores deviate from one another. Similar arguments can be made for the standard deviation.

Sixth, if there is no variability among the scores, that is, if all the scores in the distribution are identical to one another, then all quantities $(X_i - \bar{X})^2$ and their sum will be zero because X_i and $\bar{X}$ will always be identical. Therefore, when there is no variability among the scores of a distribution (if all scores have the same value), $s^2 = 0$.

Seventh, under certain conditions the variance can be partitioned and its portions attributed to different sources. This capability of being partitioned permits statisticians to ask the following type of question: A group of scores possesses a certain amount of variability. What portion of that variability can be attributed to cause A as opposed to cause B? This aspect of the variance will be taken up in more detail later, in Chapters 11 and 12.

ESTIMATION

A major task of inferential statistics, to be presented more fully in Part 2 of this book, is to estimate values in a larger group on the basis of calculations made on a small number of subjects. For example, a mean of 83 on an English exam for one classroom might be used to estimate the average performance for all classrooms in a school or even in the nation. The variance might be used in the same way.

It will be helpful now to distinguish between the values that are actually calculated and those that are estimated.

Population and sample

Frequently, a scientist performs an experiment on a relatively small group of subjects. At the conclusion of the research, however, the results are generalized to a much larger group of subjects, of which the sample is a small part.

For example, an experiment might be performed using a group of 40 students at State University, but the results are discussed in terms of all college students. The small group actually used as subjects in the experiment is known as a sample, whereas the larger group of subjects to which the researcher wishes to generalize the results is called the population.

> A **population** is a collection of subjects, events, or scores that have some common characteristic.[5]

> A **sample** is a subgroup of a population.

It is important to note that sample and population are relative terms. All students enrolled at State University might be the population from which a sample of 50 students is drawn for a given experiment, or the group might function instead as a sample of the larger population, all college students.

One obvious reason for using samples rather than populations in research is that populations are usually too large to be studied efficiently. In addition, research results that must be limited to the specific subjects studied are less interesting and less useful than those applicable to a much larger group. Therefore, the scientist designs the experiment so that generalizations from the sample to the population may be made. One aspect of this process is to **estimate** characteristics in the population by computing various measures on the sample.

Characteristics of a population are called **parameters**, while those of a sample are termed **statistics**. Therefore, the research scientist attempts to estimate parameters with statistics.

Characteristics of a good estimator

Suppose the developer of a new test of scholastic competence wanted to determine the mean and standard deviation of test scores in the population of all American high school seniors. Assume further that a random sample of 100 seniors from around the country has taken the test. The mean of that sample might be 83, and the researcher would use this to estimate the unknown population mean. Of course, this sample mean probably would not equal precisely the population mean, which in fact might be 90. How can the researcher decide whether the sample mean is a good estimator of the population mean?

---

[5]Very frequently, although not always, a population is considered to be composed of an infinite number of cases. While such a conception has certain advantages, particularly for theoretical mathematical statistics, the definition presented here will be used for pedagogical reasons.

3–4 Average of Hypothetical Sample Means for Different Numbers of Samples

Number of Samples	Average of Sample Means	Difference between Average Mean and Population Mean (90)
1	83	−7
5	96	+6
10	93	+3
50	86	−4
100	88	−2
1000	91	+1
∞	90	0

There are several criteria that a good estimator of a population parameter should meet.[6] The one that will be of major concern to us here is the quality of unbiasedness.

An **unbiased estimator** of a population parameter is one whose average over all possible random samples of a given size equals the value of the parameter.

The random sample of 100 cases that the test constructor had available is just one of a much larger number of possible samples of size 100 that could be drawn from the population of all American high school seniors. It is possible to think about selecting 5, 10, 50, 100, or even 1000 random samples of size 100, calculating the mean for each sample, and then averaging these means. Table 3–4 shows some hypothetical results of this process. The middle column reports the average of the means for 1, 5, 10, 50, 100, 1000, and an infinite number of samples (if that were possible), and the third column indicates the difference between that average mean and the (assumed) population mean of 90.

The test constructor's task is to estimate a value in the population—one that will never be calculated and thus never known exactly—with a value computed on a small sample of subjects. Special mathematical procedures exist, called expectation theory, that can tell a statistician whether an estimate is accurate or not. A hypothetical numerical example is used here in order to clarify the concepts of estimation and unbiased estimators.

[6]For more details on the characteristics of good estimators, see: W. L. Hays, *Statistics for the Social Sciences*, 2nd ed. (New York: Holt, Rinehart & Winston, 1973).

If a sample mean is an unbiased estimator of a population mean, then the average mean should tend to converge on (move toward) the population mean as the number of samples increases. Table 3–4 illustrates this tendency. The sample mean of 83 is indeed an unbiased estimator of the population mean (although that is proved by theoretical statistical procedures from expectation theory, not by such illustrations).

The fact that this course is usually called statistics and not parameters indicates that the focus is on computing measures on samples and not on populations. Consequently, for some statistics presented in this volume (such as the variance; see below), the relationship between a sample value and the corresponding parameter will be discussed, whereas in other cases only the sample statistic will be considered.

Sample and population means, variances, and standard deviations

The **sample mean** is symbolized by $\overline{X}$ and is computed with the formula

$$\overline{X} = \frac{\Sigma X}{N}$$

The **population mean** is symbolized by the Greek letter μ (read "mew"). $\overline{X}$ is frequently used as an estimator of the population mean, μ. (To facilitate distinguishing between statistics and parameters, statistics are usually abbreviated with Roman letters, while parameters have Greek designations.)

The **sample variance** is symbolized by s^2 and is calculated with the formula

$$s^2 = \frac{\Sigma\left(X_i - \overline{X}\right)^2}{N - 1}$$

The **population variance** is symbolized by the Greek letter σ^2 (read "sigma squared"). Although s^2 is known to be an unbiased estimate of σ^2, this is true only when the denominator of the formula for s^2 contains $N - 1$ rather than just N. Technically, the sample variance is defined by the formula $\Sigma(X_i - \overline{X})^2/N$, but this quantity is not an unbiased estimator of the population variance. If N is used in the denominator, the value of the sample variance does not converge on the population variance as the number of samples increases. It tends to be just a little smaller than the population value, on the average. Expectation theory has determined that dividing by $N - 1$ rather than N makes s^2 an unbiased estimator of σ^2. Since, in practice, one almost always wishes to estimate the population variance, in this book the sample variance is defined with a denominator of $N - 1$.

The **sample standard deviation** is symbolized by s and is defined to be

$$s = \sqrt{s^2}$$

The **population standard deviation** is represented by the Greek letter σ and is usually estimated by s. Although, technically, s is not an unbiased estimator of σ, the bias is so slight, especially with large samples, that we shall not worry about it further.

FORMULAS

1. Mean

$$\bar{X} = \frac{\Sigma X_i}{N}$$

The population mean is symbolized by μ.

2. Median
  a. **No duplication near the median, odd number of scores:**
     M_d is the middle score.
  b. **No duplication near the median, even number of scores:**
     M_d is the average of the two middle scores.
  c. **Duplication of scores near the median:**

$$M_d = L + \left[\frac{N/2 - n_b}{n_w} \right] i$$

where L = lower real limit of the score interval containing the median
      N = number of scores in the distribution
      n_b = number of scores falling below the lower real limit of the interval containing the median
      n_w = number of cases with the score interval containing the median
      i = the size of the score interval ($i = 1$ if the data are in whole numbers)

3. Mode
The mode, M_o, is the most frequently occurring score.

4. Range
The range is estimated by taking the largest minus the smallest score.

5. Variance

$$s^2 = \frac{\Sigma(X_i - \bar{X})^2}{N - 1} \qquad \text{(definitional)}$$

$$s^2 = \frac{N\Sigma X_i^2 - (\Sigma X_i)^2}{N(N - 1)} \qquad \text{(computational)}$$

The population variance is symbolized by σ^2.

6. Standard deviation

$$s = \sqrt{s^2} = \sqrt{\frac{\Sigma(X_i - \bar{X})^2}{N - 1}} \quad \text{(definitional)}$$

$$s = \sqrt{\frac{N\Sigma X_i^2 - (\Sigma X_i)^2}{N(N - 1)}} \quad \text{(computational)}$$

The population standard deviation is symbolized by σ.

EXERCISES

1. Compute the mean, median, and mode for each of the distributions below.

A	B	C	D
1	2	1	2
2	2	3	3
3	2	3	4
5	5	3	4
7	6	5	4
7	7	7	5
10	7	8	7
	8	8	8
	10	8	8
	11	9	
		11	

2. Show that the sum of the deviations about the mean of distribution C above is zero. Compute the sum of the squared deviations about the mean and median of distribution C. Which is less, the sum of the squared deviations about the mean or about the median?

3. Indicate for each of the following distributions which measure of central tendency would be preferred and explain why.
   a. family incomes in the United States
   b. heights of seniors in a public high school
   c. IQ scores in the third grade of a public school

4. Find the median for the following distributions.
   a. 2, 5, 6, 8, 9
   b. 0, 2, 3, 6, 8, 10
   c. 1, 2, 3, 3, 4, 5
   d. 5, 6, 6, 6, 6, 20, 100
   e. 0, 2, 3, 3, 7
   f. 3.1, 3.2, 3.3, 3.4, 3.6 (note that the score interval is .1)

5. Draw the curves for distributions in which
   a. the mean, median, and mode are identical.
   b. the mean and median are identical but the mode is different.
   c. the mean is greater than the median.
   d. the median is greater than the mean.

6. Define *variability* and compose three distributions which differ in their amount of variability.

7. Discuss the limitations of the range as a measure of variability and present some numerical examples to illustrate your point.

8. Compute with both the definitional and computational formulas the variance and standard deviation for each of the following distributions. Also compare the means of these distributions.
   a. 6, 7, 7, 8
   b. 4, 5, 9, 10
   c. 0, 1, 3, 4, 7, 7, 8, 8, 9, 9, 9, 10, 10, 10, 10

9. Discuss the properties, characteristics, and advantages of s^2 as a measure of variability. Does s have any potential advantages over s^2?

10. Why does the formula for the variance have $N - 1$ in the denominator and not N?

11. Define *unbiased estimator*.

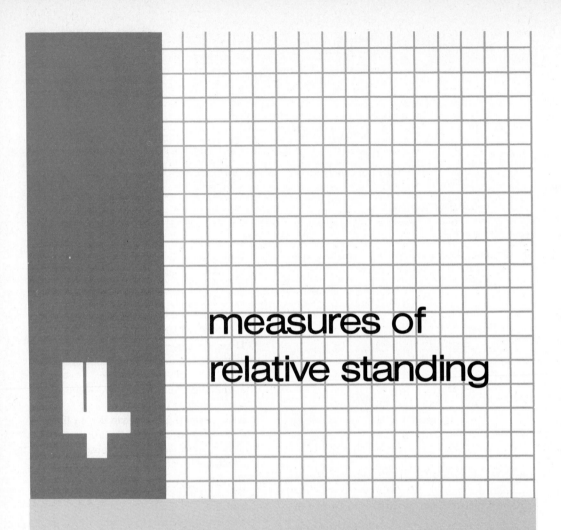

4

measures of relative standing

ALTHOUGH THE STATISTICS discussed in the previous chapter help describe an entire distribution, they do not provide direct assistance in interpreting individual scores. Suppose a teacher gives a final exam and wants to assign grades on a curve—the top 10% get A's, the next 20% get B's, and so on. (The teacher would need to determine what scores are in the top 10%, the next 20%, and so on.) Similarly, if you were a student in the class, knowing that you had a score of 88 on the exam would not provide much information on how well you did relative to other students. Knowing the mean, range, and standard deviation for the class would help somewhat, but it would also be informative to know what proportion of the students scored below you.

PERCENTILES

Percentile points and ranks provide such information about the relative standing of scores in a distribution.

> The **Pth percentile point** is the score value below which the proportion P of the cases in the distribution falls. That percentile point is said to have a **percentile rank** of P.

Thus, if a score of 88 was at the 92nd percentile rank, .92 of the people in the group scored less than the percentile point of 88; in symbols, this information would be written $P_{.92} = 88$. Notice that a percentile rank is the cumulative relative frequency of its corresponding score value. Like cumulative relative frequency, percentile rank is expressed as a proportion, which can be changed to a percentage by multiplying by 100.

The **median** is the percentile point that corresponds to a percentile rank of .50. In addition to the median, other percentile points have been given specific names. For example, the **upper quartile point** ($P_{.75}$) is the score value that separates the top .25 from the remainder of the distribution, while the **lower quartile point** ($P_{.25}$) is the score value that separates the bottom .25 from the remainder of the distribution. The **interquartile range** is defined by the score values that separate the middle .50 of the distribution from the remainder.

Computation of percentile points

Determining percentile points is similar to determining the median, which is the point corresponding to the rank of .50. Suppose the exam scores for 40 students are those presented in Table 4–1, and the teacher wants to determine the percentile point corresponding to the rank of .30. To determine the 30th

4-1 Distribution of Scores on a Statistics Examination
for a Class of 40 Students

Student No.	Score	Student No.	Score	Student No.	Score	Student No.	Score
40	97	30	80	20	72	10	56
39	93	29	79	19	72	9	55
38	92	28	78	18	72	8	55
37	91	27	78	17	71	7	55
36	88	26	78	16	70	6	55
35	85	25	78	15	65	5	51
34	85	24	76	14	61	4	49
33	84	23	75	13	60	3	49
32	83	22	75	12	59	2	48
31	82	21	74	11	58	1	46

percentile point, recall the definition of percentile point: It is that score value below which P proportion of the cases in the distribution fall. In this problem, P is .30 and the number of cases in the distribution is $N = 40$. Therefore,

$$P(N) = .30(40) = 12$$

cases must fall below the required score value. Looking at the data in Table 4-1, one can see that the twelfth student from the bottom of the distribution had a score of 59. Therefore, 12 students scored below the upper real limit of the score of 59, and consequently the percentile point corresponding to $P_{.30}$ is 59.5.

This logic is expressed in a formula similar to that introduced for the median, which will also determine percentile points in more complicated situations involving duplication of scores near the required score value.

The **percentile point** corresponding to the Pth rank is given by

$$\text{Percentile Point} = L + \left[\frac{P(N) - n_b}{n_w} \right] i$$

in which

P = the percentile rank of the required point (i.e., the proportion of cases falling below the desired point—P—ranges between 0 and 1.00)

L = the lower real limit of the score interval containing the required percentile point

N = the number of scores in the distribution

n_b = the number of cases falling below L

n_w = the number of cases falling within the score interval containing the required percentile point

i = the size of the score interval (i = 1 if the data are in whole numbers)

To illustrate, here is how the formula is used to determine the percentile point corresponding to the 20th percentile rank ($P_{.20}$) for the distribution in Table 4–1. In this case, P = .20 and N = 40, so $P(N)$ = .20(40) = 8 cases must fall below the desired score value. The eighth lowest subject scored 55, but there were four scores of 55. The lower real limit of the score of 55 is L = 54.5; there were five people who scored below 55 (n_b = 5); and four people scored 55 (n_w = 4). Thus, the 20th percentile point is given by

$$L + \left[\frac{P(N) - n_b}{n_w} \right] i$$

$$54.5 + \left[\frac{.20(40) - 5}{4} \right] 1 = 55.25$$

Computation of percentile ranks

The two illustrations just given show how one computes the score (i.e., percentile point) corresponding to a given percentile rank. However, the question can be reversed. What is the percentile rank corresponding to a given score?

A student who scored 83 might want to know what proportion of students scored lower. This information is given by the percentile rank of a score of 83. Looking at Table 4–1, one can see that the score of 83 was thirty-second in the distribution. This means that 32 people scored lower than 83.5, which is the upper real limit of 83. Similarly, 31 people scored below 82.5, which is the lower real limit of 83. Therefore, 31.5 of 40 people or 31.5/40 = .7875 of the cases recorded a score below exactly 83.0. Therefore, the score of 83 has a percentile rank of .7875; more conventionally, that score is said to be at the 79th percentile. Thus, 83 corresponds to $P_{.79}$, and 79% of the scores were below 83.

Again there is a formula that will produce this result as well as percentile ranks when duplication of scores is a problem. Note that this formula includes the same elements as the one for finding percentile points when percentile rank is known.

The **percentile rank** corresponding to a given score is determined by

$$P = \frac{n_w(X - L) + in_b}{Ni}$$

in which

P = the desired percentile rank (i.e., the proportion of cases falling below X—P ranges between 0 and 1.00)
X = the score value corresponding to the desired percentile rank
L = the lower real limit of X
n_w = the number of cases falling within the score interval containing X
n_b = the number of cases having a score lower than X
N = the number of scores in the distribution
i = the size of the score interval ($i = 1$ if the data are in whole numbers)

To determine the percentile rank for a score of 83, let $X = 83$, with the lower real limit $L = 82.5$. There is only one score of 83 in the distribution ($n_w = 1$); 31 scores are below X ($n_b = 31$); $i = 1$; and there are $N = 40$ scores in the entire distribution. Therefore,

$$P = \frac{n_w(X - L) + in_b}{Ni}$$

$$= \frac{1(83 - 82.5) + 1(31)}{40(1)} = \frac{31.5}{40}$$

$$P = .7875$$

Therefore, $X = 83$ is at the 78.75th or 79th percentile rank ($P_{.79}$).

Suppose one had to determine the percentile rank corresponding to a score of 72. In this case, $X = 72$ with a lower real limit of $L = 71.5$; three people scored exactly 72 ($n_w = 3$); and 17 people scored lower than 72 ($n_b = 17$). Consequently,

$$P = \frac{n_w(X - L) + in_b}{Ni}$$

$$= \frac{3(72 - 71.5) + 1(17)}{40(1)} = \frac{18.5}{40}$$

$$P = .4625$$

Thus, $X = 72$ is at the 46.25th or 46th percentile rank ($P_{.46}$).

CHANGING THE PROPERTIES OF SCALES

Percentiles allow us to compare the relative standing of an individual in two different distributions. For example, a student who scores at the 80th percentile on both a history and a statistics exam has performed similarly on the two tests, at least with respect to the proportion of people scoring higher or lower. In addition, percentiles help us make other comparisons between different distributions. The procedures involved require some knowledge of the effects of changing the properties of scales.

Celsius and Fahrenheit are two different scales of temperature. They differ in that they do not have the same origin (that is, 0° Celsius and 0° Fahrenheit represent different temperatures) or the same size units. The formula for converting degrees Celsius to degrees Fahrenheit is

$$°F = °C(1.8) + 32°$$

The 1.8 is the factor used to convert any given number of Celsius degrees to the corresponding number of Fahrenheit degrees. Consider the freezing and boiling points of water on the two scales. For the Fahrenheit scale, these points are 32° and 212°, respectively, whereas for the Celsius scale they are 0° and 100°, respectively. Thus, between the freezing and boiling points of water on the Fahrenheit scale there are 180 Fahrenheit degrees. Between the freezing and boiling points of water on the Celsius scale there are only 100 degrees. Therefore, the Celsius degree is larger, and it takes 1.8 Fahrenheit degrees to equal 1 Celsius degree. This is the reason the 1.8 appears in the conversion formula; it changes the Celsius units into Fahrenheit units.

However, the two scales are not yet equivalent. Even if the Celsius degrees are converted to Fahrenheit degrees, the freezing point is still at 0° and the boiling point is at (1.8)(100) = 180°: Both values are 32° short of their corresponding Fahrenheit values. To change the Celsius origin to match the Fahrenheit origin, 32° must be added.

> To change the size of the unit of measurement, multiply or divide the old values by the proper constant (conversion factor). To change the origin of a scale, add or subtract the appropriate number of units.

Most transformations common to everyday life are ones in which only the size of the unit is converted. For example:

$$\text{feet} = \frac{\text{inches}}{12} \longleftarrow \text{conversion factor}$$

$$\text{kilometers} = \frac{\text{miles}}{.6} \longleftarrow \text{conversion factor}$$

However, in research in the social sciences, transformations from one scale to another frequently involve changing both the unit size and the origin of the scale.

Effects of scale changes on the mean

What happens to the value of the mean of a distribution when the unit size and origin of the measurement scale are changed?

> The origin of a scale is changed by adding (or subtracting) a constant to every score. If a constant is added to (or subtracted from) every score in the distribution, the mean of the new distribution is the mean of the old plus (or minus) that constant. If $X' = X + c$, then

$$\overline{X}' = \overline{X} + c$$

Therefore, if the mean of a distribution is 15 and 3 is added to every score, the new mean is $15 + 3 = 18$. If 3 is subtracted from every score, the new mean is $15 - 3 = 12$. Table 4–2 (optional) presents a formal proof of this principle.

<table>
<tr><th colspan="2">4–2 Change in the Mean with a Change in Origin</th></tr>
<tr><th>Operation</th><th>Explanation</th></tr>
<tr><td>To prove
$\overline{X}' = \overline{X} + c$</td><td></td></tr>
<tr><td>1. $\overline{X}' = \dfrac{\Sigma(X + c)}{N}$</td><td>1. Definition of the mean for a distribution in which c has been added to each X.</td></tr>
<tr><td>2. $\phantom{\overline{X}'} = \dfrac{\Sigma X + \Sigma c}{N}$</td><td>2. The sum of several terms is the sum of the separate terms.</td></tr>
<tr><td>3. $\phantom{\overline{X}'} = \dfrac{\Sigma X + Nc}{N}$</td><td>3. The sum (from 1 to N) of a constant is N times that constant.</td></tr>
<tr><td>4. $\phantom{\overline{X}'} = \dfrac{\Sigma X}{N} + \dfrac{Nc}{N}$</td><td>4. Simplification.</td></tr>
<tr><td>5. $\overline{X}' = \overline{X} + c$</td><td>5. Substitution: $\dfrac{\Sigma X}{N} = \overline{X}$ and cancellation of N's.</td></tr>
</table>

OPTIONAL TABLE

The unit of a scale is changed by multiplying or dividing every score by a constant. If every score in a distribution is multiplied (or divided) by a constant, the mean of the new distribution is the mean of the old distribution multiplied (or divided) by that constant. In symbols, if $X' = cX$, then

$$\overline{X'} = c\overline{X}$$

Therefore, if the mean of a distribution is 15 and every score is multiplied by 3, the new mean is three times the old mean, or $15 \times 3 = 45$. Since dividing each score by c is the same as multiplying each score by $1/c$, if every score is divided by 3, the mean is also divided by 3: $15 \div 3 = 5$. Table 4–3 (optional) gives a general proof of this fact. In summary, if the scale of measurement is altered, the mean will change in the same manner and to the same extent as any other value on the scale.

	4–3 Change in the Mean with a Change in Unit	
Operation	**Explanation**	
To prove $\overline{X'} = c\overline{X}$		
1. $\overline{X'} = \dfrac{\Sigma cX}{N}$	1. Definition.	
2. $\quad= \dfrac{c\Sigma X}{N}$	2. The sum of a constant times a variable is the constant times the sum of that variable.	
3. $\overline{X'} = c\overline{X}$	3. Substitution: $\dfrac{\Sigma X}{N} = \overline{X}$	

OPTIONAL TABLE

Effects of scale changes on the variance and standard deviation

Transformations of scale do not have the same effect on the variance and standard deviation of a distribution as they do on the mean.

If the origin of a scale is changed by adding (or subtracting) a constant from every score in the distribution, the variance and standard deviation are not changed. In symbols, if $X' = X + c$ and s_x^2 is the variance of the original distribution, then

$$s_{x'}^2 = s_x^2$$

and

$$s_{x'} = s_x$$

If the variance of a distribution is 16 and the standard deviation is 4, then when 5 is added to every score these statistics will remain unchanged and $s_{x'}^2 = 16$ and $s_{x'} = 4$. This makes intuitive sense if one considers that s^2 and s are functions of the distance between the various score values in the distribution, not of the distance between those score values and the origin. For example, consider the distance between scores in the following distribution before and after a change of origin (5 is added to each score):

X			$X' = X + 5$		
3	7	11	8	12	16

$(X - \overline{X})$: -4 4 $(X' - \overline{X'})$: -4 4

Since s^2 and s are based upon the distances between the scores and the mean $(X - \overline{X})$ and since these distances do not change when the origin is changed, adding a constant to each score does not alter s^2 and s. The formal proof that such a change in origin leaves s^2 and s unaffected is presented in (optional) Table 4–4.

OPTIONAL TABLE

4–4 Proof that No Change Occurs in the Variance (and Standard Deviation) with a Change in Origin

Operation	Explanation
To prove $s_{x'}^2 = s_x^2$	
1. $s_{x'}^2 = \dfrac{\Sigma\left[(X + c) - (\overline{X} + c)\right]^2}{N - 1}$	1. Definition; if a constant is added to every score, the mean will also be incremented by that constant.
2. $\quad = \dfrac{\Sigma\left[X + c - \overline{X} - c\right]^2}{N - 1}$	2. Removing parentheses.
3. $s_{x'}^2 = \dfrac{\Sigma(X - \overline{X})^2}{N - 1} = s_x^2$	3. Subtraction.

Although adding or subtracting a constant does not affect s^2 and s, multiplying or dividing by a constant does change these statistics.

> If the unit of measurement is changed by multiplying (or dividing) every score in the distribution by a positive constant, the new variance will equal the old variance multiplied (or divided) by the square of that constant, and the new standard deviation will equal the old standard deviation multiplied (or divided) by that constant. In symbols, if $X' = cX$, and s_x^2 is the variance of the original distribution, then

$$s_{x'}^2 = c^2 s_x^2$$

and

$$s_{x'} = c s_x$$

Thus, if the variance of a distribution is 16 and the standard deviation is 4 and each score in the distribution is multiplied by 2, the variance of the new distribution will be $(2^2)(16) = 64$ and the new standard deviation will be $(2)(4) = 8$. The reason that multiplying or dividing by a constant changes the variance and standard deviation can be seen more clearly by examining what happens to the deviations and squared deviations used in the formula for the variance when the scale undergoes a change in unit (as when every score is multiplied by 2):

	X			$X' = 2X$	
3	7	11	6	14	22
$(X - \bar{X})$:	-4	4	$(X' - \bar{X'})$:	-8	8
$(X - \bar{X})^2$:	16	16	$(X' - \bar{X'})^2$:	64	64

Two important points are made by this example. When the unit of measurement is changed by multiplying every score by 2, (1) the distances between points are also multiplied by 2 (that is, by c), and (2) the size of the squared deviations increases by $2^2 = 4$ (that is, by c^2). In other words, because the variance uses squared deviations in its formula, the size of the variance is multiplied by c^2 when each score is multiplied by c. On the other hand, the formula for the standard deviation takes the positive square root of these squared deviations, and thus it is changed by a factor of c (assuming c is a positive constant).[1] The formal proof of this proposition can be found in Table 4–5 (optional). A numerical example illustrating all of these principles is given in Table 4–6.

---

[1] If c is negative, $s_{x'} = |c| s_x$.

4–5 Proof that the Variance Changes by c^2 and the Standard Deviation by c with a Change in Unit

Operation	Explanation
Variance **To prove** $s_{x'}^{2} = c^2 s_x^{2}$	
1. $s_{x'}^{2} = \dfrac{\Sigma(cX - c\overline{X})^2}{N - 1}$	1. Definition; if each score is multiplied by a constant the mean is multiplied by that same constant.
2. $= \dfrac{\Sigma\left[c(X - \overline{X})\right]^2}{N - 1}$	2. Factoring in the manner of $ca - cb = c(a - b)$
3. $= \dfrac{\Sigma\left[c^2(X - \overline{X})^2\right]}{N - 1}$	3. Simplification in the manner of $(ab)^2 = a^2 b^2$
4. $= \dfrac{c^2\Sigma(X - \overline{X})^2}{N - 1}$	4. The sum of a constant times a variable is that constant times the sum of the variable.
5. $s_{x'}^{2} = c^2 s_x^{2}$	5. Substitution: $\dfrac{\Sigma(X - \overline{X})^2}{N - 1} = s_x^{2}$
Standard Deviation **To prove** $s_{x'} = c s_x$	
1. $s_{x'} = \sqrt{s_{x'}^{2}}$	1. Definition.
2. $= \sqrt{c^2 s_x^{2}}$	2. Substitution: $s_{x'}^{2} = c^2 s_x^{2}$
3. $s_{x'} = c s_x$	3. Simplification in the manner of $\sqrt{a^2 b^2} = ab$

OPTIONAL TABLE

4–6 Effects of Changes in Origin and Unit on the Mean, Variance, and Standard Deviation

X		Change the Origin $Y = X + 5$		Change the Unit $W = 3X$	
X_i	X_i^2	Y_i	Y_i^2	W_i	W_i^2
3	9	8	64	9	81
5	25	10	100	15	225
6	36	11	121	18	324
10	100	15	225	30	900
$\Sigma X_i = 24$ $\Sigma X_i^2 = 170$		$\Sigma Y_i = 44$ $\Sigma Y_i^2 = 510$		$\Sigma W_i = 72$ $\Sigma W_i^2 = 1530$	

$N = 4$

$$\bar{X} = \frac{\Sigma X_i}{N} = \frac{24}{4} = 6$$

$$s_x{}^2 = \frac{N \Sigma X_i^2 - (\Sigma X_i)^2}{N(N-1)}$$
$$= \frac{4(170) - (24)^2}{4(4-1)} = 8.67$$

$$s_x = \sqrt{8.67} = 2.94$$

$N = 4$

$$\bar{Y} = \frac{\Sigma Y_i}{N} = \frac{44}{4} = 11$$

$$s_y{}^2 = \frac{N \Sigma Y_i^2 - (\Sigma Y_i)^2}{N(N-1)}$$
$$= \frac{4(510) - (44)^2}{4(4-1)} = 8.67$$

$$s_y = \sqrt{8.67} = 2.94$$

$N = 4$

$$\bar{W} = \frac{\Sigma W_i}{N} = \frac{72}{4} = 18$$

$$s_w{}^2 = \frac{N \Sigma W_i^2 - (\Sigma W_i)^2}{N(N-1)}$$
$$= \frac{4(1530) - (72)^2}{4(4-1)} = 78.00$$

$$s_w = \sqrt{78.0} = 8.83$$

Summary

	X	$Y = X + 5$	$W = 3X$
Mean	6	11	18
Variance	8.67	8.67	78.00
Standard Deviation	2.94	2.94	8.83

STANDARD SCORES AND THE NORMAL DISTRIBUTION

Standard scores

One reason for changing the origin and unit of measurement of a scale is that such transformations can be used to make two very different scales comparable. For example, suppose a teacher gives two exams to a class. A score of 88 on the first exam might mean something quite different from the same score on the second exam: It could be an extremely high score on the first test but not on the second. While percentile ranks provide some idea of relative standing, they indicate only ordinal position within a distribution. That is, percentiles show the proportion of a distribution that falls below a given score, but they do not indicate how *much* below that score the remainder of the distribution is. To illustrate, assume the results of the two tests are

$$A = (78, 81, 87, 88)$$
$$B = (59, 61, 63, 88)$$

The score of 88 has the same percentile rank within each distribution, but the 88 on test *B* represents a considerably greater achievement than the 88 on test *A* because it is so much higher than the remaining scores. Clearly, it would be desirable to be able to characterize the position of a score within a distribution more precisely than by ordinal position alone.

The interpretation of a score within a distribution may be based upon both (1) its relative standing with respect to the mean and (2) the variability of the scores within the distribution. Consider the following example:

$$A = (10, 36, 38, 40, 42, 44, 70)$$
$$B = (10, 12, 14, 40, 66, 68, 70)$$

In both distributions, the range is 60, the mean is 40, and the score of 70 deviates +30 points from the mean. But the variability is much greater in the *B* distribution. Thus, the score of 70 is somewhat less significant in group *B*, since other scores were also more likely in *B* than in *A* to deviate greatly from the mean. The interpretation of an individual score is affected by the variability of a distribution as well as by the mean.

If a single scale of measurement is adopted as standard, then all distributions under consideration, despite their different origins and unit sizes, can be transformed to that single scale for comparison. This is essentially what is done when measures are converted into standard scores (often called *z* scores). By this standardization process the relative standing of persons in one distribution may be compared with their relative standing in another distribution, even though the two distributions represent different measurement scales.

A **standard score** is defined to be

$$z_i = \frac{X_i - \overline{X}}{s_x}$$

This formula means that to transform a score (X_i) into a standard score (z_i), the mean of the original distribution $(\overline{X})$ is subtracted from the original score (X_i) and the result is divided by the standard deviation of the original distribution (s_x).

If the formula is used to standardize each raw score in a distribution, what effect does that have on the distribution's mean and standard deviation? Recall that the mean is a constant; when a constant (in this case, the mean) is subtracted from each score in a distribution, the constant will also be subtracted from the old mean: The new mean will be the old mean minus that constant. Thus, when the scores in a distribution are standardized, the new mean is the old mean minus the old mean, which equals 0. In symbols,

$$\text{New mean,} \quad \overline{z} = \overline{X} - c$$
$$c = \overline{X}$$
$$\overline{z} = \overline{X} - \overline{X}$$
$$\overline{z} = 0$$

Note that since the new mean is 0, the z scores in the new distribution will include negative as well as positive numbers.

Subtracting a constant does not alter the standard deviation of a distribution, but dividing by a constant does. When each score in a distribution is divided by a constant, the standard deviation of the new distribution will equal the original standard deviation divided by that constant. In the formula for a standard score, the constant in question is s_x, the standard deviation of the original distribution. Thus, the new standard deviation equals the old standard deviation divided by itself, or 1:

$$\text{New standard deviation,} \quad s_z = s_x/c$$
$$c = s_x$$
$$s_z = s_x/s_x$$
$$s_z = 1$$

Dividing by s_x does not change the mean in this case because the new mean is 0, and $0/s_x = 0$.

To summarize, if every score in a distribution of X_i has the mean of that distribution subtracted from it and this result is divided by the standard deviation of that distribution, the new distribution will have a mean of 0 and a standard deviation of 1.

Any single raw score may be transformed into a standard score. For example, if you had a score of 88 on an examination, the class mean was 79, and the standard deviation was 6, the standard score, z, corresponding to $X = 88$ is given by

$$z = \frac{X - \overline{X}}{s_x} = \frac{88 - 79}{6} = \frac{9}{6} = 1.50$$

This standard score is in **standard deviation units**. In the above example, the standard deviation of the X_i is 6. The z score represents the number of standard deviations that the score X is from its mean, $\overline{X}$. Since $(X_i - \overline{X})$ is 9, X_i is $\frac{9}{6} = 1.5$ standard deviations above the mean. It is in this sense that the units of the z scale are standard deviation units.

Standard scores become useful when one wishes to compare performance on several measures, each of which has a different mean and standard deviation. Suppose a teacher gives three examinations and wants to combine

4–7 Effect of Computing Standard Scores and Adding across Distributions

	Raw Scores			
Student	Test 1	Test 2	Test 3	Total
1	93	80	85	258
2	81	80	84	245
3	70	85	90	245
4	76	81	85	242
5	65	89	82	236
6	65	90	81	236
7	69	86	79	234
	$\overline{X}_1 = 74.14$	$\overline{X}_2 = 84.43$	$\overline{X}_3 = 83.71$	
	$s_1 = 10.14$	$s_2 = 4.20$	$s_3 = 3.55$	

	Standard Scores			
Student	Test 1	Test 2	Test 3	Total
1	1.86	− 1.05	.36	1.17
2	.68	− 1.05	.08	− .29
3	− .41	.14	1.77	1.50
4	.18	− .82	.36	− .28
5	− .90	1.09	− .48	− .29
6	− .90	1.32	− .76	− .34
7	− .51	.37	− 1.33	− 1.47

the scores into a single score for the purpose of grading at the end of the course. Merely adding the scores on the three tests could give a biased picture of a student's performance. Table 4–7 compares adding raw scores across tests with adding standard scores for the same data. Notice that the rank order of the students is not the same using standard scores as when raw scores are employed. This is particularly striking when students 2 and 3 are compared. They had identical raw score totals (245), but student 2 had a total z score of $-.29$ compared to 1.50 for student 3. The difference was primarily a result of student 3's scoring 90 on the third test. The 90 was an extremely high score in a distribution that otherwise did not have much variability. Therefore, that performance was weighted more in terms of standard deviation units than in terms of raw scores. A comparison of students 2 and 5 shows the case in which two students scored quite differently in terms of raw score but had the same standard score total. In general, adding scores in standard score form provides a fairer and more accurate index of performance than does adding raw scores because the means and variances of the individual distributions are equated. Standard scores take into account not only how much a student's performance deviates from the average of the group on that test but also how likely it was that other individuals in the group would score as high (or low) as that student did.

The normal distribution

Standard scores take on greater meaning when they are arranged in relation to the normal distribution. The normal distribution is different from the distributions described in the last chapter. It is a **theoretical distribution:**[2] a hypothetical or idealized distribution based on a population of an infinite number of cases. Because it is based on a population, not a sample, the normal distribution has parameters, not statistics. Thus, its mean is symbolized by μ (not $\overline{X}$), and its variance is symbolized by σ (not s^2).

The **normal distribution** is defined by

$$Y = \frac{1}{\sqrt{2\pi\sigma^2}} e^{-(X-\mu)^2/2\sigma^2}$$

in which

Y = the height of the curve at point X
X = any point along the X-axis
μ = mean of the distribution
σ^2 = variance of the distribution
π = a constant, approximately 3.1416
e = the base of Napierian logarithms, approximately 2.71828

[2]This is also called a probability distribution.

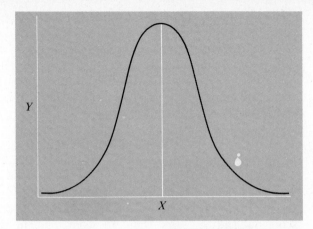

Fig. 4–1.   A normal distribution

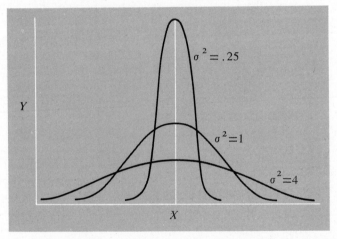

Fig. 4–2.   Three normal distributions with the same mean but different
variances

Usually, the normal distribution is described as being bell-shaped, like
the one pictured in Figure 4–1. However, because the formula includes the
population mean and variance, a different normal distribution can be drawn
for each combination of μ and σ^2. Figure 4–2 shows three normal distributions that have the same mean but different variances.

Although the extent to which the normal distribution looks bell-shaped
depends upon the values of its parameters, certain attributes are present in
every normal distribution. First, the mean, median, and mode are all the same
value, and they divide the distribution into two equal parts. Second, the
normal distribution is symmetrical about this central point, so that if the
distribution were folded over at the mean, the right and left sides would fall

exactly on top of one another. Finally, the "tails" of the distribution get closer and closer to the X-axis as they get farther from the mean, but they never touch it. These tails are said to be **asymptotic** to the X-axis, and thus the normal distribution actually stretches from minus infinity $(-\infty)$ to plus infinity $(+\infty)$ along the X-axis. However, the most important characteristics of the normal distribution can best be understood only after one transforms it into a standardized form.

The standard normal distribution   In the sections above, we saw how any distribution of X_i could be transformed or *standardized* into a distribution of z_i with a mean of 0 and a standard deviation of 1.00. Suppose the original distribution of X_i is a population distribution with a mean μ_x and a standard deviation σ_x, and suppose further that the X_i follow the normal distribution. Then, the formula

$$z_i = \frac{X_i - \mu_x}{\sigma_x}$$

converts this normal population distribution of X_i into a normal population distribution with a mean of 0 and a standard deviation of 1. This special normal distribution with parameters $\mu_z = 0$ and $\sigma_z = 1.00$ is called the **standard normal distribution**.

The real importance of the standard normal distribution derives from the fact that it is a theoretical distribution whose special characteristics allow us to determine percentile ranks and points very easily. Moreover, since normal distributions are common in nature (e.g., height, weight, intelligence) and since any normal distribution can be transformed into the standard normal distribution by employing the conversion formula above, the standard normal distribution can be used as the single reference distribution for comparing a wide variety of otherwise not comparable statistics. Notice that when different normal distributions are converted to the standard normal, each of the transformed distributions has the same mean ($\mu_z = 0$) and standard deviation ($\sigma_z = 1.00$), and thus the percentiles determined for one measure can be directly compared with those for another. For example, if percentile ranks are determined by using the standard normal, a student can be at the 87th percentile rank in math and the 87th percentile rank in English; the student is at precisely the same relative position in the math and English distributions. There is no need to worry about the fact that the original untransformed math and English distributions have different means and standard deviations because these contrasting distributions have been converted into a single distribution, the standard normal distribution.

It is important to realize that one must be able to assume that the original distribution of untransformed scores is normal in form. *Converting scores of a nonnormal distribution to standard score form does not make the standardized distribution normal.* Standardization changes only the numerical values of the mean and variance of a distribution—it does not change the distribution's form or normalize it. This can be seen in Figure 4–3. Distribution A is normal, B is not. Notice that the forms of the z-score distributions (bottom) are nearly the same as the forms of the raw-score distributions. Therefore, one must always ask if the original distribution is normal in form before converting it to the standard normal distribution.

Some of the special characteristics of the standard normal distribution can be seen by examining Figure 4–4. One can determine percentiles of the

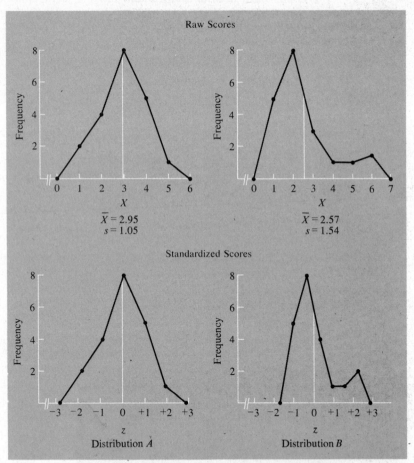

Fig. 4–3. Graphic demonstration that standardizing all scores in a distribution does not change the basic form of the distribution. Notice that -1.96 and $+1.96$ are marked off on the z-axis. They are reasonable estimates of $P_{.025}$ and $P_{.975}$ only for the normal distribution (A). Therefore, the raw-score distribution must be normal in order for the percentiles of the standard normal distribution to be accurate.

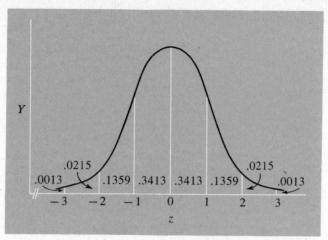

Y

.0215

.0013   .1359 .3413 .3413 .1359   .0215

.0013

-3 -2 -1 0 1 2 3

z

Fig. 4-4.   Areas under the curve of the standard normal distribution

standard normal distribution by observing that the area under the distribution curve represents the proportion of scores that occur between various
values of z. Thus, the proportion of the total area under the curve which is
located between two points along the z-axis equals the proportion of scores
falling between those two z values. For example, in the standard normal
distribution the mean, median, and mode are all located at $z = 0$. The area
under the curve between $z = 0$ and $+\infty$ is one-half the total area; thus .50 of
the cases fall above $z = 0$. Similarly, .50 of the cases fall below $z = 0$ because
the standard normal is symmetrical about $z = 0$. The important point to
remember is that *the area under the standard normal curve implies relative
frequency (or proportion of cases).*

With this in mind, certain points can be selected along the z-axis and
the proportion of the total area under the curve between these points
determined, as has been done in Figure 4-4. Here it can be seen that .3413 of
the scores are between $z = 0$ and $+1$ and another .3413 are between $z = 0$
and -1. Further, almost all the scores in a normal distribution (all but .0026)
will be included between $z = -3$ and $z = +3$. Since the z scale is in
standard deviation units, this means that in any normal distribution almost
all the scores are between three standard deviations above the mean and three
standard deviations below the mean. If the mean of a normal distribution is
55 and the standard deviation is 10, then almost all of the scores are likely to
fall between

$$\mu \pm 3(\sigma)$$
$$55 \pm 3(10)$$
$$55 \pm 30$$

25 and 85

Indeed, because the z scale is in standard deviation units, z scores from normal distributions are sometimes called **standard normal deviates**: They indicate how many standard deviations separate a score from the mean.

The few points along the z-axis that are labeled in Figure 4–4 were chosen for convenience. A list of many points along the z-axis appears in Table A of Appendix II in the back of this book. Turn to this table now. The first column, labeled z, corresponds to values along the z-axis, which have been transformed from a normal distribution of X_i by the formula

$$z_i = \frac{(X_i - \mu)}{\sigma}$$

The second column gives the proportion of area under the curve (proportion of scores) falling between the mean ($z = 0$) and the z value for that row. The third column provides the area beyond z (i.e., between the z value and $+\infty$). Thus, for $z = .35$, the proportion of area falling between the mean ($z = 0$) and $z = .35$ is .1368, and the proportion of area falling between $z = .35$ and $z = +\infty$ is .3632. Notice that although the standard normal distribution contains negative z values to the left of the mean and positive values to the right of it, the table presents only positive z values. Since the standard normal distribution is perfectly symmetrical, the area relating to each positive z value is equally appropriate for each negative z value. Thus, for example, the third column of the table implies either area to the right of a positive z value or area to the left of a negative z value.

Application of the standard normal distribution   The steps below show how certain common types of questions regarding relative standing in a distribution can be answered by using the standard normal distribution.

1. In a normal distribution with mean $\mu = 45$ and standard deviation $\sigma = 10$, at what percentile rank does a score of 58 fall?

a. Determine the z score corresponding to $X = 58$:

$$z_i = \frac{X_i - \mu}{\sigma} = \frac{58 - 45}{10} = 1.30$$

b. Draw a picture similar to Figure 4–5. Look in Appendix II, Table A, for the area between the mean ($z = 0$) and $z = 1.30$. This area constitutes .4032 of the total. Since all of the distribution to the left of the mean also falls below a z of 1.30, .5000 should be added to .4032 to obtain

$$.4032 + .5000 = .9032$$

Thus, $X = 58$ corresponds to the 90th percentile rank.

2. In the normal distribution described above, what score is at the 23rd percentile? This is the same type of problem as 1, solved in the reverse direction.

a. Draw a picture of the normal distribution, as in Figure 4–6. The question asks for the X score corresponding to the 23rd percentile. The z

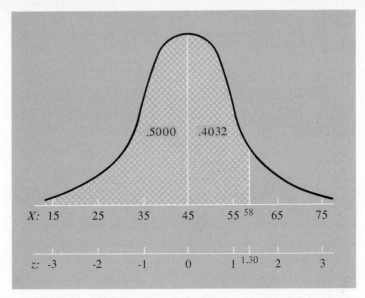

Fig. 4–5. Determining the percentile rank of a score of 58
in a normal distribution with $\mu = 45$, $\sigma = 10$

score corresponding to $P_{.23}$ will have 23% of the cases below it. The third column of the table gives the proportion of cases between z and $+\infty$. However, since the distribution is symmetrical, the column also gives the relative frequency between $-z$ and $-\infty$. Therefore, look down the third column until you find .23. The closest figure is .2296, and the z corresponding to that area is .74. Since the left side of the distribution is being worked with, the z value for the 23rd percentile is $-.74$.

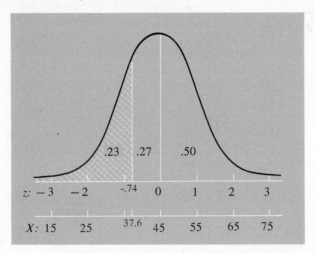

Fig. 4–6. Finding the X score at the 23rd percentile of a normal
distribution

b. Since 23% of the area falls below $z = -.74$, the problem may be solved by converting $z = -.74$ into its corresponding X value:

$$z_i = \frac{X_i - \mu}{\sigma}$$

$$-.74 = \frac{X_i - 45}{10}$$

$$X_i = 45 - 7.4$$

$$X_i = 37.6$$

Therefore, $X = 37.6$ represents the 23rd percentile.

3. What percentage of the cases in a normal frequency distribution falls between $z = -1.00$ and $z = 1.00$?

a. The problem is solved by (1) determining the areas between the mean and $z = 1.00$ and between the mean and $z = -1.00$ and (2) adding these together. The table shows that the area between the mean and $z = 1.00$ is .3413 of the total. Therefore,

$$.3413 + .3413 = .6826$$

and 68.26% of the cases fall between $z = -1.00$ and $z = 1.00$.

4. What proportion of the scores falls between $X = 88$ and $X = 95$ if the normal distribution of X_i has a mean of 85 and a standard deviation of 5?

a. Draw a picture similar to that presented in Figure 4–7 describing the problem. This diagrams the X distribution and indicates that the desired area falls between 88 and 95.

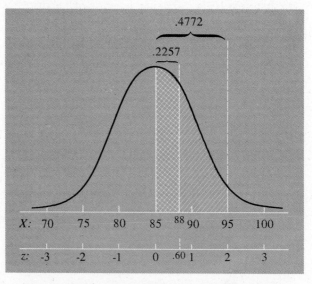

Fig. 4–7.   Determining the proportion of cases falling between $X = 88$ and $X = 95$ in a normal distribution

b. Convert 88 and 95 to z values:

$$z_i = \frac{X_i - \mu}{\sigma} = \frac{88 - 85}{5} = .60$$

$$z_i = \frac{X_i - \mu}{\sigma} = \frac{95 - 85}{5} = 2.00$$

c. Look in the table under $z = 2.00$ (corresponding to $X = 95$). The area between the mean and this point is .4772. However, the area between the mean and $z = .60$ (corresponding to $X = 88$) is not to be included in the desired answer. Therefore, since this area amounts to .2257, subtract it from the proportion of area between the mean and $z = 2.00$,

$$.4772 - .2257 = .2515$$

to obtain the required proportion.[3]

Standard scores and the standard normal distribution Because the standard normal distribution is a theoretical distribution, conversions between other normal distributions and the standard normal require μ and σ rather than their corresponding sample statistics, $\overline{X}$ and s. However, the standard normal distribution may be used even if only sample estimates of these population parameters are at hand as long as two conditions are met:

1. The population distribution from which the sample is drawn must be normal in form. One way to examine the tenability of this assumption is to plot the sample distribution and casually observe its general form. If the sample distribution does not depart severely from a normal pattern and there is no reason that it should not be normal, then this condition is satisfied. As Figure 4-3 illustrates, the percentiles of the standard normal distribution are accurate only in the case of a normal distribution (A).

2. There must be enough measurements in the sample. A reasonable number of cases is needed to determine if the sample is normal in form (to satisfy the first condition, above) and for $\overline{X}$ and s to be accurate estimators of μ and σ. For samples of approximately 30 cases or more, $\overline{X}$ and s^2 are sufficiently good estimators of their respective parameters, but it must be remembered that they are still only estimates or approximations of μ and σ.

If these two conditions are met, then one can relate the sample distribution to the theoretical standard normal distribution in the same manner as has been illustrated for normal population distributions: by using the usual formula but with μ replaced by $\overline{X}$ and σ_x replaced by s_x. Thus,

$$z_i = \frac{X_i - \overline{X}}{s_x}$$

[3]The solution presented for this type of problem assumes that the values of 88 and 95 are exactly 88.000 . . . and 95.000 . . . . . If not, then their real limits would be 87.5 to 88.5 and 94.5 to 95.5, respectively, in which case the interval of concern would be (87.5–95.5) rather than (88–95).

Although the accuracy of this procedure depends directly on the extent to which the conditions stated above are met, using the standard normal or other theoretical distributions to estimate percentiles in a sample in this manner is a very common procedure. In fact, most of the procedures of elementary statistical inference use essentially this strategy, as we will see beginning in Chapter 7. The logic of the procedure is that $\overline{X}$ and s_x are estimators of μ and σ, and therefore may be used to estimate those parameters when only sample data are available. However, the accuracy of the results obtained by using sample data and this procedure depends upon the normality of the distribution and upon the sample size, which in turn determine the accuracy of $\overline{X}$ and s_x as estimators of their corresponding parameters. Therefore, the procedure should be employed only with the knowledge that the results it yields are approximate.

Other standardized distributions

The standard normal distribution discussed in the preceding section had a mean of 0 and a standard deviation of 1 because the mean was subtracted from each score and the result divided by the standard deviation. However, these parameter values are arbitrarily designated. The distribution could just as easily be made to have a mean of 100 and a standard deviation of 20, by using the transformation

$$\text{standard score} = 20\left[\frac{X_i - \overline{X}}{s_x}\right] + 100$$

or a mean of 500 and a standard deviation of 100, by using the transformation

$$\text{standard score} = 100\left[\frac{X_i - \overline{X}}{s_x}\right] + 500$$

The Army General Classification Test of the Second World War used the former standard score, and the Graduate Record Exam has employed the latter. Thus, it is obvious that the concept of standard scores is perfectly general and that the selection of a mean of 0 and a standard deviation of 1 is convenient but arbitrary.

EXERCISES

1. Using the frequency distribution presented in Table 4–1, determine the score values corresponding to the following percentile ranks.

   a. $P_{.25}$ c. $P_{.55}$ e. $P_{.15}$ g. $P_{.2875}$
   b. $P_{.05}$ d. $P_{.875}$ f. $P_{.425}$ h. $P_{.50}$

2. Using Table 4–1, determine the percentile ranks corresponding to the following score values.

   a. 58   c. 80   e. 55   g. 85
   b. 74   d. 78   f. 60   h. 49

3. If the mean of a distribution is 80 and the

standard deviation is 5, what do the mean, variance, and standard deviation become after each of the following operations?

a. 10 is added to each score.
b. 12 is subtracted from each score.
c. Each score is multiplied by 10.
d. Each score is divided by 2.
e. Each score is divided by 8, and then 50 is subtracted from each.

4. If an original set of measurements is made in inches and has a mean of 36 and a variance of 64, what will be the mean, variance, and standard deviation if:

a. The unit is changed to feet?
b. 6 inches must be added to each measurement to correct an error?
c. The measures are converted to centimeters (2.5 centimeters = 1 inch) and then 15 centimeters is subtracted from each score?

5. What are the limitations of percentiles as measures of relative position, and how do standard scores overcome these limitations?

6. Explain and illustrate why it is necessary to consider the variance of a distribution in order to accurately reflect the relative position of a score.

7. Compute the standard score for each of the members of the following distribution. Then calculate the mean and the standard deviation of the z_i. Would you have predicted these last two values? Why?

$$X_i = (4, 5, 6, 9)$$

8. What does the area under a theoretical distribution signify?

9. What proportion of the cases of a normal distribution will fall to the left of $z = 0$? To the right of $z = 0$? To the right of $z = 1.00$? To the right of $z = -1.00$?

10. Determine the proportion of cases falling under the normal curve in the following circumstances.

a. Between $z = -1$ and $z = +1$.
b. Between $z = 1.0$ and $z = 2.0$.
c. To the right of $z = 1.5$.
d. To the left of $z = -1.96$ and to the right of $z = 1.96$.
e. To the left of $z = -2.575$ and to the right of $z = 2.575$.

11. If the mean of a normal distribution is 43 and the standard deviation is 2, within how many standard deviations of the mean would you expect almost all of the cases to fall? What score values correspond to these points?

12. In a normal distribution with mean 16 and standard deviation 10, at what percentile is a score of 21 likely to fall? $X_i = 6$? $X_i = 14$?

13. In a normal distribution with mean 40 and standard deviation 6, what score is likely to fall at $P_{.925}$? $P_{.166}$?

14. What percent of the cases are likely to fall between the values of 20 and 32 in a normal distribution with mean 25 and standard deviation 10? Between 26 and 45? Between 11 and 19?

15. If your score on a statistics test was 90, and the professor determined grades separately for each of four sections, which section would you hope was yours? Why?

a. $\bar{X} = 65, s = 13$
b. $\bar{X} = 75, s = 10$
c. $\bar{X} = 80, s = 8$
d. $\bar{X} = 85, s = 2$

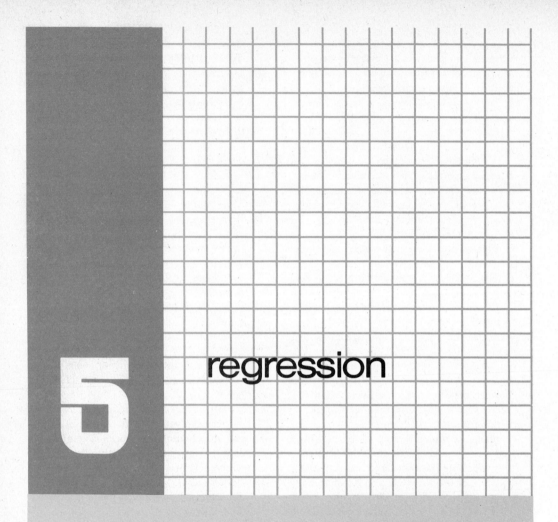

5 regression

THE TECHNIQUES that were presented in Chapters 1–4 describe (1) a set of measurements on a single variable and (2) the relative standing of individuals in the group on that single variable. We will now consider procedures that describe the relationship between two variables for a sample of observations.

For example, a researcher may be interested in the relationship between the amount of male sex hormone testosterone experimentally injected into rats and the amount of fighting subsequently displayed by the animals. Various levels of dosage might be selected, randomly or deliberately, and administered to different animals. The scientist might then time the amount of fighting each animal displays during a given period. The next task is to describe the relationship between the size of dosage of testosterone and the amount of fighting.

The above question pertains to the relationship between dosage and fighting in the group of subjects as a whole. In addition, we frequently want to be able to predict one behavior from another for a specific individual. For example, suppose an admissions officer of a college is faced with the problem of selecting a freshman class. One piece of information that is available is the Scholastic Aptitude Test scores of each applicant. What is applicant John Greenley's college grade average likely to be at a specific college if he scores 625 on the verbal scale of the SAT?

The questions of predicting one variable from another are called problems of **regression**. This term stems from the work of Sir Francis Galton, who studied the relationship between a variety of physical and behavioral characteristics in the nineteenth century. In trying to predict one trait from another, he observed that sons of tall fathers were tall, but not as much taller than average as their fathers. Conversely, sons of short fathers, though short, were also closer to the average height than their fathers. Such a tendency toward the average is called **regression toward the mean**, which refers to the fact that it is a good bet that people who are extreme on the predictor will be less extreme on the trait being predicted. Some characteristics follow this principle while others do not, but it is a good prediction strategy in most cases. In any event, the term *regression* has come to signify a variety of techniques that describe relationships between variables and predict one variable from another.

LINEAR RELATIONSHIPS

One of the simplest relationships between two variables occurs when high values on one measure are associated with high values on another variable and low values on one are associated with low values on the other. Such a relationship can be drawn on a graph, and it approximates a straight line.

109

Thus, understanding such relationships will be easier after a brief discussion of the algebra and geometry of simple linear relationships.

The equation for a straight line

Suppose a young baby sitter charges $1.00 per hour for services. Consider the relationship between money earned (labeled Y) and hours worked (labeled X). If the baby sitter does not work at all ($X = 0$), then no money is earned ($Y = 0$). If the baby sitter works two hours ($X = 2$), $2.00 is made ($Y = 2$). For four hours, $4.00 is made. A table displays these values:

Hours (X)	Dollars (Y)
0	0
2	2
4	4

The basis of this table is the relationship: an hour worked is a dollar earned. Using the symbols Y for dollars earned and X for hours worked, this statement reduces to

$$Y = X$$

This simple mathematical equation will indicate the earnings for any amount of time worked. For example, if the baby sitter works $3\frac{1}{2}$ hours ($X = 3.5$), then the amount earned is $3.50.

The same information may be expressed by graphing the equation, letting specific pairs of values for X and Y be the coordinates of a set of points. Figure 5–1 shows the X- and Y-axes and the three points from the preceding table [(0, 0), (2, 2), (4, 4)]. The line passing through them is described by the equation, $Y = X$. The line really represents an infinite number of points which indicate the amount of money earned for a given amount of time worked. For example, if the baby sitter worked $3\frac{1}{2}$ hours, a vertical line could be drawn at $X = 3.5$, which would intersect the graphed line of the relationship at the height of $Y = \$3.50$, as indicated in the figure. Hence, both the equation ($Y = X$) and the graph of that equation in Figure 5–1 describe the relationship between money earned and hours worked and provide a method of predicting one (Y) from the other (X).

A **linear relationship** is an association between two variables that may be accurately represented on a graph by a straight line.

The relationship in Figure 5–1 is obviously linear—in fact, it is perfectly linear in that all the points fall directly on the straight line. Later in this

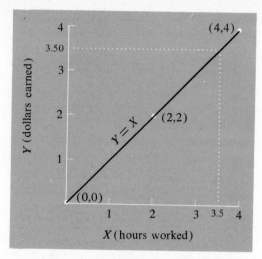

Fig. 5–1.   Graph of the line $Y = X$

chapter we will discuss relationships that are essentially linear even though some or all of the points do not fall precisely on the line. Not all relationships are linear; some are **nonlinear**. For example, the relationship between vocabulary and age between one and six years follows an S-shaped pattern, beginning upward slowly, then accelerating, and finally leveling off. This nonlinear relationship is **curvilinear**; there are other kinds of nonlinear associations as well. However, we will be concerned only with linear relationships in this text.

The linear relationship described above and graphed in Figure 5–1 is one of the simplest kinds. Other perfect linear relationships can also be described. Suppose the baby sitter is quite young and inexperienced and, as a consequence, earns only $.75 an hour. Once again, for zero hours worked ($X = 0$), pay is 0 ($Y = 0$); for one hour ($X = 1$), earnings are $.75 ($Y = .75$); and for three hours, earnings are $2.25. Putting these values into a table, one obtains

Hours (X)	Dollars (Y)
0	0
1	.75
3	2.25

The table uses the relationship: dollars earned equals $.75 times the number of hours worked, which in symbols is

$$Y = .75X$$

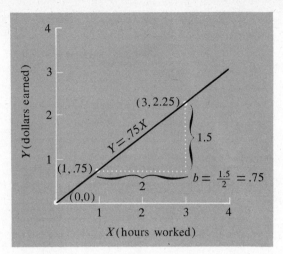

Fig. 5–2.   Graph of the line $Y = .75X$, indicating the slope of the line, b, to equal .75

Fig. 5–2 illustrates this relationship. Note that the two lines $Y = X$ (Fig. 5–1) and $Y = .75X$ (Fig. 5–2) are quite similar, differing only in tilt or slope.

The **slope** of a line is defined to be the vertical distance divided by the horizontal distance between any two points on the line.[1] In symbols, given the two points (x_1, y_1) and (x_2, y_2), the slope b of the line is

$$\text{slope} = b = \frac{\text{vertical distance}}{\text{horizontal distance}} = \frac{y_2 - y_1}{x_2 - x_1}$$

For example, two points found on the above line are (1, .75) and (3, 2.25). Call the first point (x_1, y_1) and the second point (x_2, y_2). Putting these values into the formula for the slope, one obtains

$$b = \frac{y_2 - y_1}{x_2 - x_1} = \frac{2.25 - .75}{3 - 1} = \frac{1.5}{2}$$

$$b = .75$$

The derivation of the slope is pictured in Figure 5–2. Recalling that the slope equals the vertical distance between any two points on the line divided by the horizontal distance between the same two points, one can see that this ratio of distances is .75 for the points (1, .75) and (3, 2.25). Of course, this ratio will be .75 for any two points on the line.

---

[1] Some students define the slope to be slope = rise ÷ run, remembering that in alphabetical order *rise* precedes *run*.

It is not just coincidence that the slope of this line equals .75 and the coefficient of X in the equation $Y = .75X$ is also .75. The coefficient of X in an equation of this sort is identical to the slope of the line that the equation describes. In the first illustration, $Y = X$, the coefficient of X is 1 $[Y = (1)X]$. An examination of Figure 5–1 reveals that two points on the graph are $(x_1, y_1) = (2, 2)$ and $(x_2, y_2) = (4, 4)$, yielding a slope of 1:

$$b = \frac{4 - 2}{4 - 2} = \frac{2}{2} = 1$$

The slope of a line may be negative as well as positive. Consider the line in Figure 5–3. Two points on that line are $(x_1, y_1) = (2, 3)$ and $(x_2, y_2) = (4, 2)$, which when substituted into the formula for the slope, b, yield

$$b = \frac{y_2 - y_1}{x_2 - x_1} = \frac{2 - 3}{4 - 2} = \frac{-1}{2}$$

$$b = -.50$$

To get from the point (2, 3) to (4, 2) in the graph, one must go down one unit, equivalent to going -1 vertical units, and to the right two units $(+2)$, giving a slope of

$$b = \frac{-1}{2} = -.50$$

The slope of the line determines whether the relationship is positive or negative.

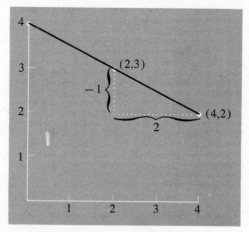

Fig. 5–3. Graph of a line with negative slope, $b = -\frac{1}{2} = -.50$. Note that low values of X are associated with high values of Y, an inverse or negative relationship.

A line with a positive slope indicates a **positive** or **direct relation-ship** and a line with a negative slope represents a **negative** or **inverse relationship**.

Notice that lines with positive slope tend to run upward from left to right on the graph, indicating that low values on one variable are associated with low values on the other and that high values on one variable are associated with high values on the other. In the baby-sitting example few hours worked earned few dollars and many hours worked earned many dollars. Since the slope in such a situation is positive, the relationship is called positive; it may also be referred to as direct because of the direct correspondence of values on the two dimensions. In contrast, if the slope is negative the line tends to run downward from left to right on the graph, indicating that low values on one dimension go with high values on the other. Because the slope is negative in these instances, the relationship is called negative; it is sometimes called inverse because high values are associated with low values (rather than high with high and low with low as in direct relationships).

As a final example of a linear relationship, suppose the baby sitter works for an agency. The agency operates in the following manner: Once parents contact and arrange for a baby sitter there is a $1.00 charge which is made regardless of whether or not the engagement is fulfilled. Baby sitting is paid at the rate of $.75 an hour in addition to the $1.00 service charge. Thus, if parents make an appointment and then have to cancel it, the baby sitter works no hours ($X = 0$) but the cost to the parents is $1.00. If the date is completed and the sitter has been occupied for three hours ($X = 3$), the charge is three hours at $.75 an hour plus $1.00 service charge for a total of $3.25. A five-hour evening would cost $5 \times$ $.75 plus $1.00 or $4.75. In tabular form,

Hours (X)	Dollars (Y)
0	1.00
3	3.25
5	4.75

These values were arrived at by multiplying .75 times the number of hours worked and adding 1.00, which can be expressed by the general equation

$$Y = .75X + 1.00$$

The plot of this equation, together with the plots that were shown in Figures 5–1 and 5–2, are presented in Figure 5–4. Observe first that the coefficients of X indicate that the line $Y = X$ has a slope of 1, while the other two lines both have a slope of .75 ($Y = .75X$ and $Y = .75X + 1$). On the graph, the line for ($Y = X$) has a clearly different tilt or slope, whereas the

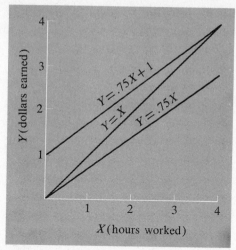

Fig. 5-4. Plot of three lines that have different slopes and intercepts

two lines that have identical slopes are parallel. But the two parallel lines differ in that one is always one unit above the other. If the Y-axis is arbitrarily selected as the place to measure this separation, the line $Y = .75X + 1$ goes through the Y-axis at 1, whereas the line $Y = .75X$ goes through the Y-axis at 0.

The point at which a line intersects the Y-axis is called the _y_-intercept, and its value is symbolied by _a_.

Note that the value of the y-intercept, or a, equals the constant that is added to the equation, as follows:

Equation	y-intercept
$Y = .75X + 1$	$a = 1$
$Y = .75X + 0 = .75X$	$a = 0$

The y-intercept can be found from the equation of the line by setting $X = 0$, which is, in fact, the point on the X scale where the Y-axis is located. Thus, to compute the y-intercept:

$$Y = .75X + 1$$

$$= .75(0) + 1$$

$$Y = +1 = y\text{-intercept}$$

In summary, one needs to know two values to determine the equation of a straight line: the y-intercept (a) and the slope (b). Further, since a provides one point and b tells one how to locate some other point on the line, a and b completely specify a particular line. Therefore,

Any straight line will have the general equation of

$$Y = bX + a$$

If $a = 1$ and $b = .75$, then the equation

$$Y = .75X + 1$$

describes the line with slope .75 and y-intercept at 1.

The usefulness of the equation of a linear relationship is that it allows one to determine a Y value given any X score. Suppose the appropriate equation is $Y = .75X + 1$, and one wishes to know the Y score corresponding to an X of 2. Substituting $X = 2$,

$$
\begin{aligned}
Y &= .75X + 1 \\
&= .75(2) + 1 \\
&= 1.5 + 1 \\
Y &= 2.5
\end{aligned}
$$

Thus, in the baby-sitting example, if the parents are to be gone for two hours, they could plan on spending $2.50 for a baby sitter through the agency.

The baby-sitting examples illustrate that graphs of linear relationships have a slope and a y-intercept and that they can provide the basis for predicting a Y value given a certain X value. However, all these cases have dealt with perfect linear relationships. That is, all of the points have fallen precisely on the line. None has deviated from it. Regrettably, most relationships observed in social science are not so precise.

A plot of an approximate linear relationship is presented in Figure 5–5. The relationship does have a positive linear trend, as indicated by the straight line drawn through it, but this line is obviously only an approximation. Yet it still might be useful to have its equation in order to make approximate predictions of Y given X. This plot represents a hypothetical relationship between Scholastic Aptitude Test (SAT) scores and freshman college grades at a highly selective college in a certain year. If the equation of the line was known, educated guesses concerning the scholastic performance of next year's applicants could be made on the basis of their SAT scores before they actually get to college and earn grades. Of course, this strategy assumes that the relationship will be much the same from year to year. Previously in this chapter, the equation of the line has been rather obvious, but this is no longer so when the relationship is not perfect. In fact, it may not be immediately clear whether the data do indeed follow a linear trend. Thus, the first step toward an equation is to check on the linear trend by constructing a scatterplot.

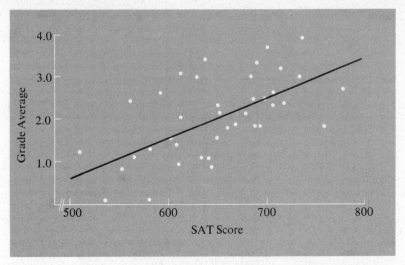

Fig. 5-5.   Approximate linear relationship between the SAT scores and
freshman grades shown in Table 5-1

Scatterplots

One way to obtain some information on the precision and linearity of a
relationship is to construct a scatterplot of it. Suppose that a college admissions officer has a group of 40 students with the SAT scores and first-year
college grades (A = 4.0) presented in Table 5-1. This collection of pairs of
scores may be treated in the same manner as the small tables of values
described in the previous section, and a point may be placed on a graph
corresponding to each pair of values.

A **scatterplot** is a graph of a collection of pairs of scores.

Figure 5-5 is an example of a scatterplot. It was constructed simply by
recording a point corresponding to each pair of scores. The variable about
which predictions are to be made (predicted variable) is always placed on the
ordinate. Thus, to plot the first pair of scores (510, 1.3) in Table 5-1, a point
is placed directly over 510 on the SAT scale and precisely to the right of a
grade average of 1.3.

Although statistical techniques exist that help in making a decision
about linearity, for our purposes an observation of the total scatterplot is
usually sufficient to determine if the trend in the data is approximately linear
or if it is nonlinear.

5–1 SAT Scores and Grades for a Sample of 40 College Students

SAT Score	Freshman Grades	SAT Score	Freshman Grades
510	1.3	659	2.1
533	0.1	663	1.7
558	0.8	670	1.8
565	2.4	678	2.1
569	1.1	679	2.9
580	0.1	680	2.3
581	1.3	687	1.8
590	2.6	688	3.2
603	1.5	693	1.8
610	1.3	698	2.4
612	0.9	700	3.6
615	2.0	710	2.5
618	3.0	710	2.2
630	2.9	718	3.1
633	1.1	724	2.3
639	3.3	734	2.9
643	1.1	739	3.8
645	0.9	750	3.9
651	1.5	767	1.7
654	2.2	778	2.6

REGRESSION CONSTANTS AND THE REGRESSION LINE

Obviously, if the scatterplot tends to be nonlinear, trying to describe it with a straight line would hardly be appropriate. If, however, it is approximately linear, the next task is to determine the equation of the line that best describes the linear relationship between the two observed variables.

Earlier in this chapter, the equation for a straight line, $Y = bX + a$, was for the case in which a perfect linear relationship existed between two available sets of measures. But suppose the relationship is linear but not perfect. How can we determine the equation for the line—that is, how can we determine the values of a and b? And suppose we want to use the observed relationship between X and Y to predict a person's Y score when all we know is his or her X score. This prediction is the fundamental task of regression, and the vocabulary of straight lines changes a bit in the new context.

> The line describing the relationship between two variables is known as the **regression line** and is expressed in the form, $\tilde{Y} = bX + a$. The variable $\tilde{Y}$ is called **Y predicted**. The values a (the y-intercept) and b (slope) are known as **regression constants**.

The task, then, is to specify the regression constants for the regression line that best describes the relationship. The criterion used to determine which line of the many possible lines is best is called the **least squares criterion**.

In order to examine the meaning of that concept it is necessary to observe a scatterplot in more detail. For purposes of illustration, the following few subjects have been selected from the original sample of 40 students. Their SAT scores (i.e., X_i) and grade averages (i.e., Y_i) are as shown.

SAT Score (X_i)	Grades (Y_i)
510	1.3
533	0.1
603	1.5
670	1.8
750	3.9

A scatterplot of these five points and two possible regression lines are presented in Figure 5–6.

For the purpose of explaining the least squares criterion for determining the regression line, several points will be reviewed, using Figure 5–6 as a

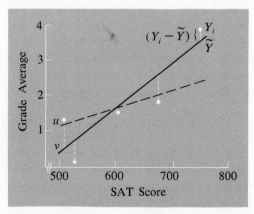

Fig. 5–6.   Plot of a selected group of points with two possible regression lines, u and v

graphic example. First, note that when a linear relationship is not perfect, some or all the points do not fall directly on the line. Second, the vertical distance between the abscissa and any point is simply that point's Y_i value (that person's grade average). Third, suppose you know a person's X score but not his or her Y score. What would the regression line, which describes the relationship between X and Y, *predict* this point to be? The predicted value of Y_i would be the vertical height of the regression line at that particular X_i value. More specifically, one would predict $\tilde{Y}$ (read "Y predicted" or "Y TIL-dah"). Fourth, $\tilde{Y}$ will probably not precisely equal the subject's actual Y_i score. That is, there will be some error in predicting actual grade average (Y_i) simply by knowing the SAT score (X_i) and the regression line. The quantity ($Y_i - \tilde{Y}$) represents the amount of this error. This is the difference between the actual and the predicted value of the person's grade average.

Now consider the two lines in Figure 5–6. Notice that in the upper right-hand corner of the graph the distance between Y_i and $\tilde{Y}$ is shown by a vertical white line. The distance between the actual point (Y_i) and its predicted value ($\tilde{Y}_i$) represents the **error** in prediction. This difference is commonly called a **residual**, since it represents distance remaining after one predicts Y_i with the regression line, $\tilde{Y}$. It is advantageous to select regression lines that make such errors (over all points in the plot) as small as possible. For example, examining the vertical distance between each point and the line in Figure 5–6, it is obvious that the points cluster more closely about line v than about line u. The distances between the points and the line might give a convenient measure of how closely the points group about the line and thus be a criterion for which line best fits the data. However, if the simple distances ($Y_i - \tilde{Y}$) are used, the positive and negative errors will cancel each other out when summed over all points, leaving unresolved the choice of which line is best. But if these error distances are squared, this will not happen. In calculating the variance, you will recall, $\Sigma(Y_i - \bar{Y})^2$ is used to measure the extent to which scores (Y_i) deviate about their mean. Similarly, $\Sigma(Y_i - \tilde{Y})^2$ provides an index of how much the points deviate from the regression line. Since it is desirable to have this total sum of squared errors as small as possible, one picks the regression line so that $\Sigma(Y_i - \tilde{Y})^2$ is a minimum.

> The **least squares criterion** for determining the regression line requires that the sum of the squared deviations between points and the regression line, $\Sigma(Y_i - \tilde{Y})^2$, is a minimum.

Since the regression line is completely determined if the y-intercept (a) and slope (b) are known, the task reduces to the selection of these regression constants in accordance with the least squares criterion.

Obtaining an equation for the regression line

Earlier in this chapter the process of determining the equation of a straight line when two points on the line were known was described. The current task is different. Now we are required to determine a and b when we do not know what the line should be and when we do not have two points that we know lie exactly on the line. We need formulas that use the (X_i, Y_i) pairs of scores to provide values for a and b that determine a line which satisfies the least squares criterion. These formulas for a and b, which can be found by calculus, are

$$b = \frac{N(\Sigma XY) - (\Sigma X)(\Sigma Y)}{N \Sigma X^2 - (\Sigma X)^2}$$

$$a = \overline{Y} - b\overline{X}$$

The expressions provide a means for computing a and b from the original scores such that when these constants are placed into the regression equation $(\tilde{Y} = bX + a)$, it will describe the line that best fits the data in the sense of having minimized the squared deviations between the data points and the line [i.e., $\Sigma(Y_i - \tilde{Y})^2$ will be a minimum].

The use of these formulas will be illustrated with the data provided by the five subjects selected previously. Of course, the admissions director mentioned in the original illustration would use the data from as many college students as possible. However, using just these few subjects will provide the reader with a simple example of the computational routine required. The procedures are presented in Table 5–2. An examination of the formulas reveals that five quantities are needed:

$$N, \Sigma X, \Sigma X^2, \Sigma Y, \text{ and } \Sigma XY$$

The N is the number of *pairs* of scores or the number of subjects, *not* the number of X and Y scores together. Recall from Chapter 1 that ΣXY means X_1 times Y_1 plus X_2 times Y_2, and so forth. That is, the values of each pair of scores are multiplied together, and these *cross products* are summed over all pairs of observations. The values of b and a are arrived at by substituting all required values into the formulas given above and reducing. The last step is to substitute the resulting constants into the regression equation, which yields

$$\tilde{Y} = .012X - 5.638$$

With the regression equation obtained, one may answer the question the admissions director is faced with. If applicant John Greenley's SAT score is 625, what would one predict his freshman grade average to be? The question

5-2 Computation of the Regression Equation

Subject	X	Y	X^2	XY
1	510	1.3	260100	663.0
2	533	.1	284089	53.3
3	603	1.5	363609	904.5
4	670	1.8	448900	1206.0
5	750	3.9	562500	2925.0

$$\Sigma X = 3066 \quad \Sigma Y = 8.6 \quad \Sigma X^2 = 1919198 \quad \Sigma XY = 5751.8$$
$$(\Sigma X)^2 = 9400356$$
$$\overline{X} = 613.20 \quad \overline{Y} = 1.72$$
$$N = 5$$
$$b = \frac{N(\Sigma XY) - (\Sigma X)(\Sigma Y)}{N\Sigma X^2 - (\Sigma X)^2} = \frac{5(5751.8) - (3066)(8.6)}{5(1919198) - 9400356}$$
$$b = \frac{2391.4}{195634} = .012$$
$$a = \overline{Y} - b\overline{X} = 1.72 - (.012)(613.20) = -5.638$$
$$\tilde{Y} = bX + a$$
$$\tilde{Y} = .012X - 5.638$$

is answered by substituting $X = 625$ into the regression equation and computing $\tilde{Y}$:

$$\tilde{Y} = .012(625) - 5.638$$
$$\tilde{Y} = 1.862$$

On the graph this amounts to drawing a vertical line through $X = 625$ parallel to the Y-axis and determining the Y value at the point of intersection of that vertical line and the regression line.

Plotting the regression line

It is sometimes necessary to plot a regression line once you have obtained its algebraic equation. Two points determine a straight line, and the regression constants, a and b, will provide those two points. First, mark off a set of axes. Remember, the variable doing the predicting (variable X) goes on the horizontal axis (X-axis), and the variable being predicted (variable Y) goes on the vertical axis (Y-axis). Since the value of a is the y-intercept, simply place a point on the Y-axis at the point corresponding to the value calculated for a. In this case $a = -5.638$, which would be a point on the ordinate 5.638 units below 0.

Now the task is to determine any second point on the line. Recall that the slope represents the amount of vertical distance relative to horizontal distance between any two points. Thus, start with the point a that you just made on the Y-axis and move up vertically b units. From there, move horizontally one X unit to the right. That is, if the slope is .75, move up .75 units on the Y-axis for every 1 unit you move right on the X-axis. If the slope is $-.75$, move down .75 units on the Y-axis for every 1 unit on the X-axis. The second point on the regression line may be determined more simply if the means of the X and Y distributions are known, because $(\overline{X}, \overline{Y})$ is always a point on the line. Thus, $(\overline{X}, \overline{Y})$ and a are two points on the line.

Linearity

The procedures described in the preceding sections are appropriate for data having an underlying linear relationship in the population. A linear relationship probably exists in the population when the form of the scatterplot of the sample data approximates a straight line. Obviously, if it is doubtful that a linear relationship exists, it is foolish to attempt to fit a straight line to data which would be more appropriately described by some curvilinear trend.

The range of X

When predicting Y from a specific X score, one must be careful to restrict predictions to those X values that fall within the range of X values available in the original data. It is this original set of data that one uses to determine if the relationship is linear. Since one's confidence in the assumption of linearity is based only upon the original range of X values, predicting for more extreme X_i exceeds the information on the linearity of the relationship. Consider an example in which rats were placed under food deprivation for from 0 to 48 hours prior to performing ten trials in a two-choice problem. The solid line in Figure 5–7 represents the hypothetical relationship between hours of deprivation and percentage of responses correct. The equation of this hypothetical line is

$$\tilde{Y} = .015X + .02$$

To predict performance for 31 hours of deprivation, simply substitute $X = 31$ into the equation, obtaining a predicted figure of 48.5% correct responses. If $X = 72$, which is outside of the range of original X values, then the regression equation would predict that 110% of the trials would be correct! Clearly, the

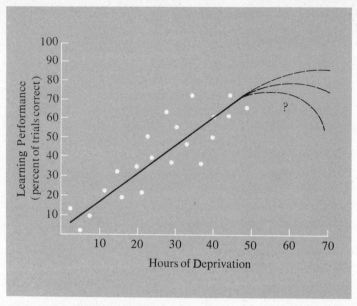

Fig. 5–7.   A hypothetical relationship between hours of deprivation and
learning performance (solid line) and the ambiguity of the form
of the relationship for X values outside the tested range
(dashed lines)

relationship between hours of deprivation and learning performance is not
linear over such an extensive range of values. Furthermore, performance
cannot continue to improve with each extra hour of deprivation, because
ultimately the animal will die. Therefore, it is appropriate to predict only
from X values falling within the range of the original values because information on the linearity of the relationship at more extreme values is not
available.

The second regression line

All the discussion so far has been directed at predicting Y from X. It would
appear that if two variables are measured on each person in a group, the
relationship between them should be the same regardless of whether one
predicts Y from X or X from Y. It is true that the *degree* of relationship is the
same regardless of the direction of prediction, as will be demonstrated in the
next chapter on correlation. However, the values of the regression constants
will be different depending upon which variable is being predicted. This fact
becomes more obvious if one considers the SAT–grades example in terms of
the y-intercept. When trying to predict grades from SAT scores, the y-inter-
cept of the line was measured in the units of the predicted variable, namely,

− 5.638 grade points. If the direction of prediction is reversed, making SAT scores the predicted dimension and placing those scores along the ordinate (*Y*-axis), then the *y*-intercept for the regression line should be in SAT-score points, not grade points. Obviously, the value of *a* will be different in these two cases because the measurement scale constituting the *Y*-axis is different.

The fact that the regression constants are different, depending upon which variable is being predicted, may be understood further by examining the manner in which *a* and *b* were derived. When predicting grades from SAT scores, Σ(grade − predicted grade)^2 is minimized. In contrast, when SAT scores are being predicted, the Σ(SAT − predicted SAT)^2 must be minimized. This is diagramed in Figure 5–8. It happens that a line that minimizes one of these sums does not necessarily minimize the other sum. Therefore, there are usually two regression lines, one for predicting *Y* and one for predicting *X*. However, if a relationship is perfectly linear, so that all the points fall on the regression line, or if both the *X* and *Y* distributions are standardized, there is only one regression line.

The formulas given above for *a* and *b* were for predicting *Y* from *X*. This case is sometimes referred to as the **regression of Y on X**. In contrast, when *X* is predicted from knowledge of *Y* it is called the **regression of X on Y**. Although different formulas exist for calculating the regression constants for the second regression line, they are rarely used. All you need to remember is that for any computational situation the variable being predicted is labeled *Y* while the variable doing the predicting is labeled *X*. Then, use the formulas as presented for the regression of *Y* on *X*.

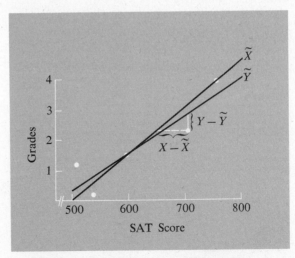

Fig. 5–8.   The two regression lines for the sample data, *Y* on *X* ($\tilde{Y}$) and *X* on *Y* ($\tilde{X}$). The lines connecting two of the points to the regression lines indicate the distances that, when squared, are minimized in determining $\tilde{Y}$ (solid lines) and $\tilde{X}$ (dashed lines).

STANDARD ERROR OF ESTIMATE

The regression procedures described above determine the straight line that predicts Y from X. Except in rare cases, the actual data points will not all fall precisely on the regression line—that is, there will be some error in estimating Y scores, even though the regression line is selected to minimize such error.

Sometimes the scatterplots for two sets of data show regression lines that are essentially the same, but with a larger amount of variability of points about the regression line, or error, in one plot than in the other. Figure 5–9 portrays such a case. The points in graph A cluster more closely around the line than do those in graph B. Therefore, the sum of the squared deviations between the points and the line—the squared errors of prediction—will be less in graph A than in B. As a result, the typical prediction of Y for an individual X score will be more accurate in A than in B. The standard error of estimate is a measure of the amount of error in prediction.

> The **standard error of estimate**, $s_{y \cdot x}$ (read "$s\ y$ dot x") is defined to be the square root of the sum of squared deviations of Y_i about the line ($\tilde{Y}$) divided by $(N - 2)$. In symbols,
>
> $$s_{y \cdot x} = \sqrt{\frac{\Sigma(Y_i - \tilde{Y})^2}{N - 2}}$$

The standard error of estimate is abbreviated $s_{y \cdot x}$ in order to distinguish it from the standard deviation of the Y distribution, which is written s_y (s_x for the X distribution), and to indicate that it is appropriate for the regression of Y on X. Consider the logic of this index.

First, recall the discussion in Chapter 3 of variability of points about the mean. In that case, an index of variability about a constant value, the mean, was developed by taking the sum of the squared deviations of points about

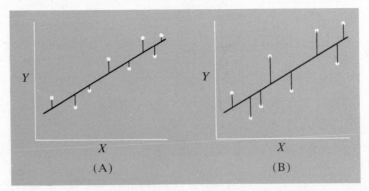

Fig. 5–9.   Two regression lines with different amounts of error in prediction. Graph B has larger errors than does graph A.

that mean and dividing them by $N - 1$ to obtain the variance and its square root, the standard deviation:

$$s_y = \sqrt{\frac{\Sigma(Y_i - \bar{Y})^2}{N - 1}}$$

The present task is similar: to index the variability of the Y scores, not about their mean, but about the regression line, $\tilde{Y}$. If the sum of these squared deviations, $\Sigma(Y - \tilde{Y})^2$, is divided by $N - 2$, one has

$$s_{y \cdot x} = \sqrt{\frac{\Sigma(Y_i - \tilde{Y})^2}{N - 2}}$$

Conceptually, this statistic is very similar to the standard deviation, except that it reflects variability about the regression line ($\tilde{Y}$) rather than about the mean ($\bar{Y}$). Note that while the denominator for the standard deviation was $N - 1$ in order to improve the estimation of σ, the denominator for the standard error of estimate is $N - 2$ to improve $s_{y \cdot x}$ as an estimator of its corresponding population parameter, $\sigma_{y \cdot x}$.

The formula for $s_{y \cdot x}$ given above is a definitional formula. To use it for computation would be tedious since a $\tilde{Y}$ would have to be computed for every X. Therefore, an alternative formula is used which will give the identical result but is easier to compute, despite its forbidding appearance:

$$s_{y \cdot x} = \sqrt{\left[\frac{1}{N(N-2)}\right]\left[N\Sigma Y^2 - (\Sigma Y)^2 - \frac{[N\Sigma XY - (\Sigma X)(\Sigma Y)]^2}{N\Sigma X^2 - (\Sigma X)^2}\right]}$$

For the five subjects used for illustrative purposes in Table 5–2, we have the following quantities:

$$N = 5$$

$\Sigma Y^2 = 22.4$ (obtained by squaring each Y_i and adding these five squares)

$(\Sigma Y)^2 = 73.96$ (obtained by squaring ΣY_i)

$\Sigma XY = 5751.8$

$\Sigma X = 3066$

$\Sigma Y = 8.6$

$\Sigma X^2 = 1919198$

$(\Sigma X)^2 = 9400356$

Substituting into the formula

$$s_{y \cdot x} = \sqrt{\left[\frac{1}{N(N-2)} \right] \left[N\Sigma Y^2 - (\Sigma Y)^2 - \frac{\left[N(\Sigma XY) - (\Sigma X)(\Sigma Y) \right]^2}{N\Sigma X^2 - (\Sigma X)^2} \right]}$$

we have

$$s_{y \cdot x} = \sqrt{\left[\frac{1}{5(5-2)} \right] \left[5(22.4) - (73.96) - \frac{\left[5(5751.8) - (3066)(8.6) \right]^2}{5(1919198) - 9400356} \right]}$$

$$= \sqrt{\left[\frac{1}{15} \right] \left[38.04 - \frac{[2391.4]^2}{195634} \right]}$$

$$= \sqrt{\left[\frac{1}{15} \right] [8.8079]}$$

$$s_{y \cdot x} = .77$$

The standard error of estimate provides a measure of the error in the prediction of Y. Notice that it is essentially an average error over the entire scatterplot; it will *not* be an accurate estimate of the error in predicting Y for each particular value of X. Actually, $s_{y \cdot x}$ will overestimate the error when predicting from X scores near the mean of X but underestimate the error when predicting from X scores that deviate substantially from the mean of X. Special formulas exist that can adjust for this bias.[2]

It is also necessary to observe that $s_{y \cdot x}$ expresses the average error in prediction in the units of the Y variable. In the example above, $s_{y \cdot x}$ is expressed in grade points. Obviously, then, the standard error of estimate cannot be used to compare the error in prediction for two different Y variables. Therefore, $s_{y \cdot x}$ is useful when the same Y variable is being predicted from two different predictors. For example, perhaps grade averages are predicted for a group of applicants first from the verbal scores and then from the quantitative scores on an entrance exam. The two standard errors of estimate can be used to reflect the relative error in predictions for these two cases. The use of $s_{y \cdot x}$ to compare the accuracy of prediction technically

[2] If X_p is the predictor X_i score, then the estimated standard error of $\tilde{Y}$ for X_p is

$$s_{y \cdot x} = \sqrt{\frac{1}{N} + \frac{\left(X_p - \bar{X} \right)^2}{\sum\limits_{i=1}^{N} \left(X_i - \bar{X} \right)^2}}$$

requires that the variance of the predicted variable (i.e., of Y) be the same in each scatterplot. This can be achieved by standardizing the variables (see pages 95–97) before computing regression constants and standard errors. Then standard errors can be used to compare the accuracy of predictions across relationships involving different standardized variables.

However, another method of comparing relationships between two variables that can be applied to different pairs of measures is to calculate the correlation coefficient, as described in the next chapter.

FORMULAS

1. Geometry of a straight line

$$Y = bX + a$$

Slope $(b) = \dfrac{y_2 - y_1}{x_2 - x_1}$ for points (x_1, y_1) and (x_2, y_2) on the line

y-intercept $(a) = Y$ value at $X = 0$

2. Regression of Y on X
   a. Regression equation

$$\tilde{Y} = bX + a$$

b. Slope

$$b = \frac{N(\Sigma XY) - (\Sigma X)(\Sigma Y)}{N\Sigma X^2 - (\Sigma X)^2}$$

c. y-intercept

$$a = \overline{Y} - b\overline{X}$$

d. Standard error of estimate

$$s_{y \cdot x} = \sqrt{\frac{\Sigma(Y_i - \tilde{Y})^2}{N - 2}} \quad \text{(definitional formula)}$$

$$s_{y \cdot x} = \sqrt{\left[\frac{1}{N(N-2)}\right]\left[N\Sigma Y^2 - (\Sigma Y)^2 - \frac{[N(\Sigma XY) - (\Sigma X)(\Sigma Y)]^2}{N\Sigma X^2 - (\Sigma X)^2}\right]}$$

(computational formula)

EXERCISES

1. Compute the regression constants for the regressions of Y on X, W on X, and W on Y. Also write the regression equation for each of these relationships.

X	Y	W
4	2	8
6	5	4
7	5	4
9	6	1
2	0	7
3	5	6

2. Using the equations from exercise 1, what would you predict in the following cases?
   a. $\tilde{Y}$ for $X = 6$
   b. $\tilde{W}$ for $X = 3$
   c. $\tilde{W}$ for $Y = 3$
   d. $\tilde{Y}$ for $X = 12$
3. Determine the standard errors of estimate, $s_{y \cdot x}$, $s_{w \cdot x}$, and $s_{w \cdot y}$, in exercise 1 above.

4. The squared deviations $\Sigma(Y_i - \tilde{Y})^2$ are called squared errors of prediction or residuals. Explain why they are "errors" and describe their role in the least squares criterion. Why is the regression line the line of "best fit"?
5. If a salesperson receives a base pay of $400 per month and a 6% commission on sales, what is the equation relating sales and income for this person?
6. Compare the standard error of estimate with a simple standard deviation. How are the two measures similar and how are they different?
7. Why are there usually two regression lines?
8. Indicate whether each of the following sets of facts could or could not exist simultaneously. If not, why not?
   a. $a = 5$, $b = 1.00$, the regression line crosses the X-axis at 5
   b. $a = 10$, $b = .5$, $s_y = 4$
   c. $b = -.4$, $s_y = 15$, $s_{y \cdot x} = 20$

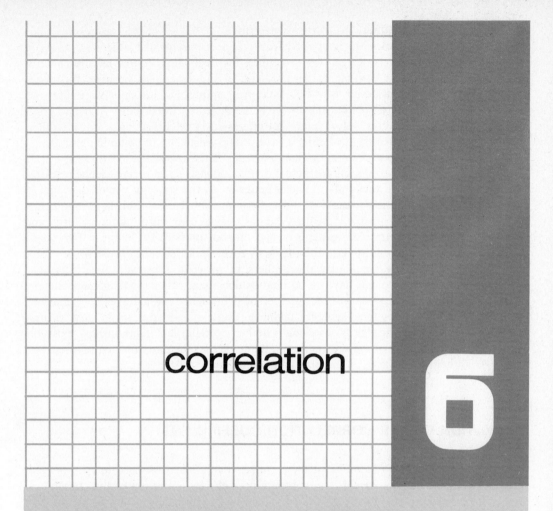

correlation

6

THE REGRESSION METHODS described in the last chapter are useful when one variable is specified as the predictor of the other. In terms of research design (see Prologue), regression procedures require that the independent variable be the predictor of the dependent variable. But sometimes such a design is not possible or desirable. For example, we might wish to know if a relationship exists between tested intelligence (IQ) and some measure of creativity. We are not interested in predicting one from the other; we are interested in whether a relationship exists and in obtaining an index of its direction and strength. Such an index is the **correlation coefficient**.

The correlation coefficient is used extensively in social science research, especially in the study of social behavior, in the study of personality, and in testing. In these areas it is frequently impossible for researchers to manipulate the lives of people. Scientists cannot decide who will produce children and thereby control the genetic composition of individuals; nor can they make some people introverts and others extroverts; nor can they cause some people to be more intelligent than others in order to observe the effects such differences might have on occupational success, educational accomplishment, social facility, happiness, interpersonal influence, or many other behaviors. Instead, they must settle for taking a sample of individuals who vary on these characteristics and observing whether correlational relationships exist between pairs of these variables, in the hope that the pattern of correlations will help them (and us) understand complex human behaviors.

DERIVATION OF THE CORRELATION COEFFICIENT

To develop a formula for the correlation coefficient we need to keep in mind that the square of the correlation coefficient represents the amount of variability in the Y_i that is associated with differences in the variable X. For example, suppose a reading test is given to a group of children upon their completion of first grade. Not every child obtains the same score on this reading test, and the extent to which the scores are dissimilar from child to child is reflected in the variability of the scores (their variance, s^2). Suppose that an intelligence test is also administered at the end of first grade, so that the mental age of each child is available. Since the children will not likely all have the same mental age, there will be variability in the distribution of mental ages as well. Now, what proportion of the difference (i.e., variability) in the children's reading scores is reflected in the variability of the mental ages of those subjects? That proportion is the square of the correlation coefficient. We now will consider the details underlying this statement.

Recall that the concept of variability expresses the extent to which individual scores differ from one another. We index the amount of variability with the variance, s^2, which statistic is based on the squared differences between each score and the mean of the distribution. The strategy below breaks up this variability into two parts; one aspect of variability is associated with another variable, and the other aspect is not. In the population, the total variability in reading score (i.e., in the Y_i) is symbolized by $\sigma_y{}^2$. The square of the standard error of estimate, $\sigma_{y \cdot x}{}^2$, reflects the amount of variability in reading which is *not* linearly related to mental age, since that square is the variability in Y_i *after* one predicts Y from its relationship to mental age (X). Therefore, $\sigma_y{}^2 - \sigma_{y \cdot x}{}^2$ represents the amount of variability in Y which is indeed associated with Y's relationship to X. When this quantity is expressed as a proportion of the total variability, $(\sigma_y{}^2 - \sigma_{y \cdot x}{}^2)/\sigma_y{}^2$, one has the square of the population correlation coefficient.

To examine the logic more concretely, consider Figure 6–1, which presents an imperfect relationship between mental age and reading score for five pupils. The score for one pupil (with mental age 110 months) and its deviations from the mean and from the regression line have been illustrated in detail. Call this pupil Johnny. Locate three points in Figure 6–1 for him: Y_i, Johnny's actual observed reading score; $\overline{Y}$, the mean of all five reading scores; and $\tilde{Y}_i$, the reading score which would be predicted for Johnny on the basis of the relationship between reading and mental age (i.e., the regression line at $X = 110$). To begin, consider how you might go about predicting Johnny's score if you had no knowledge of the relationship between reading and mental age. In the absence of any additional information, a good estimate of Johnny's reading score would be the mean score for the entire

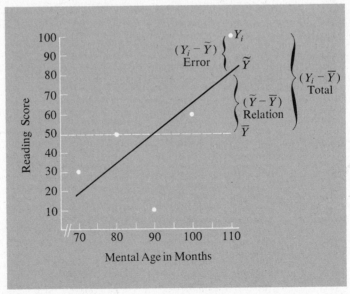

Fig. 6–1. Scatterplot showing three deviations: $(Y_i - \overline{Y})$, $(Y_i - \tilde{Y})$, and $(\tilde{Y} - \overline{Y})$

group of children in the sample, since the squared deviations of scores about their mean is a minimum. In the case of the data presented in Figure 6–1, that prediction would be $\overline{Y} = 50$ (depicted by the dashed white line in Figure 6–1). However, since Johnny actually scored 100 on the reading test ($Y_i = 100$), the prediction of 50 would be in error by

$$\left(Y_i - \overline{Y}\right) = 100 - 50 = 50$$

Now consider the score you would predict for Johnny if you knew that he had a mental age of 110 months and if you also knew the linear relationship between mental age and reading score. Given this information, you would predict Johnny's reading score to be the point on the regression line at $X = 110$, namely, $\tilde{Y} = 80$. Again, there would be error in this estimation:

$$(Y_i - \tilde{Y}) = 100 - 80 = 20$$

Notice that in this case, although there is still an error, it is less than when $\overline{Y}$ was used to make the prediction. Specifically, the reduction in error is given by

$$\left(Y_i - \overline{Y}\right) - (Y_i - \tilde{Y}) = 50 - 20 = 30$$

Simplifying the left side of the above expression, one obtains

$$\left(Y_i - \overline{Y}\right) - (Y_i - \tilde{Y}) = Y_i - \overline{Y} - Y_i + \tilde{Y}$$

$$Y_i - \overline{Y} - Y_i + \tilde{Y} = (\tilde{Y} - \overline{Y})$$

$$(\tilde{Y} - \overline{Y}) = 80 - 50 = 30$$

That is, as shown in Figure 6–1, the distance between the point and the mean of the Y_i can be partitioned into two parts:

$$\left(Y_i - \overline{Y}\right) = (\tilde{Y} - \overline{Y}) + (Y_i - \tilde{Y})$$

Since $(Y_i - \tilde{Y})$ reflects error still remaining after the prediction based on the regression line, $(\tilde{Y} - \overline{Y})$ must represent that segment of $(Y_i - \overline{Y})$ that is associated with X.

However, measures of variability and proportions of variability usually involve squared deviations summed over all subjects. It happens that when

$$\left(Y_i - \overline{Y}\right) = (\tilde{Y} - \overline{Y}) + (Y_i - \tilde{Y})$$

is squared and summed for all subjects, all the cross-product terms [i.e., $(\tilde{Y} - \overline{Y})(Y_i - \tilde{Y})$] drop from the expression and the result is

$$\Sigma\left(Y_i - \overline{Y}\right)^2 = \Sigma(\tilde{Y} - \overline{Y})^2 + \Sigma(Y_i - \tilde{Y})^2$$

This states that the total squared deviations of points about their mean may be partitioned into two parts. Notice that the quantity $\Sigma(Y_i - \tilde{Y})^2$ represents the squared deviations that still remain after prediction has been made with the regression line. Therefore, $\Sigma(\tilde{Y} - \overline{Y})^2$ must represent the segment of the total squared deviations in Y_i that indeed is associated with X. Thus the conceptual translation from

$$\Sigma(Y_i - \overline{Y})^2 \quad = \quad \Sigma(\tilde{Y} - \overline{Y})^2 \quad + \quad \Sigma(Y_i - \tilde{Y})^2$$

to

↓	↓	↓
Total squared deviations	= squared deviations associated with X	+ squared deviations associated with error in prediction using X

Since $\Sigma(\tilde{Y} - \overline{Y})^2$ represents the squared deviations in Y associated with X and $\Sigma(Y_i - \overline{Y})^2$ represents the total squared deviations in Y, their ratio

$$\frac{\text{Squared deviations associated with } X}{\text{total squared deviations}} = \frac{\Sigma(\tilde{Y} - \overline{Y})^2}{\Sigma(Y_i - \overline{Y})^2}$$

constitutes the proportion of squared deviations in Y associated with differences in X. Because variability is defined by squared deviations, this ratio is often said to express the **proportion of variability** in Y that is associated with differences in X.[1] If the value of the ratio was .55 for the mental age–reading example described above, it would mean that 55% of the total variability in reading scores is associated with differences in mental age.

Although the ratio indeed provides an index of degree of relationship by indicating the proportion of variability in Y that is associated with differences in X, it does not indicate the direction of the relationship. Is it positive or negative? That is, are high scores on one variable associated with high scores on the other (positive or direct relationship), or do high scores on one measure relate to low scores on the other (negative or inverse relationship)? Regrettably, since this proportion is a ratio of two positive quantities, it will always be positive. However, the square roots of a number in fact include both a positive and a negative root. Hence, there are two results of taking the square root of a^2, $+a$ and $-a$, because both $(+a)^2$ and $(-a)^2$ equal a^2. Therefore, one could take the square root of the proportion of variability, accepting the positive root if the relationship between X and Y is positive or the negative root if it is negative. This root of the proportion of variability in Y attributable to X is the correlation coefficient.

---

[1]Strictly speaking, such an interpretation requires that the conceptual variable X be measured without error. However, in practice this assumption is often disregarded.

The **Pearson product-moment correlation coefficient**, named after Karl Pearson and symbolized by *r*, is defined as

$$r = \sqrt{\frac{\Sigma(\tilde{Y} - \overline{Y})^2}{\Sigma(Y_i - \overline{Y})^2}}$$

This index, or coefficient of correlation, was developed by Karl Pearson, one of the fathers of modern statistics. It is called a coefficient, not because it is used to multiply other quantities, but because it is a unitless index. It is a number ranging from − 1.00 through .00 to + 1.00 that reflects the extent of a linear relationship. It is unitless because it is not expressed in the units of measurement—it is just a quantity that varies with the direction and degree of linear relationship. It is called a product-moment correlation because it is defined in terms of the product of the moments of the two variables, moment being a concept in mathematical statistics.

Other indices of correlation exist. One, the **Spearman rank-order correlation**, is presented in Chapter 13. In addition, other correlation coefficients can be used to express a wide variety of relationships, such as the degree of nonlinear association between two variables (called **eta** and pronounced "ATE-ah"), the degree of relationship between several predictor variables and one criterion (i.e., predicted) variable (**multiple correlation**), and the correlation between two variables after each variable's relationship with a third variable has been removed (**partial correlation**). These other correlation coefficients are described in advanced textbooks on statistics.

Once again the formula that defines a statistic is inconvenient to use in computational work. To use this expression requires that $\tilde{Y}$ be computed for every value of *X*. However, by substituting the equation ($\tilde{Y} = bX + a$) into the above formula and then substituting the raw-score formulas for both *b* and *a*, the above quantity becomes

$$r = \frac{N(\Sigma XY) - (\Sigma X)(\Sigma Y)}{\sqrt{[N\Sigma X^2 - (\Sigma X)^2][N\Sigma Y^2 - (\Sigma Y)^2]}}$$

The distinct advantage of this formula, in addition to its ease of computation when large numbers of subjects are involved, is that since the formula for *b*, the slope of the regression line, was used in its derivation, one does not have to be concerned about selecting the positive or negative root to indicate the direction of the relationship. The slope is positive for positive relationships and negative for negative relationships; therefore, since this formula includes the slope, it provides the appropriate sign for *r* without further labor.

Table 6–1 presents a numerical example of both formulas; one can see that they yield identical results. The regression line for the data is $\tilde{Y} = .75X + 1.25$. $\tilde{Y}$ has been computed for each pair of scores. Columns five through

6–1 Calculation of the Correlation Coefficient

X	Y	$\bar{Y}$	$\tilde{Y}$	$(Y_i - \bar{Y})^2$	$(Y_i - \tilde{Y})^2$	$(\tilde{Y} - \bar{Y})^2$	X^2	Y^2	XY
9	10	5	8.0	25	4.00	9.00	81	100	90
7	6	5	6.5	1	.25	2.25	49	36	42
5	1	5	5.0	16	16.00	.00	25	1	5
3	5	5	3.5	0	2.25	2.25	9	25	15
1	3	5	2.0	4	1.00	9.00	1	9	3
25	25			46	23.50	22.50	165	171	155

Definitional Formula

$$r^2 = \frac{\text{variability associated with } X}{\text{total variability}} = \frac{\Sigma(\tilde{Y} - \bar{Y})^2}{\Sigma(Y_i - \bar{Y})^2} = \frac{22.50}{46} = .49$$

$$r = \sqrt{r^2} = \sqrt{.49} = .70$$

Computational Formula

$$r = \frac{N\Sigma XY - (\Sigma X)(\Sigma Y)}{\sqrt{[N\Sigma X^2 - (\Sigma X)^2][N\Sigma Y^2 - (\Sigma Y)^2]}}$$

$$= \frac{5(155) - (25)(25)}{\sqrt{[5(165) - (25)^2][5(171) - (25)^2]}}$$

$$r = \frac{150}{\sqrt{(200)(230)}} = \frac{150}{\sqrt{46000}} = \frac{150}{214.476} = .70$$

$$r^2 = \text{proportion of variance associated with } X = (.70)^2 = .49$$

seven yield $\Sigma(Y_i - \bar{Y})^2$, $\Sigma(Y_i - \tilde{Y})^2$, and $\Sigma(\tilde{Y} - \bar{Y})^2$. The square of the correlation coefficient is defined as the proportion of the total variability in Y that is associated with differences in X. In this case $r^2 = .49$. The correlation coefficient, r, is the square root of this value, or .70. In the last three columns of Table 6–1 are the values X^2, Y^2, and XY for use in the computational formula. Again, N is the number of subjects or pairs of scores. These values are substituted into the computational formula, which gives the result $r = .70$. The square of r is .49, which is the proportion of variance in Y that is associated with differences in X.

PROPERTIES OF THE CORRELATION COEFFICIENT

The range of r

The correlation coefficient may assume values from -1 to $+1$. Consider first the case of a perfect linear relationship between X and Y in which the points all fall precisely on a line of nonzero slope (i.e., the line is not parallel to the X-axis). In this situation each point Y_i is identical to the point $\tilde{Y}$, therefore $\Sigma(\tilde{Y} - \bar{Y})^2$ and $\Sigma(Y_i - \bar{Y})^2$ are identical, and

$$r = \sqrt{\frac{\Sigma(\tilde{Y} - \bar{Y})^2}{\Sigma(Y_i - \bar{Y})^2}} = \sqrt{1} = \pm 1.00$$

Thus, in the case of a perfect relationship in which all points fall on the regression line ($b \neq 0$), the correlation coefficient will equal $+1.00$ or -1.00, depending upon whether the relationship is positive or negative.

Suppose now that there is no relationship between X and Y. The scatterplot may appear to be a circular clustering of points, and the regression line will be parallel to the X-axis and will be the same line as $\bar{Y}$. Since the two lines, $\tilde{Y}$ and $\bar{Y}$, are identical, then $\Sigma(\tilde{Y} - \bar{Y})^2$ must be equal to zero, leaving

$$r = \sqrt{\frac{\Sigma(\tilde{Y} - \bar{Y})^2}{\Sigma(Y_i - \bar{Y})^2}} = \sqrt{\frac{0}{\Sigma(Y_i - \bar{Y})^2}} = .00$$

Therefore, when there is no linear relationship, $r = .00$.

Can r ever be greater than 1.00 or less than -1.00? No, because it can be shown that $\Sigma(\tilde{Y} - \bar{Y})^2$ is always less than or equal to $\Sigma(Y_i - \bar{Y})^2$, which means that the fraction

$$\frac{\Sigma(\tilde{Y} - \bar{Y})^2}{\Sigma(Y_i - \bar{Y})^2}$$

will always be less than or equal to 1.00, and therefore so will r.

Hence, the correlation coefficient ranges in value between -1.00 and $+1.00$. It is ± 1.00 if all the points fall precisely on a line of nonzero slope, and it is .00 if there is no linear relationship at all. If there is no variability in Y_i (i.e., they are all the same value, as shown in Figure 6–3D, p. 146), r is not defined.

Variance interpretation of r^2

Examine the preceding facts in terms of the proportion of variability of the Y_i that is attributable to X (assuming that X is measured without error). Consider the heights of school children from kindergarten to sixth grade. Obviously, although the distribution of heights has a mean, most of the scores are not precisely equal to the mean value. They deviate from the mean, and the extent to which they deviate constitutes the total variability in Y_i (in this case, heights). But, careful thought suggests that at least some of the variability in the heights of children is attributable to the fact that the children span a large age range, and there is certainly a relationship between the age of the children and their heights. It might be of interest to ask what proportion of the variability in heights is associated with differences in age. If the numbers used in Table 6–1 were appropriate for this example, $r^2 = .49$ would mean that 49% of the variability in the heights of the children in the school is attributable to the fact that the children are of different ages. The higher this proportion, the greater the degree of linear relationship between age and height.

Suppose the baby sitter from the example in Chapter 5 has recorded both the hours worked and the money earned on sitting jobs over the last year. Obviously, the baby sitter did not earn the same amount of money for each job. That is, there is variability in the per-job return on baby sitting. What proportion of that variability is attributable to the hours worked on each job? If the pairs of points (hours, earnings) were plotted, they would all fall on a straight line with slope equal to the hourly rate. Since the regression line perfectly predicts earnings (Y) from hours worked (X), there is no variability in Y_i that is not predictable from X. Therefore, 100% ($r^2 = 1.00$) of the variability in the Y_i is attributable to X.

Suppose one attempts to relate height and IQ. There is almost no relationship between these variables. The regression line is parallel to and $\overline{Y}$ units above the X-axis (i.e., $\tilde{Y} = \overline{Y}$). Of course there would be variability in the heights of the subjects, but none of this variability would be attributable to the fact that the subjects differed in IQ. The fact that 0% of the variability in height is associated with IQ is reflected in the correlation of .00 ($r^2 = .00$, $r = .00$).

The relation between r and r^2

It has been said that the square of the correlation coefficient (r^2) may be interpreted as the proportion of variance in Y_i attributable to differences in X. The correlation coefficient is the square root of this proportion with the algebraic sign indicating the direction of the relationship. Since r is obviously not the same as r^2, one must be careful to interpret the terms properly. For example, consider the following table:

Correlation: r	Proportion of Variance: r^2
.10	.01
.20	.04
.30	.09
.40	.16
.50	.25
.60	.36
.70	.49
.80	.64
.90	.81
1.00	1.00

Note that a correlation of from .10 to .30 suggests that not very much variance in Y_i is associated with differences in X (1%–9%). In fact, a correlation of .50, which is frequently considered high in psychological and educational research, implies that only 25% of the variance in Y_i is associated with X. That means that 75% of the variability in Y_i is associated with factors other than X. One needs a correlation of .71 before one can say that half of the variability in Y_i is attributable to X. The implication is that in terms of proportion of variance in Y, the *unsquared* correlation coefficient (r) gives the impression of indicating a higher degree of relationship with X than does the proportion of variance interpretation.

The effect of scale changes on r

In Chapter 4 it was shown how adding a constant to every score or multiplying every score by a constant affects the value of the mean and variance of that distribution. It was also demonstrated that adding and multiplying (or subtracting and dividing) really amounts to changing the origin and the unit of the measuring scale. An extension of that exercise is to ask what happens to the correlation coefficient if the unit and/or origin of either scale of measurement (X and/or Y) is altered.

> The correlation coefficient does not change if every score in either or both distributions is increased or multiplied by a constant ($c \neq 0$). Thus, r is not altered by changes in the origin and unit of the measurement scale.

The algebraic proof of this assertion is given in Table 6–2 (optional). Adding or subtracting, multiplying or dividing, or both adding (or subtracting) and multiplying (or dividing) X and/or Y by the same or different nonzero constants does not alter the value of r. Therefore, r does not change when one or both scales of measurement are changed.

This result has important implications for the use of the correlation coefficient: Whether measurement is in feet or inches, minutes or seconds, units or dozens, the correlation between the variables will be the same. If r indexes the degree of relationship between two variables (e.g., age and height), it must be the same regardless of whether age is measured in months or years and height is measured in inches or centimeters. Indeed, the fact that r remains the same, even though the origin and/or unit of measurement for one or both of the variables is changed, gives this statistic a large range of applications.

The relation between correlation and regression

Slope and correlation   By certain algebraic manipulations it can be shown that

$$r = b_{yx}\left(\frac{s_x}{s_y}\right)$$

This expression states that the correlation coefficient is a joint function of the slope of the regression line (b_{yx}) and the standard deviations of the two variables (s_x and s_y). Consequently, the size of r is not equivalent to the slope of the regression line alone. However, if the standard deviations of the X and Y variables are made identical by converting both the X and Y distributions to standard scores (so that they both have a standard deviation of 1.00), then

$$r = b_{z_y z_x}\left(\frac{s_{z_x}}{s_{z_y}}\right) = b_{z_y z_x}\left(\frac{1}{1}\right)$$

$$r = b_{z_y z_x}$$

in which the subscripts z_y and z_x indicate values determined on the standardized X and Y distributions.

In short, when both X and Y distributions are in standard score form (and thus have the same standard deviation), the correlation coefficient is precisely the slope of the regression line. Note, however, that the slope of the line computed with raw scores and the slope computed with standardized scores will not likely be the same value. That is, ordinarily $b_{yx} \neq b_{z_y z_x}$.

r_{xy} and r_{yx}   Usually there will be two regression lines, one for the regression of Y on X and the other for X on Y. However, the correlation coefficient for these two cases is the same. That is, $r_{xy} = r_{yx}$. The truth of this

6–2 Proof that No Change Occurs in the Correlation Coefficient with a Change in Origin or Unit

Operation	Explanation
Change in Origin (Adding a Constant)	
1. $r = \dfrac{\Sigma(X-\bar{X})(Y-\bar{Y})}{\sqrt{\Sigma(X-\bar{X})^2\Sigma(Y-\bar{Y})^2}}$	1. Definition of r (see next section).
2. $= \dfrac{\Sigma\big[(X+c)-(\bar{X}+c)\big]\big[(Y+k)-(\bar{Y}+k)\big]}{\sqrt{\Sigma\big[(X+c)-(\bar{X}+c)\big]^2\Sigma\big[(Y+k)-(\bar{Y}+k)\big]^2}}$	2. Substituting $(X+c)$ for X and $(Y+k)$ for Y; as a result, the means will be $\bar{X}+c$ and $\bar{Y}+k$.
3. $= \dfrac{\Sigma(X+c-\bar{X}-c)(Y+k-\bar{Y}-k)}{\sqrt{\Sigma(X+c-\bar{X}-c)^2\Sigma(Y+k-\bar{Y}-k)^2}}$	3. Removing parentheses.
4. $r = \dfrac{\Sigma(X-\bar{X})(Y-\bar{Y})}{\sqrt{\Sigma(X-\bar{X})^2\Sigma(Y-\bar{Y})^2}}$	4. Subtracting the c's and k's within parentheses leaves the formula for r unchanged.

Change in Unit (Multiplying by a Constant)

1. $r = \dfrac{\Sigma(cX - c\bar{X})(kY - k\bar{Y})}{\sqrt{\Sigma(cX - c\bar{X})^2 \Sigma(kY - k\bar{Y})^2}}$

    1. Definition of r with cX and kY substituted for X and Y; as a result, the means will be $c\bar{X}$ and $k\bar{Y}$.

2. $= \dfrac{\Sigma c(X - \bar{X})k(Y - \bar{Y})}{\sqrt{\Sigma[c(X - \bar{X})]^2 \Sigma[k(Y - \bar{Y})]^2}}$

    2. Factoring in the manner of $ab - ac = a(b - c)$.

3. $= \dfrac{ck\Sigma(X - \bar{X})(Y - \bar{Y})}{\sqrt{c^2 k^2 \Sigma(X - \bar{X})^2 \Sigma(Y - \bar{Y})^2}}$

    3. In the numerator, $\Sigma cW = c\Sigma W$, and in the denominator, $(ab)^2 = a^2 b^2$.

4. $= \dfrac{ck\Sigma(X - \bar{X})(Y - \bar{Y})}{ck\sqrt{\Sigma(X - \bar{X})^2 \Sigma(Y - \bar{Y})^2}}$

    4. $\sqrt{a^2 b^2 WZ} = ab\sqrt{WZ}$

5. $r = \dfrac{\Sigma(X - \bar{X})(Y - \bar{Y})}{\sqrt{\Sigma(X - \bar{X})^2 \Sigma(Y - \bar{Y})^2}}$

    5. Cancellation, leaving r unchanged.

can be readily seen by examining one form of the formula for r:

$$r = \frac{\Sigma(X - \bar{X})(Y - \bar{Y})}{\sqrt{\left[\Sigma(X - \bar{X})^2\right]\left[\Sigma(Y - \bar{Y})^2\right]}}$$

If the positions of X and Y are reversed, the formula retains the same meaning.

From an intuitive standpoint this is as it should be. If r is a measure of the degree of relationship between X and Y, the two variables should be related to the same extent regardless of the direction of prediction. Conversely however, the regression lines predict in score units, and if X and Y are not identical scales, then it is reasonable that the regression lines should be different for the two directions of prediction.

$s_{y \cdot x}$ **and** r The relationship between the square of the correlation coefficient and the square of the standard error of estimate is approximately

$$r^2 = 1 - \frac{s_{y \cdot x}^2}{s_y^2}$$

This equation states that the size of the relationship as reflected in the square of the correlation coefficient is a function of the ratio of the variability of points about the regression line (as expressed by $s_{y \cdot x}^2$) to the total variability of the Y scores (s_y^2).[2] More specifically, r^2 and r will be higher if the fraction

$$\frac{s_{y \cdot x}^2}{s_y^2}$$

is relatively small. In words, the correlation will be high if the variability of the points about the regression line is small relative to their total variability.

Moreover, since the magnitudes of both $s_{y \cdot x}^2$ and s_y^2 are reflected in a scatterplot, it is possible to judge the relative size of different correlations by observing their graphs. Consider the graphs in Figure 6–2. Plots A and B have approximately the same variability in Y_i but differ in the extent to which the points cluster about the line. Since the $s_{y \cdot x}$ in A is less than in B, the fraction

$$\frac{s_{y \cdot x}^2}{s_y^2}$$

is smaller, and thus the correlation is larger, for A than for B. The converse situation is presented in graphs C and D. The variability of Y_i is greater in D

---

[2]In the population the relationship is exact, but it is only approximate for samples.

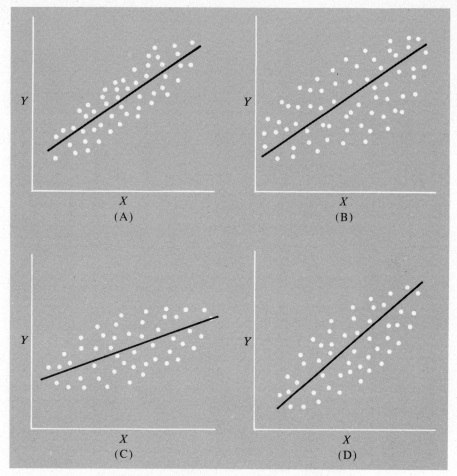

Fig. 6–2. Hypothetical scatterplots. In graphs A and B the s_y are equivalent, but since the $s_{y \cdot x}$ is less in A its correlation is higher. In graphs C and D, the $s_{y \cdot x}$ are comparable, but since the s_y is greater in D the correlation is higher.

than in C, while the $s_{y \cdot x}$ is approximately the same in each. Therefore, the ratio

$$\frac{s_{y \cdot x}^{2}}{s_{y}^{2}}$$

is smaller in D than in C, and consequently the correlation is higher in D.

Figure 6–3 provides several examples of different scatterplots with their respective correlation coefficients. Scatterplots A, B, and C reflect the concepts just discussed. Scatterplots D, E, and F represent some interesting

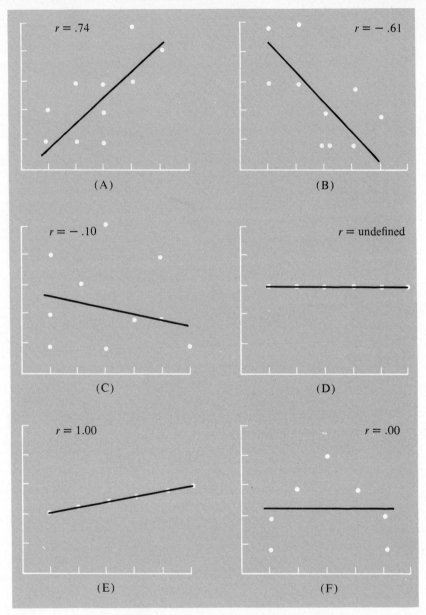

Fig. 6–3.   Sample scatterplots with their respective regression lines and correlations

special cases. Plot D shows a perfect relationship in the sense of clustering, but one which is maximally imperfect with respect to the variability of Y_i. Here the deviations $\Sigma(\tilde{Y} - \overline{Y})^2$ and $\Sigma(Y_i - \overline{Y})^2$ are 0, making $r = \frac{0}{0}$, which is best stated **r is undefined**. This result makes intuitive sense also, because prediction of Y_i is not improved by knowing the correlation, since regardless of the X value one always predicts Y_i to be $\overline{Y}$. Hence, knowledge of a person's X score does not assist prediction over using $\overline{Y}$, and the correlation is undefined. However, Plot E shows that if all points fall precisely on a regression line that does not have a slope of 0 (i.e., is not parallel to the X-axis), then the correlation is 1.00. Plot F depicts the case of a perfect but nonlinear relationship. Depending upon the nature of the curvilinearity, one can obtain r's of various sizes. Thus, it is not valid to conclude that if $r = .00$ there is no relationship between X and Y; rather, the conclusion should be that there is no *linear relationship* between X and Y. These facts punctuate the advice to examine the scatterplot before going on with procedures of linear relationship.

SAMPLING FACTORS THAT CHANGE THE CORRELATION COEFFICIENT

Up to this point the discussion of correlation has been concerned with developing a measure of the degree of linear relationship between two variables in a sample. However, in practice one usually wishes to use the sample correlation, r, to estimate the correlation that exists in the larger population. This parameter is symbolized by ρ (in English, read "rho"). The topics discussed in this section are concerned with practical problems that arise when one wants to estimate ρ with r. Obviously, the accuracy of any statistic as an estimate of a population value depends upon how representative the sample is of the population. The illustrations below describe in detail how r may be a distorted estimate of the population value when certain biases exist in the sample.

A restricted range

As discussed in a previous section, the size of r is a function of the relative values of $s_{y \cdot x}^2$ and s_y^2 such that r becomes large as $s_{y \cdot x}^2$ becomes small relative to s_y^2. Therefore, if the degree of clustering about the regression line is fairly constant over all segments of the line, then as the range and (usually) thus the variance of Y_i is reduced, the correlation is reduced.

Consider the following example. Suppose a new test of language skills is given to some pupils in grades one through six. Further, the mental age (MA)

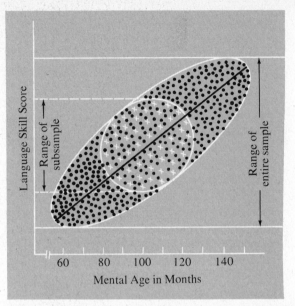

Fig. 6–4.   Hypothetical scatterplot showing relatively equal clustering of points about the regression line for a group of pupils and for a subsample of third graders, but a marked reduction in the range of the Y_i for the third graders. The correlation between language skills and mental age is much larger for the total group than for the subsample.

from a standardized IQ test is also available for each youngster, and the correlation between MA and language skills is computed for the entire sample. A hypothetical plot of this relationship is presented in Figure 6–4. Suppose the correlation is .76. In other words, for children in grades one through six performance on the test of language skills is rather closely related to MA. However, now consider the degree of relationship between these two tests for the third graders alone. The scores of the third graders are indicated in Figure 6–4 with x's rather than dots. Notice that for this subsample the variability of the Y_i (amount of vertical dispersion of the scores) is markedly reduced relative to the entire sample. However, the dispersion about the regression line for the subsample is about equal to that of the entire group of pupils. Since

$$r = \sqrt{1 - \frac{s_{y \cdot x}^2}{s_y^2}}$$

and restricting the range tends to reduce the size of s_y^2 relatively more than the size of $s_{y \cdot x}^2$, the fraction $s_{y \cdot x}^2 / s_y^2$ tends to become larger and thus r will become smaller. The correlation for just the third graders might be .28.

Simply put, the size of a correlation is a function of the variance of the Y_i relative to the standard error of estimate. Usually when the sample of scores is restricted, the correlation is less than it would be if the complete range were sampled. The safest course to follow is to limit the interpretation of a correlation to the population from which the sample was drawn. Thus, the correlation of .76 between the test of language skills and MA is appropriate for children in grades one through six, but that figure may not reflect the degree of relationship for children in a portion of that sample, such as third graders.

The limitation is sometimes overlooked when a given test is applied in a new context. Suppose that a test of reading ability is given to children in grades one through six, and these scores are correlated with some other criterion of reading competence. A high degree of relationship between the test and the criterion would suggest that the test is a valid indicator of reading ability. Suppose the correlation for the entire sample is .85. On the basis of this validity information, someone may propose to use the test on all second graders in order to single out those students who need special reading instruction. However, when the sample is restricted to second graders, the test may not be nearly as valid as it was for the entire sample. Perhaps, the correlation is only .15 for this subgroup, a figure that would certainly discourage using the test for the purpose suggested.

Extreme groups

The absolute size of r is usually increased when researchers select extreme groups of subjects. For example, a researcher might select very good and very poor readers in an elementary school in order to identify personality factors that may distinguish the groups. In the course of the research an IQ test might be given, and the researcher might wish to know the degree of relationship between IQ and reading achievement as measured by a standard reading test. Suppose the two tests are correlated $r = .84$. However, since the subjects were either very good or very poor readers, the scores would tend to be in two groups as displayed by the dots in Figure 6–5. All the cases which would have fallen between these two groups (x's) were eliminated; if they had been included, r would equal .66, not .84.

Why should selecting extreme groups on one variable increase the size of r over what would be obtained with more random sampling? The answer is seen more easily if the following formula for r is considered:

$$r = \frac{\Sigma(X - \bar{X})(Y - \bar{Y})}{\sqrt{\Sigma(X - \bar{X})^2 \Sigma(Y - \bar{Y})^2}}$$

The numerator is composed of the sum of products of the deviation of an X

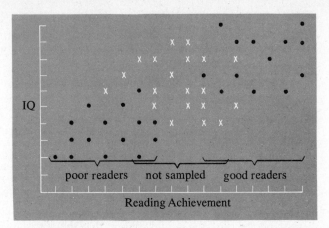

Fig. 6-5. Scatterplot for a sample made up of two extreme groups, poor
           and good readers (dots). The r = .84. The x's indicate scores
           of students of average reading ability, who were not sampled.
           The r over all subjects is .66. Selection of extreme groups
           increases r.

value from its mean and the deviation of a Y value from its mean. Therefore,
r becomes large when there are many subjects whose X and Y scores both
deviate markedly from their respective means. Selecting extreme groups
eliminates subjects whose scores would be near the means (which would be
located between the two extreme groups), leaving those subjects who have
large $(X_i - \bar{X})(Y_i - \bar{Y})$ values to predominate in the group. As a result, the
correlation coefficient is likely to be larger than if random sampling had been
employed.

Combined groups

One must also be cautious when a correlation between two variables is
computed for subjects from two groups that differ in their mean values on
one or both of the variables. For example, suppose the correlation between
mental age and fear of dying is approximately .10 for a group of first graders
but $-.40$ for sixth graders. If the two groups of children are combined into
one, the correlation reverses to approximately $+.52$. How is that possible?

Figure 6–6 illustrates what may happen when groups that differ in
mean values are combined for purposes of correlation. The first graders have
lower mental ages and are also less concerned with death, and therefore
points for them cluster in the lower left corner of the scatterplot. The
correlation between mental age and fear of death within that group is very
low, $r = .10$. Conversely, the sixth graders have higher mental ages and show
considerably more concern about death. Points for them therefore cluster in

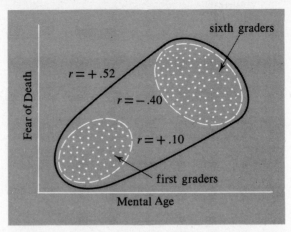

Fig. 6–6.   Illustration of how the relationship between mental age and fear of death may be approximately +.10 for first graders and −.40 for sixth graders, but +.52 for the two groups combined

the upper right corner of the plot. Within this group there is a moderately negative association between mental age and fear of death, $r = -.40$. But the scatterplot for these two groups together shows a positive relationship, extending from lower left to upper right with two extreme groups to enhance the correlation to +.52, which is a highly unrealistic representation of the true state of affairs.[3]

It is quite possible to have any combination of positive and negative correlations between disparate subgroups and the r of their combined group. Figure 6–7 presents some of these possibilities. If the group means are not very different, then combining these groups, while it may change the value of r somewhat, will not do so merely because of the combination itself. If, on the other hand, the group means are different, the r for the combined sample is not likely to faithfully represent the situation.

An extreme score

Last, consider the problem of an extreme case in a sample. In Figure 6–8, most of the scores (dots) cluster in a circular array, but there is one extreme case, x. Without x the correlation is .05, but with it $r = .48$. It is interesting to

[3]Although the data on mental age and fear of death are fictitious, they reflect some general trends found in research reported in the following papers: F. C. Jeffers, C. R. Nichols, and C. Eisdorfer, "Attitudes of Older Persons to Death," *Journal of Gerontology* 16 (1961): 53–56; A. Mauer, "Adolescent Attitudes toward Death," *Journal of Genetic Psychology* 105 (1964): 75–90; J. M. Natterson and A. G. Knudson, "Children and Their Mothers: Observations Concerning the Fear of Death in Fatally Ill Children," *Psychosomatic Medicine* 22 (1960): 456–65.

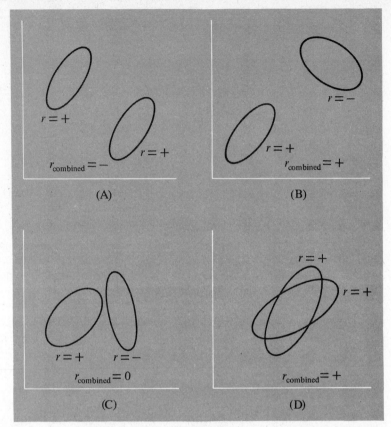

Fig. 6–7.   Graphs A, B, and C show effects on r of combining two groups
of subjects that have different means. Graph D shows that if the
two groups have comparable means the r may change somewhat, but not merely because of the regrouping.

note that if one of the dots, rather than x, were dropped, the correlation
would not change a great deal. Only cases that deviate markedly from the
general cluster have a very large effect. Because the numerator of r contains
the expression $(X_i - \overline{X})(Y_i - \overline{Y})$, a score deviating from the means $\overline{X}$ and $\overline{Y}$
will increase or decrease the numerator of r and therefore alter the size of a
correlation a great deal.

Ordinarily, if sufficient numbers of cases are sampled in a random
manner, such a situation does not occur. However, if the sample is small,
extreme cases can play a significant role in determining the size of r. In this
situation some researchers use other types of correlation coefficients, which
are not so sensitive to an extreme score (see Chapter 13).

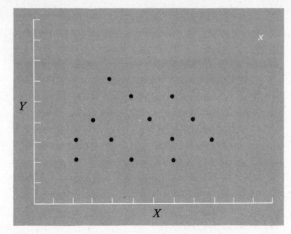

Fig. 6–8. Illustration of the effects of a single extreme case, x. Without x the correlation is .05, but with it $r = .48$.

CAUSALITY AND CORRELATION

The correlation coefficient represents the degree of observed linear association between two variables, not the extent of their causal relationship. Although having a runny nose correlates with having a cold, one would hardly suggest that the runny nose *causes* the cold. We may say that the cold causes the runny nose, but the correlation between runny nose and cold is the same regardless of which way one states it. Because the sun comes up when we wake in the morning does not prompt the megalomaniacal delusion that our getting up causes the sun to rise.

If A correlates with B, three possible causal relationships exist:

> A causes B,
> B causes A, or
> C causes both A and B.

Of course, C may be quite remote, with long causal chains interposed before A and B actually occur, but the point is that there is always the possibility that a third variable may produce an observed relationship between A and B. Consequently, one can never infer causality between two variables solely on the basis of their correlation. For example, the physical abuse of children is correlated with the frequency of various problem behaviors in adolescence and adulthood (e.g., alcohol and drug abuse, suicide, prostitution). However, many children who are abused are also unloved, rejected, and neglected. It is possible that the physical abuse per se, despite its horror, does not cause these problems but that the other factors, usually correlated with abuse, actually lead to these unhappy outcomes. Correlation suggests covariation but not necessarily causality.

6–3 Complete Computational Example for Major Statistics in Chapters 2 through 6

Raw Data

SAT	Grades
$\Sigma X = 3066$	$\Sigma Y = 8.60$
$(\Sigma X)^2 = 9400356$	$(\Sigma Y)^2 = 73.96$
$\Sigma X^2 = 1919198$	$\Sigma Y^2 = 22.40$
$N = 5$	$N = 5$

$$\Sigma XY = 5751.8$$

Intermediate Quantities

(I) $N\Sigma X^2 - (\Sigma X)^2 = 5(1919198) - 9400356 = 195634$

(II) $N\Sigma Y^2 - (\Sigma Y)^2 = 5(22.40) - 73.96 = 38.04$

(III) $N\Sigma XY - (\Sigma X)(\Sigma Y) = 5(5751.8) - (3066)(8.6) = 2391.40$

Means, Variances, and Standard Deviations

$$\overline{X} = \frac{\Sigma X}{N} = \frac{3066}{5} = 613.20$$

$$s_x^2 = \frac{N\Sigma X^2 - (\Sigma X)^2}{N(N-1)} = \frac{(I)}{N(N-1)} = \frac{195634}{5(4)} = 9781.70$$

$$s_x = \sqrt{s_x^2} = \sqrt{9781.70} = 98.90$$

$$\overline{Y} = \frac{\Sigma Y}{N} = \frac{8.6}{5} = 1.72$$

$$s_y^2 = \frac{N\Sigma Y^2 - (\Sigma Y)^2}{N(N-1)} = \frac{(II)}{N(N-1)} = \frac{38.04}{5(4)} = 1.90$$

$$s_y = \sqrt{s_y^2} = \sqrt{1.90} = 1.38$$

6–3 continued

Regression of Y on X

$$b = \frac{N(\Sigma XY) - (\Sigma X)(\Sigma Y)}{N\Sigma X^2 - (\Sigma X)^2} = \frac{(III)}{(I)} = \frac{2391.4}{195634} = .012$$

$$a = \overline{Y} - b\overline{X} = 1.72 - .012(613.20) = -5.638$$

$$\tilde{Y} = bX + a = .012X - 5.638$$

If $X = 620$, $\tilde{Y} = .012(620) - 5.638 = 1.80$

$$s_{y \cdot x} = \sqrt{\left[\frac{1}{N(N-2)}\right]\left[N\Sigma Y^2 - (\Sigma Y)^2 - \frac{[N\Sigma XY - (\Sigma X)(\Sigma Y)]^2}{N\Sigma X^2 - (\Sigma X)^2}\right]}$$

$$= \sqrt{\left[\frac{1}{N(N-2)}\right]\left[(II) - \frac{(III)^2}{(I)}\right]}$$

$$= \sqrt{\frac{1}{5(3)}\left[38.04 - \frac{(2391.4)^2}{195634}\right]} = \sqrt{.5873}$$

$$s_{y \cdot x} = .77$$

Correlation

$$r = \frac{N(\Sigma XY) - (\Sigma X)(\Sigma Y)}{\sqrt{[N\Sigma X^2 - (\Sigma X)^2][N\Sigma Y^2 - (\Sigma Y)^2]}} = \frac{(III)}{\sqrt{(I)(II)}}$$

$$= \frac{2391.40}{\sqrt{(195634)(38.04)}}$$

$$r = .88$$

COMPUTATIONAL PROCEDURES

This and previous chapters contain several computational examples for statistics; the computations have been presented as illustrations of isolated techniques. In practice, it is often the case that many of these statistics are computed within the context of a single problem. When that is done, certain computational conveniences are available. Therefore, Table 6–3 presents an integrated example that displays the calculation of several statistics. It also illustrates a shortcut for the computational labor: The task is facilitated if three subquantities are computed first.

$$\textbf{(I)} \quad N\Sigma X^2 - (\Sigma X)^2$$
$$\textbf{(II)} \quad N\Sigma Y^2 - (\Sigma Y)^2$$
$$\textbf{(III)} \quad N(\Sigma XY) - (\Sigma X)(\Sigma Y)$$

The table demonstrates how these three subquantities can be used to simplify the computation of the variance, slope, standard error of estimate, and correlation coefficient.

FORMULAS

1. Pearson product-moment correlation

$$r = \sqrt{\frac{\Sigma(\tilde{Y} - \bar{Y})^2}{\Sigma(Y - \bar{Y})^2}} \qquad \text{(definitional formula)}$$

$$r = \frac{N(\Sigma XY) - (\Sigma X)(\Sigma Y)}{\sqrt{\left[N\Sigma X^2 - (\Sigma X)^2\right]\left[N\Sigma Y^2 - (\Sigma Y)^2\right]}} \qquad \text{(computational formula)}$$

r^2 = proportion of Y variance associated with differences in X

2. Correlation and regression

$$r = b_{yx}\left(\frac{s_x}{s_y}\right) \qquad \text{(}r\text{ in terms of slope and variance)}$$

$$r = b_{z_y z_x} \qquad \text{(standard score version of relationship between } r \text{ and slope)}$$

$$r = \sqrt{1 - \frac{s_{y\cdot x}^2}{s_y^2}} \qquad \text{(}r\text{ in terms of the variability about the line, } s_{y\cdot x}^2, \text{ relative to the total variability, } s_y^2, \text{ in } Y\text{)}$$

EXERCISES

1. Below are three scores for each of nine subjects. Compute the correlation between A and B, A and C, and B and C. Add 5 to each score in distribution A and then multiply each score by 2. Recompute the correlation between A and B. Explain the effect of changing scales on the correlation.

Subject	A	B	C
1	3	1	7
2	1	8	4
3	7	9	5
4	3	2	2
5	8	3	1
6	0	7	6
7	5	3	10
8	6	1	7
9	9	0	3

2. Compute the correlation for the following data. Then add the score pair (12, 8) and recompute. Why does adding one score change the correlation so much? Can you think of other pairs of scores which would alter the situation in a different direction? Illustrate.

X	Y
0	3
2	2
1	0
2	4
3	2
5	4
4	5

3. If children are divided into two groups containing just the top 50 and bottom 50 children on a First Grade Readiness Test in a school of 500 children, why might the relationship between IQ and leadership potential as determined by a specially designed test given to these two extreme groups be unreasonably high relative to that obtained on the complete group of 500?

4. What squared deviations are said to compose the variability in Y associated with differences in X? Explain the logic of the reasoning.

5. Why are there usually two regression lines but only one correlation for a pair of variables, X and Y?

6. If the point $(\overline{X}, \overline{Y})$ is (15, 25), why will a pair of scores such as (50, 85) probably influence the regression and correlation between X and Y more than a pair of scores on the order of (10, 22)?

7. Explain why the correlation coefficient is identical to the slope of the regression line if the scores are in standard score form but not if raw scores are used.

8. In what way is the ratio of the variability about the regression line to the total variability in the Y_i related to the value of r?

9. Discuss and explain whether or not the following combinations of values are possible or impossible:
   a. $N = 25$, $b = .43$, $r = .55$
   b. $s_y = 9$, $r = .83$, $b = -.74$
   c. $s_y = 12$, $b = .70$, $s_{y \cdot x} = 15$
   d. $r = .15$, $b = .80$, $s_y = 12$, $s_x = 12$
   e. $N = 20$, $r = 1.00$, $s_{y \cdot x} = 2.5$
   f. $r = b_{z_x z_y}$

INFERENTIAL STATISTICS

topics in probability

THE FIRST SIX CHAPTERS of this book have been concerned with **descriptive statistics**—procedures that describe and summarize groups of measurements. Attention is now turned to **inferential statistics**, which includes techniques for making inferential decisions when only partial information is available.

The basis of inferential statistics is probability, and this chapter presents an introduction to the traditional topics of probability. The next chapter focuses on the application of the concept of probability to the processes of statistical inference. Although an introductory knowledge of formal probability will be helpful to students who plan to attend graduate school in the social sciences or who will study statistics again at a more advanced level, the material in this chapter is not a prerequisite for understanding the remainder of the text. Therefore, this chapter may be considered optional in some courses.

SET THEORY

Sets and relations among sets

A discussion of probability is greatly facilitated if it can be couched in terms of set theory. Therefore, some very elementary definitions and operations of set theory will be presented first.

A **set** is a well-defined collection of things, typically objects or events.

The ordinary concept of a set, such as a set of drinking glasses or a set of carving knives, is quite similar to the mathematical notion of a set. The critical factor in the definition is that there be some quality that the objects in the set possess or some rule that they follow which defines them as members of the set. For example, husbands are a set. Every human male, if he is married, is a member of the set of husbands. If he is not married, if he is a she, if he is a giraffe, then he is not a member of the set of husbands. Thus, a set is a collection of objects or events that are distinguishable from all other objects on the basis of some particular characteristic or rule. It is customary to label a set by some capital letter, such as A, B, or C.

An **element** of a set is any one of its members.

For example, if Jim Simpson is married, then he is an element of the set H of husbands.

A very important concept for the study of probability is that of subset.

162

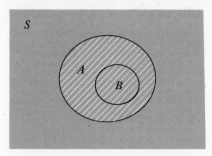

Fig. 7–1.   Set–subset relationship. S is all college students (universal set), A is college students at State (set), and B is a sample of the students at State (subset).

If every element in set A is also an element of set B, then A is a **subset** of B.

Since the phrase *is a subset of* is rather long and cumbersome, it is customary to write it with the symbol $\subseteq$. A subset, like a set, is represented by a capital letter.

There are two special sets that should be mentioned, the universal and the empty set.

The **universal set** includes all things to be considered in any one discussion. It is symbolized by S.

The **empty set** (**null set**) contains no elements. It is symbolized by $\varnothing$.

The universal set is simply an inclusive set that defines the general type of objects or events being discussed. A set is a subset of this universal set; that set in turn may have subsets and elements. For example, one of the primary applications to statistics of the concepts of set and subset is their analogy to the concepts of population and sample. A sample is a subset of the population, or universal set. In Figure 7–1, one might consider the entire space, S (the universal set), as the population composed of all college students. Perhaps the sample is composed only of college students attending State University and is symbolized by A. Further, the researcher randomly selects only a few college students at State, symbolized by B. Thus, $A \subseteq S$, $B \subseteq S$, and $B \subseteq A$.

If S is the entire space and A is a subset of S, then the symbol A', read "not A," denotes the set of all elements in S which are *not* in A. A' is called the **complement** of A.[1]

[1]Other symbols that are sometimes used for the complement of A include $\bar{A}$, $\tilde{A}$, A^c and $\sim A$.

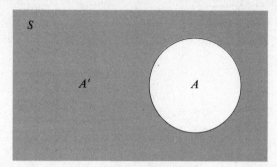

Fig. 7–2.   Complementation. S is the universal set, A is a subset of S, and
A' (not A) is the complement of A, since it contains all the
elements in S that are not in A.

If S includes all the integers from 1 to 10, and A includes 1, 4, 5, 8, and 9,
then A' includes 2, 3, 6, 7, and 10. The complement relationship is displayed
in Figure 7–2.

> If every element in A is also an element of B and if every element in
> B is also an element of A, then A **equals** B (if $A \subseteq B$ and $B \subseteq A$,
> then $A = B$).

If $A = \{1, 2, 5, 9\}$ and $B = \{5, 1, 9, 2\}$, then $A = B$. Note that the order in
which the elements of A and B are stated is irrelevant.

Operations

An important operation is the union of two sets.

> Given two sets, A and B, the **union** of A and B is the set of all
> elements that **(1)** are in A, **or (2)** are in B, **or (3)** are in both A and
> B.

The word *union* is symbolized by $\cup$, so that the union of A and B is written
$A \cup B$ and read "A union B."
    In Figure 7–3, the white-lined area represents the union of A and B,
symbolized $A \cup B$. The crucial word to remember about the concept of
union is *or*. The criterion for including any element in the **union** of A and B is
whether that element is contained in *A* **or** *B*. For example, if A includes $\{1, 2,
3, 4, 5\}$ and B includes $\{3, 4, 5, 6, 7, 8\}$, then $A \cup B = \{1, 2, 3, 4, 5, 6, 7, 8\}$.
Notice that numbers contained in both A and B are not represented twice in
$A \cup B$. That is, $A \cup B$ is *not* equal to $\{1, 2, 3, 4, 5, 3, 4, 5, 6, 7, 8\}$.

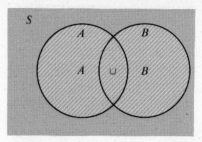

Fig. 7–3.   Union. *A* union *B* (*A* ∪ *B*) is represented by the white-lined area. *A* ∪ *B* contains all elements in *A*, in *B*, or in both *A* and *B*.

Another operation on sets is that of intersection.

Given two sets, *A* and *B*, the **intersection** of *A* and *B* contains all elements that are in **both *A* and *B***, but not in *A* or *B* alone.

The symbol for intersection is ∩, and "*A* intersection *B*" is written *A* ∩ *B*. In Figure 7–4, which illustrates the intersection of *A* and *B*, *A* ∩ *B* is the white-lined portion of the diagram. Thus, if *A* = {1, 2, 3, 4, 5} and *B* = {3, 4, 5, 6, 7, 8}, then *A* ∩ *B* = {3, 4, 5}.

Here the emphasis is on the word *and*, because in order for an element to be a member of the **intersection** of *A* and *B*, it must be contained in **both *A* and *B***. The intersection represents the common portion of two sets, or the elements shared by two sets.

It is useful to be more explicit about the criteria for union (∪) and intersection (∩). An element belongs to the **union** of *A* and *B* if it qualifies under *any one* of the following criteria: (1) If it is a member of *A*, (2) If it is a member of *B*, or (3) If it is a member of both *A* and *B*. However, in determining whether an element qualifies for inclusion in the **intersection** of *A* and *B*, (3) is the only criterion. An element must be a member of both *A* and

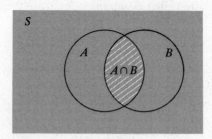

Fig. 7–4.   Intersection. *A* ∩ *B* is represented by the white-lined area, which includes all elements that are members of both *A* and *B*.

7-1 Terminology of Set Theory

1. Set: A well-defined collection of things.
2. Element: A member of a set.
3. Universal Set, S: The set of all things under discussion.
4. Empty Set, ∅: A set containing no elements.
5. Subset: If every element in A is also in B, A is a subset of B ($A \subseteq B$).
6. Complement: Set A' is the complement of set A if it contains every element in S that is not in A.
7. Equality: Set A equals set B if every element in A is also in B and every element in B is also in A. (If $A \subseteq B$ and $B \subseteq A$, then $A = B$.)
8. Union: The union of sets A and B, $A \cup B$, is the set of all elements that are (1) in A, (2) in B, or (3) in both A and B.
9. Intersection: The intersection of A and B, $A \cap B$, is the set of all elements that are in both A and B.
10. Disjoint Sets: Two sets are disjoint if they share no common elements ($A \cap B = \emptyset$).

B. The difference between the important words in the definitions also highlights the distinction between union and intersection. $A \cup B$ contains elements that are in A or B or in both A and B; whereas $A \cap B$ contains only those elements that are in both A and B simultaneously.

There is a special case of intersection that defines a particular relationship between two sets. Suppose $A \cap B$ contains no elements, that is, $A \cap B = \emptyset$. Such a condition says that A and B share no common elements. No element in A is also in B and no element in B is also in A. If $A \cap B = \emptyset$ then A and B are called **disjoint sets**, because there is no common element to join

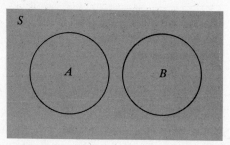

Fig. 7-5.   Disjoint sets. $A \cap B = \emptyset$, that is, A and B have no common elements.

them together. Figure 7–5 illustrates a pair of disjoint sets. Another pair of disjoint sets is A and its complement, A'. Since A' contains all elements not in A, it is clear that for any set A, A and A' are disjoint.

A summary of the terms and concepts discussed in this section is presented in Table 7–1.

SIMPLE CLASSICAL PROBABILITY

Probability usually involves thinking about an **idealized experiment** in which a given phenomenon is repeatedly observed an indefinite number of times under ideal conditions. For example, when someone says the chance of flipping a coin and obtaining a head is $\frac{1}{2}$, this means that over an uncountable number of flips of that coin under fair and ideal conditions, heads will turn up half the time. This imagined flipping of a coin over and over again is the idealized experiment.

In probability, the collection of all the elements of an idealized experiment is called the **sample space**, which corresponds to the universal set and is symbolized by S. The elements of this set are known as **elementary events**, or **outcomes**. An **event** is any subset of elements in S. Therefore, an event may be an elementary event, a subset of elementary events, or all elementary events (i.e., S). Thus, in the idealized experiment of rolling a single die, the sample space consists of the set of six elementary events (or outcomes): {1, 2, 3, 4, 5, 6}. Examples of an event in this idealized experiment include a roll of a deuce {2}, a roll of three or less {1, 2, 3}, and a roll of an even number {2, 4, 6}.

Simple classical probability involves making probabilistic statements about a sample space whose elementary events are all equally likely to occur. That is why discussions of probability frequently employ the flipping of coins, drawing of cards, and rolling of dice, because each of these experiments consists of a set of equally likely elementary events.

In a sample space containing equally likely elementary events, the **probability** of a given event A is defined to be the number of outcomes in A divided by the total number of outcomes in the sample space, S. If $\#(A)$ signifies the number of outcomes in A and $\#(S)$ symbolizes the number of outcomes in S, then $P(A)$, the "probability of event A," is given by

$$P(A) = \frac{\#(A)}{\#(S)}$$

In tossing a coin, the sample space consists of the two equally likely outcomes, heads and tails, and the probability of a head is

$$P(\text{head}) = \frac{\#(\text{head})}{\#(\text{possible outcomes})} = \frac{1}{2}$$

In the idealized experiment of throwing a die, the sample space consists of the equally likely alternatives $\{1, 2, 3, 4, 5, 6\}$, and the probability of throwing a two is $\frac{1}{6}$. The probability of a five is also $\frac{1}{6}$. The probability of a four or less is $\frac{4}{6}$.

It is important to remember that in any situation, the probability of an event can neither be less than .00 nor greater than 1.00. A probability value is always between .00 and 1.00, inclusive.

Conditional probability

Conditional probability refers to a situation in which the probability of a given event is dependent upon the occurrence or nonoccurrence of another event.

Consider the following situation. Suppose one takes the 12 picture cards from a deck of playing cards (4 jacks, 4 queens, and 4 kings) and shuffles them. Define event A as drawing a king from this deck. Since there are four kings in the deck of 12 cards, the probability of event A is

$$P(A) = \frac{\#(A)}{\#(S)} = \frac{4}{12} = \frac{1}{3}$$

Suppose that whatever card is drawn on this trial is *not* put back into the deck before making a second draw. Such a procedure is known as **sampling without replacement**, because cards are not replaced into the deck after their selection. Now consider the probability of event B, obtaining a king on the second draw. This probability depends upon the result of the first draw and thus is an instance of conditional probability. If a king was drawn first and not replaced, then for the second draw there would be three kings left in the deck, which now consists of 11 cards. Thus, $P(B) = \frac{3}{11}$. However, if a card other than a king was drawn on the first trial, all four kings would remain in the deck of 11. Then, $P(B) = \frac{4}{11}$.

> **Conditional probability** is the probability that an event will occur given that some other event has already occurred. The probability that event B will occur given that event A has already occurred is written
>
> $$P(B|A)$$
>
> which is read "the probability of B given A."

In the example above, the probability of B (drawing a king on the second draw without replacement) is conditional upon the occurrence of event A (drawing a king on the first draw). Given that A has indeed occurred, the probability of B is written

$$P(B|A) = \frac{3}{11}$$

since three kings are left in the deck of 11 remaining cards after one king has been selected on the first trial.

Dependent and independent events

The above example also illustrates the concept of dependent events.

> Two events are said to be **dependent** if the occurrence of one event alters the probability of the occurrence of the other.

Since the probability of obtaining a king on the second draw (event B) is either $\frac{3}{11}$ if a king was drawn on the first trial (event A) or $\frac{4}{11}$ if a non-king was picked first, events A and B are said to be dependent: The probability of B depends on the occurrence or nonoccurrence of event A.

> Two events are said to be **independent** if the occurrence of one event does not alter the probability that the other event will occur. This can be expressed in terms of conditional probability as

$$P(B|A) = P(B)$$

When $P(B|A) = P(B)$, this implies that the probability of B is not affected by (i.e., not conditional upon) the occurrence of A.

In the example given above for conditional probability, cards are drawn from a 12-card deck without replacement. But suppose cards are replaced after a selection is made, a circumstance called **sampling with replacement**. Now consider the probability of event B, obtaining a king on the second draw, as a function of the occurrence of event A, obtaining a king on the first draw. If a king is selected first and put back into the deck before the second draw, the probability of getting a king on the second draw is $\frac{4}{12}$, since there will be four kings in the complete deck of 12 cards. If a king is not found on the first draw and the card is replaced, there are again four kings in 12 cards. Thus, the probability of getting a king on the second draw is independent of (i.e., not conditional upon) the outcome of the first event. In symbols,

$$P(B|A) = P(B) = \frac{4}{12}$$

which means that the probability of B given that A has occurred is simply the probability of B, ignoring what has happened with respect to event A.

An interesting application (or lack of application) of the concept of independence is known as the **gambler's fallacy**. The gambler's fallacy is the failure to appreciate the independence of some sequential events, and it is epitomized by a gambler who has lost 10 straight times at cards and feels that there is a better than average chance of winning the next game because of this long string of misfortunes. Another example is the parent who feels that after five boys in a row, surely the probability of having a girl as a sixth child is greater than .50. Actually, the theoretical probabilities of these events do not depend upon previous events. The probability of having a girl (or a boy) is approximately .50 for any given birth, and the sex of a forthcoming child is probably independent of the sex of the previous children. Therefore, the probability of having a girl is approximately .50 whether it is one's first child or whether the child will have 10 brothers (or sisters) upon its arrival. So, too, for the gambler. It is no more likely that the gambler will be a winner after a bad losing streak than during a fantastic winning streak.

Games of chance are very likely to be independent in actuality as well as theoretically. On the other hand, if a basketball team loses 10 games in a row before facing its major rival, a school of approximately equal ability, the probability of winning the game may be more than .50 if the team is high-spirited and determined to salvage an otherwise poor season. Of course, the probability of winning could be less than .50 if the 10 losses have so demoralized the team that it can hardly face the humiliation of another game. Thus, although the independence of events is an important characteristic in assessing probabilities, in practice it is sometimes difficult to determine whether events are independent or not.

Mutually exclusive events

Two events may also be mutually exclusive.

> Two events, A and B, of the same idealized experiment are considered to be **mutually exclusive** if they share no elementary events, that is, if A and B are disjoint sets ($A \cap B = \varnothing$).

When $A \cap B = \varnothing$, this implies that if one event has occurred, the other has not. Consider the event A of drawing a king on a single draw from that special deck composed only of four jacks, four queens, and four kings. Abbreviate the cards by giving them two initials, the first for the denomination and the second for the suit (e.g., KS for the King of Spades). Since event A is drawing a king, set A contains four elements (all the kings):

$$A = \{KS, KC, KH, KD\}$$

Let event B be drawing a queen in a single draw. Set B also contains four elements:

$$B = \{QS, QC, QH, QD\}$$

Note first that there is no element in A that is also in B. Therefore, A and B are disjoint sets and $A \cap B = \emptyset$. Note also that if on a single draw A occurs, B has not occurred; if you draw a king you do not also draw a queen. In this case, drawing a king and drawing a queen are mutually exclusive events.

In contrast, consider now the events of drawing a king (event A) and drawing a red card (event B) from this special deck:

$$A = \{KH, KD, KS, KC\}$$
$$B = \{KH, KD, QH, QD, JH, JD\}$$

These events are not mutually exclusive because they have two common elements, $\{KH, KD\}$; therefore $\#(A \cap B) = 2$, not 0, and A and B are not disjoint. Further, it is not necessarily true that if A occurs, B does not occur. If the king of hearts or diamonds is drawn, then both event A and event B have simultaneously occurred. Therefore, two events are mutually exclusive if their sets are disjoint, and when this is true, the occurrence of one event implies that the other event did not occur.

Although mutually exclusive events may appear to be very similar to independent events, the two concepts should not be confused. The difference between them is that *mutually exclusive* refers to the sharing or not sharing of *elements of sets*, whereas *independent* is defined in terms of the *probabilities* of two events. Therefore:

$$\text{Mutually exclusive events means } A \cap B = \emptyset$$
$$\text{Independent events means } P(B|A) = P(B)$$

PROBABILITY OF COMPLEX EVENTS

Probability of intersection, $A \cap B$

What is the probability of $A \cap B$?

The probability that two events, A and B, will both occur is

$$P(A \cap B) = P(A)P(B|A)$$

This means that the probability that both A and B will occur is the probability of A times the probability of B given that A has occurred.

In the special case when A and B are independent events, $P(B|A) = P(B)$—which is the definition of independent events. In such a case, substituting $P(B)$ for $P(B|A)$ in the above expression, we have the following:

The probability that two *independent* events, A and B, will both occur is

$$P(A \cap B) = P(A)P(B)$$

For example, recall the deck of 12 face cards used in the previous example and consider determining the probability of drawing 2 kings in succession without replacement. Restated, this amounts to finding the probability of obtaining a king on the first draw (event A) *and* obtaining a king on the second draw (event B). Thus, the task is to compute the probability that both A and B will occur, which can be symbolically stated in terms of their intersection:

$$P(A \cap B)$$

The formula given above states that the required probability will be the product of the probability of A times the conditional probability of B given that A has occurred. The probability of drawing a king on the first trial is $\frac{4}{12}$, since there are 4 kings in the deck of 12 cards. The conditional probability of drawing a king on the second trial given that a king has already been drawn and not replaced is $\frac{3}{11}$ (three kings would remain in the deck, which now contains 11 cards). Thus, the required probability is

$$P(A \cap B) = P(A)P(B|A)$$

$$P(A \cap B) = \frac{4}{12} \cdot \frac{3}{11} = \frac{12}{132} = .09$$

Thus, on the average one could expect to draw two consecutive kings without replacement only 9 times in 100 such two-card draws.

Why are the probabilities multiplied? Consider the solution of the above problem in more detail. The task is to determine the probability of $A \cap B$, which by the laws of classical probability should be given by the number of ways $A \cap B$ can occur divided by the total number of two-card sequences in the sample space. In symbols,

$$P(A \cap B) = \frac{\#(A \cap B)}{\#(S)}$$

First, how many ways can $A \cap B$ occur? Simply stated, there are four ways a king could be obtained on the first draw (one way for each king in the deck), and *for each one of those four ways* there exist three possible ways a king could be selected on the second trial given that one was picked on the first

7–2 Number of Ways of Drawing Two Consecutive Kings $(A \cap B)$

First Draw	Second Draw	Number of Ways
King of spades	King of clubs King of hearts King of diamonds	3
King of clubs	King of spades King of hearts King of diamonds	3
King of hearts	King of spades King of clubs King of diamonds	3
King of diamonds	King of spades King of clubs King of hearts	3
$[\text{First draw } (A)]$	$\times\ [\text{second draw } (B\|A)]$	$=[\text{total } (A \cap B)]$
4	$\times$ 3	$=$ 12

draw. Therefore, there are four threes, or 4×3 ways of getting two consecutive kings without replacement. The 4 is the $\#(A)$ and the 3 is the $\#(B|A)$, so

$$\#(A \cap B) = \#(A) \times \#(B|A) = 4 \times 3 = 12$$

These 12 ways of obtaining two kings are listed in Table 7–2.

Second, how many total possible outcomes are there in this situation? The logic is the same: There are 12 possible outcomes on the first draw (one for each card in the deck), and *for each one of these 12 outcomes* there are 11 possible results on the second draw (one for each of the remaining 11 cards). Thus,

$$\#(S) = \#(S_A) \times \#(S_{A|B}) = 12 \times 11 = 132$$

where S_A and $S_{B|A}$ are the sample spaces for events A and $B|A$, respectively.

The probability in question is found by dividing the number of specified outcomes by the total number of possible outcomes in the idealized experiment. As determined above, this amounts to

$$P(A \cap B) = \frac{\#(A \cap B)}{\#(S)} = \frac{\#(A) \times \#(B|A)}{\#(S_A) \times \#(S_{B|A})} = \frac{4 \times 3}{12 \times 11} = \frac{12}{132} = .09$$

$$P(A \cap B) = P(A) \times P(B|A) = \frac{4}{12} \times \frac{3}{11} = \frac{12}{132} = .09$$

To summarize, the first part of the preceding expression is the probability of event A, $P(A)$, and the second part is the conditional probability of B given A, $P(B|A)$. The probability of the intersection of two events equals the probability of A times the conditional probability of B given A. In the case being illustrated, the probability of drawing two consecutive kings is $\frac{4}{12} \times \frac{3}{11}$ $= \frac{12}{132} = .09$. This means that if you were to repeatedly draw two consecutive cards from the special deck, on the average you might expect that nine of every 100 such pairs would be pairs of kings.

Turning to a different example, suppose you arrange to play a game of racquetball with a classmate. Your partner warns that arranging a 5 P.M. game is chancy, since afternoon laboratory classes run late about 50% of the time. Moreover, the probability is only .60 that a court will be available at 5 P.M. What is the probability that the two of you play your game at 5 P.M?

Define event A to be "your partner is able to play at 5 P.M.," and event B to be "a court is available." Thus, the probability of the complex event, having a partner *and* a court free, is

$$P(A \cap B) = P(A)P(B|A)$$

Since the probability of a free court is not altered by whether your partner can play, A and B are independent, and $P(B|A) = P(B)$, which in this case is .60. $P(A)$ is the probability that your partner can play, which is .50, so

$$P(A \cap B) = P(A)P(B|A) = P(A)P(B) = (.50)(.60) = .30$$

In words, the likelihood that you can play a racquetball game with your partner at 5 P.M. is .30, or on 3 of every 10 afternoons.

Please note that one needs to remember only that

$$P(A \cap B) = P(A)P(B|A)$$

This formula allows for the independence of A and B, in which case $P(B|A) = P(B)$. The real problem in using this expression is translating the verbal statement of the problem into events A and B, and then determining if the intersection of A and B is required. Most often, the intersection of A and B is needed if the problem states (or could be restated to say) that *both A and*

B must occur to satisfy the complex condition. The key word is *and*:

1. You must draw a king on the first try *and* on the second.
2. Your partner must finish in time *and* the court must be free.

The word *and* signifies that the joint occurrence of *A* and *B* is required, and thus $A \cap B$ is an appropriate description of the problem.

The formula

$$P(A \cap B) = P(A)P(B|A)$$

also provides an expression for the conditional probability of *B* given *A*. Dividing by $P(A)$ and transposing, we have

$$P(B|A) = \frac{P(A \cap B)}{P(A)}$$

Therefore, the **conditional probability of *B* given *A*** equals the probability that *A* and *B* will both occur divided by the probability of *A*. In symbols,

$$P(B|A) = \frac{P(A \cap B)}{P(A)}$$

If the probability that your team will win a two-game basketball tournament is .15 $[P(A \cap B) = .15]$ and the probability of winning your first game is .50 $[P(A) = .50]$, then before the tournament begins you can speculate that the probability that you win the tournament given that you win the first game is

$$P(B|A) = \frac{P(A \cap B)}{P(A)} = \frac{.15}{.50} = .30$$

Probability of union, $A \cup B$

The probability of the occurrence of either one of two events, *A* or *B*, that is, $A \cup B$, is

$$P(A \cup B) = P(A) + P(B) - P(A \cap B)$$

This states that the probability that either *A* or *B* will occur equals the probability of *A* plus the probability of *B* minus the probability of both *A* and *B*. However, the probability of $A \cap B$ is zero if *A* and *B* are mutually exclusive. (This is because $A \cap B = \varnothing$ and $P(\varnothing) = .00$). Therefore, if *A* and

B are mutually exclusive events, the preceding formula is simplified:

If *A* and *B* are mutually exclusive, the probability that either *A* or *B* will occur is

$$P(A \cup B) = P(A) + P(B)$$

This states that if *A* and *B* are mutually exclusive the probability that either *A* or *B* will occur is the sum of the probabilities of *A* and *B*.

Return to the special deck of 12 face cards. What is the probability of selecting either a king or a red card in a single draw? If event *A* is drawing a king and event *B* is drawing a red card, then the question is: What is the probability of *A or B*, which can be represented as the probability of the union of these events, $P(A \cup B)$.

The longer formula given above requires three probabilities. The probability of getting 1 of the 4 kings in a deck of 12 cards is $P(A) = \frac{4}{12}$. Since half the cards are red, $P(B) = \frac{1}{2}$. The probability of $A \cap B$ is simply the likelihood that both a king and a red card turn up, which will happen if either the king of hearts or king of diamonds is selected—2 of the 12 cards. Thus, $P(A \cap B) = \frac{2}{12}$. Therefore, the required probability is

$$P(A \cup B) = P(A) + P(B) - P(A \cap B)$$
$$= \frac{4}{12} + \frac{1}{2} - \frac{2}{12} = \frac{8}{12}$$
$$= .67$$

But what is the rationale of this formula? Why does one add the probabilities of *A* and *B*, and why must their intersection be subtracted? Consider the example in more detail. The problem is to determine the probability of the union of *A* and *B*, and according to the laws of classical probability this should equal the number of outcomes in $A \cup B$ divided by the total number of outcomes on a single draw:

$$P(A \cup B) = \frac{\#(A \cup B)}{\#(S)}$$

Using the same abbreviations as before, the possible outcomes for events *A* (a king) and *B* (a red card) are

$$\#(A) = (KH, KD, KS, KC) = 4$$
$$\#(B) = (KH, KD, QH, QD, JH, JD) = 6$$

In order for an outcome to qualify for $A \cup B$ it must be in either *A* or in *B*. If one simply adds the outcomes in *A* to those in *B*, some outcomes will be counted twice—namely, those that are in both *A* and *B*, that is, in $A \cap B$. In this example, the outcomes in the intersection of *A* and *B* are *KH* and *KD*. If these shared outcomes are subtracted from the sum, the correct number of elements in $A \cup B$ is produced. Specifically,

$$\#(A) \quad + \quad \#(B) \quad - \#(A \cap B)$$

$$\overbrace{(KH + KD + KS + KC)} + \overbrace{(\cancel{KH} + \cancel{KD} + QH + QD + JH + JD)} - \overbrace{(\cancel{KH} + \cancel{KD})}$$

Simplifying, one finds that

$$KH + KD + KS + KC + QH + QD + JH + JD$$

are the eight outcomes that are either a king or a red card. Thus,

$$\#(A \cup B) = \#(A) + \#(B) - \#(A \cap B) = 4 + 6 - 2 = 8$$

The probability of $A \cup B$ is

$$P(A \cup B) = \frac{\#(A \cup B)}{\#(S)}$$

$$= \frac{\#(A) + \#(B) - \#(A \cap B)}{\#(S)}$$

$$= \frac{\#(A)}{\#(S)} + \frac{\#(B)}{\#(S)} - \frac{\#(A \cap B)}{\#(S)} = \frac{4}{12} + \frac{6}{12} - \frac{2}{12} = \frac{8}{12} = .67$$

$$\downarrow \qquad\qquad \downarrow \qquad\qquad \downarrow$$

$$P(A \cup B) = P(A) + P(B) - P(A \cap B)$$

Two further examples. First, suppose a student applies to two graduate schools, A and B, and assesses the chances of getting into them as $P(A) = .10$ and $P(B) = .25$. What is the probability of getting into at least one of the schools? Since either A or B satisfies the required outcome, the probability is given by

$$P(A \cup B) = P(A) + P(B) - P(A \cap B)$$
$$= .10 + .25 - (.10)(.25) = .325$$

Second, what is the probability of getting an odd number or a six in a single roll of a die? If A is the event of getting an odd number and B the event of getting a six, then the required probability is

$$P(A \cup B) = P(A) + P(B) - P(A \cap B) = \tfrac{3}{6} + \tfrac{1}{6} - 0 = \tfrac{4}{6}$$

Note that $P(A \cap B) = 0$ because A and B are mutually exclusive. As before, it is not necessary to change the formula for cases in which A and B are mutually exclusive. Instead, one merely sets $P(A \cap B)$ equal to zero.

The word *or* in a verbal statement of a problem tends to imply that the union of two or more events is required, just as the word *and* signals the intersection of two or more events. Thus, the above discussion on the probability of the union of two events dealt with (1) either a king *or* a red card, (2) getting accepted at either school *A or* school *B*, and (3) rolling an odd number *or* a six.

Although the two probabilistic laws described above have been stated only in terms of two events, the formulas can be generalized to cases involving more than two events. For example, if *A*, *B*, and *C* are *independent*, then

$$P(A \cap B \cap C) = P(A)P(B)P(C)$$

If *A*, *B*, and *C* are *mutually exclusive*, then

$$P(A \cup B \cup C) = P(A) + P(B) + P(C)$$

Consider a last example, which involves a combination of union and intersection. What is the probability of selecting either an ace or a king on the first draw and a red card or a face card (*J, Q, K*) on the second draw from an ordinary deck of 52 cards if selection is performed with replacement? Let

$$A = \text{an ace on the first draw}$$
$$B = \text{a king on the first draw}$$
$$C = \text{a red card on the second draw}$$
$$D = \text{a face card on the second draw}$$

Reducing the statement of the problem to symbols, the required complex event is [(*A* or *B*) and (*C* or *D*)], and thus the probability is

$$P[(A \cup B) \cap (C \cup D)]$$

Because the selection is performed with replacement, the two events $(A \cup B)$ and $(C \cup D)$ are independent. Further, *A* and *B* are mutually exclusive, whereas *C* and *D* are not. Therefore, expanding the above expression,

$$P[(A \cup B) \cap (C \cup D)] = [P(A) + P(B)]$$
$$\times [P(C) + P(D) - P(C \cap D)]$$

The necessary quantities are:

$$P(A) = \tfrac{4}{52} = \tfrac{1}{13}$$

$$P(B) = \tfrac{4}{52} = \tfrac{1}{13}$$

$$P(C) = \tfrac{26}{52} = \tfrac{1}{2}$$

$$P(D) = \tfrac{12}{52} = \tfrac{3}{13}, \text{ and}$$

$$P(C \cap D) = P(C)P(D) = \left(\tfrac{1}{2}\right)\left(\tfrac{3}{13}\right) = \tfrac{3}{26}$$

Hence, the result is

$$\left[\tfrac{1}{13} + \tfrac{1}{13}\right]\left[\tfrac{1}{2} + \tfrac{3}{13} - \tfrac{3}{26}\right] = \left(\tfrac{2}{13}\right)\left(\tfrac{8}{13}\right) = \tfrac{16}{169} = .095$$

METHODS OF COUNTING

The classical definition of probability rests on the principle of dividing the number of elementary events in A by the number of elementary events in S. However, determining how many different basketball teams can be put on the court if each team has 10 players who can play any position is very tedious if one has to write out all 252 five-player teams. Therefore, it is desirable to be able to assess the number of outcomes in a complex event in some other manner than by listing each of them. Certain counting methods are available to facilitate this task.

Permutations

A **permutation** of a set of objects or events is an ordered sequence of the elements from that set. The number of ordered sequences of r objects which can be selected from a total of n objects is symbolized by

$$_nP_r$$

which is read "the number of permutations of n things taken r at a time."

If one has four objects, A, B, C, and D, then $ABCD$, $ACBD$, $ADBC$, $ADCB$ are some of the 24 possible permutations of the four objects taken four at a time. If these four objects are taken two at a time, then AB, BA, AC, CA, AD, DA, BC, CB are some of the 12 permutations of four objects taken two at a time. Note that the definition states "ordered sequence." That means that AB and BA are two different permutations; that is, order makes a difference.

The number of permutations of n things taken r at a time, $_nP_r$, equals

$$_nP_r = \frac{n!}{(n-r)!}$$

The symbol $n!$, read "n factorial," is defined as

$$n! = n(n-1)(n-2)(n-3)\ldots \quad (1)$$

Thus, $5! = (5)(4)(3)(2)(1) = 120$. Remember that $0! = 1$.

If $r = n$, the number of permutations of n things (taken n at a time) equals

$$_nP_n = \frac{n!}{(n-n)!} = \frac{n!}{0!} = n!$$

Let us examine the logic behind these expressions.

Suppose one has five objects, A, B, C, D, and E. Consider first the number of permutations of these five objects taken five at a time, $_nP_n = {}_5P_5$. Think of the task as having to fill five positions. There are five possible objects with which to fill the first position. *For each one of those five selections*, the second position may be filled with any one of the four remaining objects, since an ordered sequence does not permit selecting with replacement. This means that there are $(5)(4) = 20$ ways of filling the first two positions. Note that one multiplies because *for each one* of the first five possibilities, there exist four ways to fill the second. The third position may be filled with any one of the three remaining objects; there are two ways to fill the fourth; and only one object (or way) remains for the last position. Therefore, the total number of ordered sequences or permutations of five things is

$$(5)(4)(3)(2)(1) = 5! = 120$$

To generalize, there are $n!$ permutations of n things (taken n at a time),

$$_nP_n = n!$$

Now suppose you have five things but need to know the number of permutations of only three of the five elements at a time. Table 7–3 lists the 60 permutations of five things (A, B, C, D, and E) taken three at a time. The 60 permutations are arranged into 10 groups, each containing the 6 permutations of three elements.

The number of permutations of n things taken r at a time, $_nP_r$, in this case $_5P_3$, is determined with the following logic. There are five ways to fill the first position, four ways to fill the second position, and three ways to fill the third position, and that is all the positions required. Thus one multiplies to arrive at the answer:

$$_nP_r = n(n-1)(n-2)\ldots(n-r+1) = (5)(4)(3) = 60$$

The same result is arrived at by dividing $n!$ by $(n-r)!$:

$$_nP_r = \frac{n!}{(n-r)!} = \frac{(5)(4)(3)(2)(\cancel{1})}{(2)(\cancel{1})} = (5)(4)(3) = 60$$

7–3 The Sixty Permutations and Ten Combinations (*Underlined*)
of Five Things Taken Three at a Time

<u>ABC</u>	<u>ABD</u>	<u>ABE</u>	<u>ACD</u>	<u>ACE</u>
ACB	ADB	AEB	ADC	AEC
BCA	BAD	BAE	CAD	CAE
BAC	BDA	BEA	CDA	CEA
CAB	DAB	EAB	DAC	EAC
CBA	DBA	EBA	DCA	ECA
<u>ADE</u>	<u>BCD</u>	<u>BCE</u>	<u>BDE</u>	<u>CDE</u>
AED	BDC	BEC	BED	CED
DAE	CBD	CBE	DBE	DCE
DEA	CDB	CEB	DEB	DEC
EAD	DBC	EBC	EBD	ECD
EDA	DCB	ECB	EDB	EDC

Thus, the formula for the number of permutations of *n* things taken *r* at a time is

$$_nP_r = n(n - 1)(n - 2) \ldots (n - r + 1) = \frac{n!}{(n - r)!}$$

Consider the problem of picking the first-, second-, and third-place winners in a race involving seven horses of unknown, and thus presumably equal, speed. What is the probability of success at this task? There is only one way to pick the three horses and assign them to the proper places (event *A*), but there are $_7P_3$ ways of ordering seven horses in groups of three. The required probability is

$$P(A) = \frac{\#(A)}{\#(S)} = \frac{1}{_7P_3} = \frac{1}{\dfrac{7!}{(7 - 3)!}} = \frac{1}{\dfrac{(7)(6)(5)(\cancel{4})(\cancel{3})(\cancel{2})(\cancel{1})}{(\cancel{4})(\cancel{3})(\cancel{2})(\cancel{1})}} = \frac{1}{(7)(6)(5)}$$

$$P(A) = \frac{1}{210} = .0048$$

Therefore, in a crowd of 10,000 people, none of whom had any information on the horses and all of whom guessed randomly, on the average 48 people would pick correctly, in order, the first three horses to finish.

Combinations

A **combination** is any set or subset of objects or events, regardless of their internal order. The number of groups of r objects that can be selected from n objects is symbolized[2] by $_nC_r$ and given by

$$_nC_r = \frac{n!}{(n-r)!r!}$$

When the number of combinations of n things taken n at a time is desired, only one combination exists. Note also that while AB and BA are two different *permutations* of A and B, they are the same *combination*; permutations are ordered sequences, whereas combinations are not defined with respect to the order of their elements. Consider the logic of determining $_nC_r$.

Recall that the number of permutations of n objects taken r at a time is

$$_nP_r = \frac{n!}{(n-r)!}$$

Thus, if $n = 5$ and $r = 3$,

$$_5P_3 = \frac{5!}{2!} = 60$$

The number of combinations, however, is much less than 60, because many of the permutations are simple reorderings of a single combination. Table 7–3 presents the 60 permutations of five things taken three at a time, but they are arranged into groups corresponding to the 10 combinations. It can be seen that there are six permutations *for every one combination* (underlined in the table). In general, if objects are to be taken r at a time, there will be $r!$ permutations for every combination. Thus, since there are three elements in each group, there are $3! = (3)(2)(1) = 6$ permutations for each combination as indicated by the groupings of six in Table 7–3. One divides the number of permutations by $r!$ in order to obtain the number of combinations, since there are $r!$ permutations for every combination. Therefore, the expression for combinations is

$$_nC_r = \frac{_nP_r}{r!} = \frac{\dfrac{n!}{(n-r)!}}{r!} = \frac{n!}{(n-r)!r!}$$

To illustrate, suppose a football coach has seven guards on the team. Assuming that it makes no difference who plays the right and the left positions, how many different pairs of guards can be fielded? Since order is

[2]Some other books use the symbol $\binom{n}{r}$.

unimportant, this amounts to asking: How many combinations are there of seven people taken two at a time?

$$_7C_2 = \frac{7!}{5!2!} = \frac{(7)(6)(\cancel{5})(\cancel{4})(\cancel{3})(\cancel{2})(\cancel{1})}{(\cancel{5})(\cancel{4})(\cancel{3})(\cancel{2})(\cancel{1})(2)(1)} = 21$$

What is the probability of being dealt a five-card poker hand containing all spades? If A is the event of being dealt all spades, any one of the many combinations of 13 cards in that suit taken five at a time will do. The sample space S includes the number of five-card hands in a deck of 52 cards. The required probability is

$$P(A) = \frac{_{13}C_5}{_{52}C_5} = \frac{\dfrac{13!}{8!5!}}{\dfrac{52!}{47!5!}} = \frac{(13)(12)(11)(10)(9)}{(52)(51)(50)(49)(48)}$$

$$P(A) = .0005$$

Binomial probability

A special application of determining the number of combinations occurs when there are only two possible outcomes—success/failure, heads/tails—for each of several trials or occasions. For example, in six tosses of a coin, what is the probability of obtaining exactly two heads? It is important to notice that the question does not specify the order in which the heads and tails must appear. It only requires that exactly two of the six tosses be heads. Therefore, the solution to this problem may be achieved by first ascertaining the probability of any one specific sequence of two heads and four tails and then determining how many such sequences are possible.

One potential sequence is (H, H, T, T, T, T). The probability of a head on any single flip is $\frac{1}{2}$ and the probability of a tail is also $\frac{1}{2}$. The probability of getting two heads in succession is $(\frac{1}{2})(\frac{1}{2})$, of two heads followed by a tail is $(\frac{1}{2})(\frac{1}{2})(\frac{1}{2})$, and of (H, H, T, T, T, T) is $(\frac{1}{2})(\frac{1}{2})(\frac{1}{2})(\frac{1}{2})(\frac{1}{2})(\frac{1}{2}) = \frac{1}{64}$.

Further, the probability of the sequence (T, T, T, H, H, T) is also $(\frac{1}{2})(\frac{1}{2})(\frac{1}{2})(\frac{1}{2})(\frac{1}{2})(\frac{1}{2}) = \frac{1}{64}$. Indeed, the probability of *any* particular sequence of two heads and four tails is $\frac{1}{64}$.

How many such sequences of two heads and four tails are there? It happens that the number of sequences of two heads in six tosses is the same as the number of combinations of six things taken two at a time:

$$_6C_2 = \frac{6!}{2!\,(6-2)!} = 15$$

Since the event of two heads in six flips can occur in 15 different ways, each with a probability of $\frac{1}{64}$, the required answer is given by the sum of these 15 mutually exclusive probabilities, or

$$\tfrac{1}{64} + \tfrac{1}{64} + \cdots + \tfrac{1}{64} = 15\left(\tfrac{1}{64}\right) = \tfrac{15}{64}$$

This problem is an example of **binomial probability**, and the process of finding its solution may be formalized into a general expression.

> In a sequence of n independent trials that have only two possible outcomes (arbitrarily call them "success" and "failure") with the probability p of a success and probability q of a failure (note that $q = 1 - p$), the probability of exactly r successes in n trials is
>
> $$_nC_r\, p^r q^{n-r}$$
>
> or
>
> $$P(r, n; p) = \frac{n!}{r!\,(n - r)!} p^r q^{n-r}$$

To illustrate further, suppose an urn contains one red and three green balls. If a ball is returned to the urn after each drawing, what is the probability of selecting three reds in five tries? A success is defined as the selection of a red ball. Since on any one trial only one of the four balls in the urn is red, the probability of a success is one-fourth ($p = \frac{1}{4}$). Since $q = 1 - p$ and since three of the four balls available on any one trial are failures, $q = \frac{3}{4}$. The event under consideration is to obtain three red balls ($r = 3$) in five draws ($n = 5$). Therefore, the desired probability is

$$P(r, n; p) = \frac{n!}{r!\,(n - r)!} p^r q^{n-r}$$

$$P\left(3, 5; \tfrac{1}{4}\right) = \frac{5!}{3!\,(5 - 3)!}\left(\frac{1}{4}\right)^3\left(\frac{3}{4}\right)^{5-3} = \frac{90}{1024} = .088$$

FORMULAS

1. **Classical probability**

   a. $P(A) = \dfrac{\#(A)}{\#(S)}$

   b. $P(\varnothing) = \dfrac{\#(\varnothing)}{\#(S)} = \dfrac{0}{\#(S)} = 0$

   c. $P(S) = \dfrac{\#(S)}{\#(S)} = 1.00$

2. **Probability of intersection**
   a. $P(A \cap B) = P(A)P(B|A)$
   b. If A and B are independent [i.e., $P(B|A) = P(B)$], then $P(A \cap B) = P(A)P(B)$.
3. **Conditional probability**

$$P(B|A) = \frac{P(A \cap B)}{P(A)}$$

4. **Probability of union**
   a. $P(A \cup B) = P(A) + P(B) - P(A \cap B)$
   b. If A and B are mutually exclusive (i.e., $A \cap B = \varnothing$), then $P(A \cup B) = P(A) + P(B)$.
5. **Permutations**
   a. ${}_nP_r = \dfrac{n!}{(n-r)!}$

   b. ${}_nP_n = n!$
6. **Combinations**

$$_nC_r = \frac{n!}{(n-r)!r!}$$

7. **Binomial probability, the probability of r successes in n independent trials**

$$P(r, n; p) = \frac{n!}{(n-r)!r!}p^r q^{n-r}$$

in which

$$p = \text{the probability of a success,}$$
$$q = \text{the probability of a failure, } (q = 1 - p)$$

EXERCISES

1. Determine which of the following events are mutually exclusive:
   a. flipping a head, flipping a tail with one toss of a fair coin.
   b. rolling an even number, rolling less than a three with one turn of a die.
   c. drawing a jack, drawing an ace in one selection.
2. Determine which of the following events are independent:
   a. flipping a head on the first toss and a tail on the second toss of a coin.
   b. selecting an ace and then selecting a red card when drawing is done with replacement; when drawing is done without replacement.
   c. having a boy as a first child and having a girl as a second child.
3. What is the relationship between conditional probability and independence?
4. Determine the following probabilities:
   a. In a deck of 12 cards (4 jacks, 4 queens,

4 kings), what is the probability of drawing a king? a jack or a queen? a red card or a jack?

b. What is the probability of obtaining either (1) at least a four or (2) less than a three in a single roll of a die? of obtaining at least three or an odd number in a single roll of a die?

c. What is the probability of rolling two successive sixes with a die?

d. If an urn contains two red and three green balls, what is the probability of selecting a green ball given that a red ball has been drawn on the first selection and not replaced? and replaced? If drawing is without replacement, what is the probability of selecting two successive green balls? a red and then a green ball? a red and a green ball in any order?

5. Determine the following permutations:
  a. $_5P_5$ **b.** $_5P_2$ **c.** $_7P_3$

6. If a room has four chairs, how many different seating arrangements are there if six people come in?

7. Determine the following combinations:
  a. $_4C_4$ **b.** $_6C_4$ **c.** $_7C_3$

8. How many different relay teams of four swimmers could a coach put together with eight swimmers to choose from? If the order of swimming made a difference, then how many teams could be put together?

9. What is the probability of being dealt a three-card hand composed of all clubs from the special deck of 12 cards (exercise 4a)? a three-card hand composed of one suit?

10. What is the probability of naming the three top horses (regardless of specific position) in a field of nine? What is the probability of correctly designating first, second, and third positions from the field of nine? Given that you have picked the top three horses, what is the probability of correctly assigning them to the first three positions?

11. Suppose seven infants are familiarized with a given visual stimulus. Later they are permitted to look at that familiar pattern and two other stimuli that represent two degrees of similarity to the familiar one. The discrepancy theory predicts that looking time should be an inverted-U function of magnitude of discrepancy and the probability of this ordering of response is $\frac{1}{3}$. What is the probability that exactly five of the seven infants should respond in this way? five or more of the seven infants?[3]

12. In two rolls of a die, what is the probability of obtaining either the event of a six and then a number less than three or the event of a number at least three and then an even number?

[3]Inspired by R. B. McCall and J. Kagan, "Stimulus-Schema Discrepancy and Attention in the Infant," *Journal of Experimental Child Psychology* 5 (1967): 381–90.

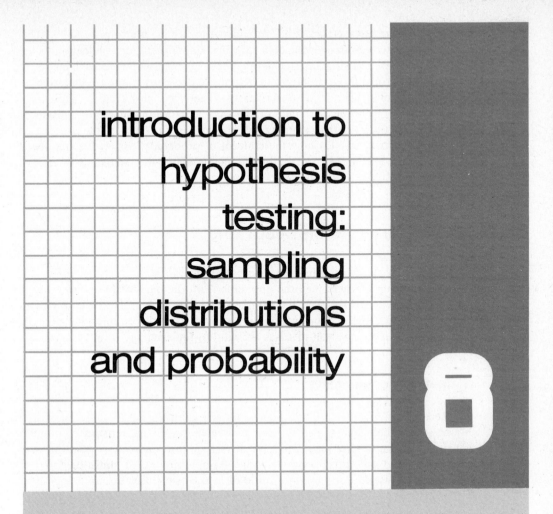

introduction to hypothesis testing: sampling distributions and probability

8

THE PREVIOUS CHAPTER has provided an introduction to probability. This chapter emphasizes the concept of probability and its application to the task of drawing statistical inferences. Statistical inference consists of invoking probability statements to help make decisions about a population on the basis of a sample of observations. Therefore, this chapter presents a discussion of (1) the adequacy of the sample as a faithful representation of the population, (2) the sampling distribution as an index of the amount of sampling error, and (3) how the concept of probability as relative frequency can be used to make probability statements about the likelihood of certain events.

SAMPLING AND SAMPLING DISTRIBUTIONS

The task of statistical inference is to permit decisions about characteristics of a large group of cases (i.e., the population) on the basis of observations of a small subset of cases (i.e., the sample). Since estimates of population characteristics are made from information provided by the sample, the methods used to select this subset of cases are vital to the accuracy of inference.

Methods of sampling

Recall from Chapter 3 that a **population** is a complete set of subjects, events, or scores that have some common characteristic. Characteristics of populations are known as **parameters**. A subset or portion of a population is a **sample**, and quantities computed on a sample are called **statistics**. Often a statistic is used to estimate the value of a parameter. For example, a pollster might select a sample of 1000 cases with which to estimate the percentage of the American adult population that supports the President.

A statistic used to estimate a parameter is no better than the sample upon which it is computed. Of the many ways to select a sample, one of the commonest is simple random sampling.

In **simple random sampling**, all elements of the population have an equal probability of being selected for the sample.

Suppose you want to have a random sample of students at your university for an opinion poll on the quality and appropriateness of their educational experience. You might obtain a list of all students in the school, go to a **random number table**, and select a sample of 50. A random number table (Table J) is located in Appendix II of this book; it consists of rows and columns of random numbers. The numbers are random in the sense that for any single digit position, each of the 10 numbers from 0 to 9 has an equal chance of occupying that position. This means that every digit was selected independently of every other digit. Further, not only are all the single digits random, but all two-, three-, or N-digit numbers are also random.

To use the table for getting a sample of 50 students from a list of 3000, assign each student a number between 1 and 3000. Then go to the random number table and mentally block the table off into four-digit columns. Read down the columns until you obtain 50 four-digit numbers that fall between 0001 and 3000 inclusive. The students assigned numbers corresponding to these 50 numbers constitute a randomly selected sample.

If you read down a one-digit column in the table, you may feel that the numbers are not random at all. For example, if you were to write what you thought was a random sequence of 0s and 1s, you might write 01001001011101, etc. Most people are very hesitant to put more than three of a kind in succession (e.g., 0000 or 1111). Yet, in any random selection of four 0s or 1s, the probability of 0000 or 1111 is .125 or 125 in every 1000 such sets of four. Look at the first column of numbers in the random number table and observe such a statistically random sequence by letting even numbers be called 0 and odd numbers be called 1. The point of this explanation is that random sampling refers to the process of *how* a sample is selected, not to which cases actually are selected. In the long run, large random samples will be **representative** of all aspects of the population, but any single randomly selected sample could be quite atypical.

In most cases, random sampling means that each subject is selected independently from other subjects. **Independence** in sampling means that the selection of any element (i.e., subject) for inclusion in the sample does not alter the likelihood of drawing any other element of the population into the sample. In almost all cases in which random sampling is required in this text, the implication is that the elements have been independently selected.[1]

Since independent random sampling is the basis for most of the procedures of statistical inference discussed in this text, it would be wise to consider the characteristics of samples that actually are taken in research. For example, suppose one wants to sample the population of a given town by selecting every two-hundredth name in the telephone book. Would this be a random sample of townsfolk, i.e., does each person in the town have an equal likelihood of being selected? No, this would not be a random sample of

---

[1]Sometimes it is important for an investigator to ensure that certain segments of the population are represented in the sample. Political pollsters, for example, attempt to sample the voting public in such a way that each area of the country, each ethnic group, each religious group, etc., is appropriately represented. Thus, if 10% of the population is Catholic, 10% of the sample would be randomly selected from Catholic voters. Such a sample is called a **proportional stratified random sample**. The sampling is random and independent within these groups but not between them. Since all of the procedures outlined in this text are appropriate for simple random samples, stratified sampling and its associated statistical procedures will not be considered further.

townsfolk, because people who do not own a phone and people with unlisted phone numbers would be systematically excluded. There are two things one could do to correct this situation. First, a list of all people in the town could be obtained from the city government or census bureau, and a random-number table could then be employed. But this might be a very time-consuming task, and it assumes the availability of the necessary list. The second possibility is to change the definition of the population in accordance with the nature of the sample. Since you cannot get a random sample of all town people, change the population to all town people listed in the telephone book and discuss the results and conclusions in terms of this group rather than in terms of all residents. The important point is for the researcher to be aware of precisely what population the sample was selected from and to limit conclusions to that population.

The tediousness of obtaining a truly random sample is so inhibiting that in practice researchers frequently do not select random samples. Nevertheless, statistically, they treat their samples as if they were random. For example, a group of rats provided by an animal supplier is not a random selection of rats. Rats are raised in cages set on tiers, some of which are closer to the light than others, and the amount of illumination in the rat's rearing experience can influence some types of later behavior. Further, when the rats arrive at the laboratory and are assigned to experimental groups, it is sometimes tempting to place the first 10 in one group, the second 10 in the next, and so on. But it happens that the more curious and active rats frequently come over to the side of the shipping box when it is opened. These rats are more accessible and easier to pick up. They are thus selected first and go into the first group if the above procedure is carried out. It is a good idea to use a random number table to determine group assignments; one should at least alternate by assigning one rat to the first group, the next to the second, etc.

Another form of bias in sampling occurs when volunteers are used. College students who volunteer for experiments are probably somewhat different with respect to academic ability, motivation, etc., from students who do not volunteer. Who volunteers can be an interesting issue in its own right, and it determines what kind of research can and cannot be done. For example, some developmental psychologists are dependent upon parents to volunteer their infants for observation. It is likely that highly educated parents are more receptive to science's "need" for subjects than are less educated parents. Thus, the sample one obtains is not random because it may contain a preponderance of infants from highly educated families.

How can you safeguard against such bias in your sample? The best way is simply to be cautious and aware of what you are doing. In addition, it is advisable to measure the sample you have selected on several dimensions (e.g., age, education, "normality" etc.) appropriate to the research (as long as the measurement of these traits does not influence the subjects in any way) so that readers can judge whether the sample has characteristics similar to the population you want to discuss. The goal is to ensure that inferences and generalizations are made to the appropriate population.

Sampling distributions and sampling error

Even when the sample is random and appropriate, it is likely to be different from the population because it is a sample of considerably fewer cases than are contained in the population and because individuals differ.

Suppose there are 20 students in your statistics class and the professor springs a surprise quiz of 10 questions. The scores of the 20 students comprise the population of raw scores and are presented in the left-hand column of Table 8–1. Now suppose a random sample of $N = 4$ students is taken for the purpose of estimating the population mean (the mean of the 20 students, which is actually 3.90). The first such sample of four students provides the scores (1, 5, 9, 0), which have a mean of 3.75, as indicated in Table 8–1. But

8–1 Population Distribution of Raw Scores, Ten Observed Sample Distributions, and an Empirical Sampling Distribution of the Means

Population Distribution of Raw Scores		10 Observed Sample Distributions $(N = 4)$	Empirical Sampling Distribution of the $\bar{X}$'s
6	2	(1, 5, 9, 0)	3.75
9	5	(0, 3, 1, 5)	2.25
0	1	(5, 8, 3, 0)	4.00
3	2	(1, 5, 0, 7)	3.25
1	1	(7, 6, 1, 3)	4.25
5	2	(3, 2, 1, 7)	3.25
7	7	(2, 0, 3, 5)	2.50
7	8	(1, 2, 1, 1)	1.25
1	1	(2, 7, 1, 7)	4.25
3	7	(9, 7, 6, 2)	6.00
$\mu = 3.90, \sigma = 2.88$			Mean of $\bar{X}$'s = 3.48 Standard deviation of $\bar{X}$'s = 1.31

suppose another random sample of $N = 4$ is selected and its mean is 2.25. Looking down the right-hand column in Table 8–1, which shows the means for 10 randomly selected samples from the population, one can see that randomly selected samples from the same population vary in the value of their means. Moreover, not one sample yields a mean that precisely equals the population mean. Why do samples differ from one another?

In a random sample each subject in the population has an equal *opportunity* to be drawn into the sample. But this does not say that the sample itself will faithfully reflect the population characteristics. One could randomly select a sample that happened to contain many exceptionally bright students. One reason random samples differ one from another is simply that they are composed of different individuals.

Returning to Table 8–1, we see that the means computed on the 10 samples could themselves be considered scores in a distribution—a distribution of means rather than of raw scores. Such a distribution has a special name and function in statistics.

> The distribution of a statistic calculated on each one of a collection of samples of size N drawn from a given population is called a **sampling distribution**.

Thus, the distribution of the 10 sample means in Table 8–1 is a sampling distribution.

For purposes of clarity, the table shows an **empirical sampling distribution**. The word *empirical* signifies "experienced" or "observed," and these 10 means are observations presumably made by collecting 10 samples and computing $\overline{X}$ for each group. In contrast, a **theoretical sampling distribution** is a theoretical distribution of a statistic, and its characteristics are determined mathematically rather than by repeated observations.

The sampling distribution is introduced in this chapter and should be carefully distinguished from the two raw-score distributions introduced earlier. The type of distribution of central concern throughout Part 1 is the **sample distribution**, which is a collection of scores obtained from a subgroup of a population. The middle column of Table 8–1 shows 10 sample distributions. Also introduced in Part 1 (Chapter 3) is the concept of a **population distribution**, which is the full array of raw scores that includes the sample distribution. Now we have another type of distribution. A **sampling distribution** is a distribution of a statistic, not of raw scores. Sampling distributions may be empirical (based upon actual samples and observations), like the one in the right-hand column of Table 8–1, or they may be theoretical (pertaining to the population).

Note that a sampling distribution differs from both a sample distribution and a population distribution in that it is a collection of statistics rather than a collection of raw scores. The statistic whose sampling distribution is shown in the third column of Table 8–1 is the mean, but one can also obtain sampling distributions of other statistics. For example, one could compute the standard deviations of the ten sample distributions shown in the table and thus obtain a sampling distribution of the standard deviation.

Just as a distribution of raw scores has certain characteristics, such as a mean and a standard deviation, so too does the sampling distribution of a statistic. The sampling distribution shown in Table 8–1, for example, itself has a mean of 3.48 and a standard deviation of 1.31. Therefore, it will be necessary to have terms and symbols to represent the mean, standard deviation, and variance of the several distributions. Table 8–2 summarizes these terms and symbols. First, notice at the left that distributions are composed either of raw scores or of statistics, and in the latter case they are called sampling distributions. In addition, a distribution may be based on a sample or a population. Distributions are also either empirical (i.e., based on actual observations) or theoretical (i.e., defined in terms of their characteristics—mean, variance, shape—and not actual observations). Samples are

8–2 Terms and Symbols for the Mean, Standard Deviation, and Variance in Different Types of Distributions

Distributions of Raw Scores	Mean	Standard Deviation	Variance
Sample	$\overline{X}$	s_x or s	s_x^2 or s^2
Population	μ_x or μ	σ_x or σ	σ_x^2 or σ^2

Sampling Distribution of the Mean	Mean	Standard Error of the Mean	Square of the Standard Error of the Mean
Sample	$\overline{X}_{\bar{x}}$	$s_{\bar{x}}$	$s_{\bar{x}}^2$
Population	$\mu_{\bar{x}}$ or μ	$\sigma_{\bar{x}}$	$\sigma_{\bar{x}}^2$

Equivalences

$$\mu_x = \mu_{\bar{x}} = \mu$$

$$s_{\bar{x}} = \frac{s_x}{\sqrt{N}} \text{ and } \sigma_{\bar{x}} = \frac{\sigma_x}{\sqrt{N}}$$

usually empirical, and populations are usually theoretical, but this is not always the case; for example, see the empirical population in Table 8–1.

Each of the four kinds of distributions listed in Table 8–2 has a mean, standard deviation, and variance, but they have different names and symbols. When these quantities pertain to samples or empirical distributions they are symbolized by the letters $\overline{X}$, s, and s^2; when they pertain to populations or theoretical distributions, they are symbolized by the Greek letters μ, σ, and σ^2. A subscript designates the distribution to which each quantity refers.

As will be explained below, some characteristics of the various distributions are equivalent in value, and thus the symbols used to represent their means and standard deviations are sometimes used interchangeably. In order to understand the logic behind the equivalences, however, you must be able to identify the distribution that each symbol for the mean or standard deviation refers to. Use Table 8–2 for reference as the discussion proceeds.

Standard error of the mean As the table shows, the mean of a theoretical sampling distribution of the mean is symbolized by $\mu_{\overline{x}}$. The Greek μ indicates that this mean is a parameter, not a statistic, and the subscript $\overline{x}$ signifies that it is a mean of means. The symbol $\mu_{\overline{x}}$ can be distinguished from μ_x, which represents the mean of a population distribution of raw scores. As it happens, however, these two means have the same value:

$$\mu_{\overline{x}} = \mu_x = \mu$$

The symbol μ without a subscript is customarily used to indicate this value. In addition, recall that μ can be estimated by the sample mean $\overline{X}$ from a single sample: $\overline{X}$ is used to estimate $\mu_{\overline{x}}$ as well as μ_x.

Besides a mean, a theoretical sampling distribution of means has a standard deviation.

> The standard deviation of a sampling distribution of a statistic is called the **standard error** of that statistic. Consequently, the standard deviation of the sampling distribution of the mean is known as the **standard error of the mean**.

Because of the great importance of the concept of a standard error in inferential statistics, it is crucial for you to have a firm grasp of its meaning. Consider the population standard error of the mean, $\sigma_{\overline{x}}$. We have said that the mean of one sample of scores will not likely equal the mean of another sample of scores, even if both samples are randomly selected from the same population. The fact that means differ in value from one sample to another simply testifies to the existence of variability in such a distribution of means. Since the standard deviation is a numerical index of variability, the standard deviation of the sampling distribution of means, $\sigma_{\overline{x}}$, is a numerical index of the extent to which means vary from one sample to another.

More generally, the standard error of the mean is an index of the amount of error that results when a single sample mean is used to estimate a population mean; that is, it is an index of **sampling error**. Thus, if $\sigma_{\bar{x}} = 5$ for samples of 20 males on a reading test but $\sigma_{\bar{x}} = 10$ for samples of 20 females on the same test, there is less sampling error for males than for females. This implies that the more-or-less random variation between means due to differences in the nature of one sample and another is less for males than for females.

It was stated above that the mean of the theoretical sampling distribution of the mean is identical to the mean of the population distribution of raw scores, $\mu_{\bar{x}} = \mu_x = \mu$. In contrast, the standard deviation of the sampling distribution of the mean (the standard error of the mean) is not identical to the standard deviation of the population distribution of raw scores, but it is related to it.

The standard error of the mean, $\sigma_{\bar{x}}$, equals the standard deviation of the population of raw scores divided by the square root of the size of the sample upon which the means are based:

$$\sigma_{\bar{x}} = \frac{\sigma_x}{\sqrt{N}}$$

If the standard deviation of the population of raw scores is $\sigma_x = 2.88$ and the sample size is $N = 4$, the theoretical standard error of the mean is

$$\sigma_{\bar{x}} = \frac{\sigma_x}{\sqrt{N}} = \frac{2.88}{\sqrt{4}} = 1.44$$

The same relationship between the standard deviation of raw scores and the standard error of the mean holds for samples as for populations:

$$s_{\bar{x}} = \frac{s_x}{\sqrt{N}}$$

Estimating $\sigma_{\bar{x}}$ with $s_{\bar{x}}$ Just as s_x is used to estimate σ_x, so can $s_{\bar{x}}$ be used to estimate $\sigma_{\bar{x}}$:

The population standard error of the mean, $\sigma_{\bar{x}}$, may be estimated by

$$s_{\bar{x}} = \frac{s_x}{\sqrt{N}}$$

which can also be expressed in terms of the raw scores of the sample as

$$s_{\bar{x}} = \sqrt{\frac{N\Sigma X_i^2 - (\Sigma X_i)^2}{N^2(N-1)}}$$

in which N is the size of the sample of X's.

Two things should be noticed about the standard error of the mean as expressed by $s_{\bar{x}} = s_x/\sqrt{N}$. First, all that is required to calculate $s_{\bar{x}}$ is the standard deviation and the N from a single sample of cases, yet $s_{\bar{x}}$ represents an estimate of the amount of variability (or sampling error) in means from all possible samples of size N from the population of raw scores. Thus, it is not necessary to select several samples in order to estimate the population sampling error of the mean; $s_{\bar{x}}$ estimates $\sigma_{\bar{x}}$, and all that is required to calculate $s_{\bar{x}}$ is s_x and N from a single sample of raw scores.

Second, observe that the formula for $s_{\bar{x}}$ states that the standard deviation of the sample must be divided by the square root of N: $s_{\bar{x}} = s_x/\sqrt{N}$. Therefore, the variability of means from sample to sample will always be smaller than the variability of raw scores. Also note that as N becomes larger, $s_{\bar{x}}$ becomes smaller. Thus, the variability of sample means decreases as the size of the sample increases. Consequently, for large samples one expects $\bar{X}$, the sample estimator of the population mean μ, to be less variable from sample to sample, and thus a more accurate estimate of μ, than if the sample size were smaller. In short, when parameters must be estimated, it is a good idea to have as large a sample as possible.

Other standard errors   The sampling distribution and standard error of the mean have been discussed in detail, but a sampling distribution and a standard error exist for any statistic. In each case the logic is the same. Random samples differ in their characteristics, and any statistic will vary somewhat from sample to sample. The theoretical sampling distribution is the distribution of a particular statistic determined for all possible samples of size N, and the standard deviation of that statistic's sampling distribution is its standard error. Therefore, one can imagine standard errors for the median, the variance, and even for the difference between two sample means.

Sampling distributions and normality   From this point forward, *sampling distribution* will refer to a *theoretical* sampling distribution. *Empirical* sampling distributions are not used in statistics except to help students understand the concept of a distribution of a statistic. However, the symbol $s_{\bar{x}}$ is used to indicate that $\sigma_{\bar{x}}$ is being estimated by sample observations.

Many of the procedures described in this and later chapters rest on the assumption that the sampling distribution of means is normal in form. This is the case if one of two conditions is met.

Given random sampling, the sampling distribution of the mean
1. is a normal distribution if the population distribution of the raw scores is normal.
2. approaches a normal distribution as the size of the sample increases.

If the population distribution of raw scores is normal, the sampling distribution of the mean will also be normal. However, since you rarely have the population at your disposal, how can you know if the population distribution is normal? One way to make an educated guess is to determine whether a random sample from the population is normally distributed. Some variables are usually not normally distributed because there are a few extreme scores. For example, IQs of all 21-year-olds are probably normally distributed but IQs of all 21-year-old college students are not because low or extremely low scores are not represented as frequently in college groups as are extremely high scores. Family income, the latency for a rat to move out of a startbox in a maze, and percent correct on a relatively easy exam are variables that are not usually normally distributed. Notice that these variables are bounded on one end of their scales (e.g., $0 income, 0 seconds, 100% correct). The scores will tend to fall near the bounded end of the scale, and the distribution is likely to be skewed. Fortunately, many of the variables measured in social sciences can be assumed to be normally distributed. When variables are not normal, the statistical techniques described in Chapter 13, rather than those described below, may be used.

A second way to obtain a normal sampling distribution of the mean is to select a large enough sample of raw scores. The sampling distribution of the mean will approach a normal distribution as the sample size increases, *even though the population distribution of raw scores on the variable of interest is not normal.* Just how many cases constitute a sufficiently large sample depends upon many factors, one of which is the extent of the departure from normality of the population distribution. If the population distribution does not deviate too much from normality, samples of size $N = 2$ might produce a sampling distribution of the mean that is quite normal, whereas if the nonnormality in the population is severe, N's of 20, 30, or several hundred might be necessary. In short, the sampling distribution of the mean will approach a normal form as the size of the samples of raw scores increases.

The following fact is one of the reasons normality is necessary in order for the statistical procedures to be described:

If the population distribution of raw scores is normal and the observations are independent and randomly selected, the sample mean and variance (and standard deviation) are independent of one another across samples.

Any two variables (including the statistics $\overline{X}$ and s^2) are independent if they are unrelated to each other. In this case, if many samples are obtained from a normal population, the distribution of variances for samples having a mean of 10, for example, will be approximately the same as for samples having a mean of 12 or some other value. The independence of the mean and variance of a normal distribution will be important later.

Let's review. Most of the statistical procedures to be described subsequently are dependent upon several principles. These include: (1) Samples are composed of randomly and independently selected observations. (2) A sampling distribution is the distribution of a statistic; the standard deviation (standard error) of a sampling distribution is an index of the extent to which the statistic varies from one sample to another (i.e., of sampling error). (3) The sampling distribution of the mean is normal if the population distribution of raw scores is normal or the size of the sample is large. (4) The mean and standard deviation of a random sample from a normal population are independent.

PROBABILITY AND ITS APPLICATION TO HYPOTHESIS TESTING

The purpose of inferential statistics is to assist in making inferences and judgments about what exists on the basis of only partial evidence. This is accomplished by using probability. In a way, most people use subjective probability every day. You want to go to the football game this afternoon, but someone has warned that it is going to rain. You look outside an hour before the game and, while there are clouds, it is not threatening. Yet, the weather reporter says there is a 70% chance of rain. You make a judgment about going to the game based on the statement of probability: an inference that relies only on partial evidence.

In scientific research as well, probability statements are used to make inferences and judgments on the basis of only partial information. For example, a social psychologist was interested in the extent to which people could be made to inflict pain on a colleague by the application of social pressure to conform.[2] The social psychologist's experiment was structured so that a male subject joined three other men in what was described as an important learning experiment. In this experiment the subject and two other men were to teach the fourth person a simple task by administering electric shock to the "learner" for his incorrect responses. Actually, though unknown to the subject, the three other people were collaborators of the experimenter. In fact the entire situation—shock, learning, etc.—was staged for the subject. Forty subjects were urged by their "colleagues" to turn up the purported shock level as high as possible because it was "essential to the experiment." The "learner" grimaced and squirmed in accordance with the level of shock the subject thought he was administering. Another 40 subjects were not pressured to elevate the presumed shock level. The results showed that the subjects who were not admonished by their colleagues to turn up the shock set their shock controls at an average intensity reading of approximately 3.5 arbitrary units, while the subjects who were pressured averaged a shock level

[2] S. Milgram, "Group Pressure and Action against a Person," *Journal of Abnormal and Social Psychology* 49 (1964): 137–43.

of approximately 14 arbitrary units, even though the "learner" screamed and gyrated in response to that much "shock." This experiment showed that for the subjects tested, social pressure seemed to result in the administration of four times as much "shock" as when no social pressure was applied.

How much confidence can we place in the results of one such study? Was this outcome simply a fluke, an unlikely chance result that would not happen again if the experiment were repeated? Would other groups of men in the population from which this sample was taken have responded essentially in the same way? There is no certain method of determining the answers to these questions, short of testing every person in the population. Therefore, we must quantify our uncertainty about what truly exists in the population and adopt some conventions for deciding what to conclude. The fundamental tool in this process is **probability**, the quantification of uncertainty.

More specifically, the task of statistical inference in the above experiment is to decide whether the mean of 14 is sufficiently greater than the mean of 3.5 to warrant the conclusion that the social pressure actually had an effect. Although 14 is substantially more than 3.5, it still could be a chance result. Not every member of the pressured group raised the purported shock level higher than every member of the non-pressured group, and some members of the non-pressured group may have set the shock level higher than even the average of 14 observed for the experimental condition. After all, suppose one just sampled two groups of 40 subjects and did not pressure either of the groups, so that subjects from both groups were treated in precisely the same way. Even in this case we would not expect the averages of these two groups to be exactly the same, because of sampling error. The job of probability in statistical inference is to quantify the likelihood of just such a possibility—that is, probability tells us how often, in 100 experiments in which two groups of 40 subjects have been treated exactly alike, we would expect a difference as large as 14 versus 3.5. If the answer is not very often, then it is probably unreasonable to suppose that this was simply a chance result, and we would conclude that the social pressuring probably caused the subjects to increase the amount of shock they administered. On the other hand, if a difference of 14 versus 3.5 is fairly likely, even if the social pressure had no effect whatsoever, then one would have to conclude that the experiment provides no evidence that social pressure affected the amount of shock the men gave their colleagues.

Scientists prefer to use as an index of the likelihood of events probability rather than their personal impressions or subjective feelings. The numerical probability is public knowledge (all scientists can observe and understand it); the probability of one event can be easily compared with the probability of a different event; and certain conventions can be adopted about how high or low the probability must be in order to justify one decision or another. Therefore, a knowledge of the concept of probability is essential to understanding the process of statistical inference and decision making in science.

Probability and relative frequency

As was explained in Chapter 7, the determination of a simple probability implies an idealized experiment. In figuring the chances of a head or a tail when flipping a coin, one actually assumes an experiment in which the coin is tossed over and over again. It is assumed that both results are equally likely. On any single throw, either a head or a tail will occur, but over the long range of such flips of a coin, the ratio of heads to all possible outcomes will approach $\frac{1}{2}$, or .50. Thus,

> **Probability** is theoretical relative frequency, the relative frequency of score values in a theoretical distribution based upon an unlimited number of cases.

As this definition implies, the probability of different events can be determined from theoretical relative frequency distributions.

For example, suppose Figure 8–1 represents the theoretical relative frequency distribution of the IQs of 10-year-old American children. Now consider the probability that a single child selected at random from this

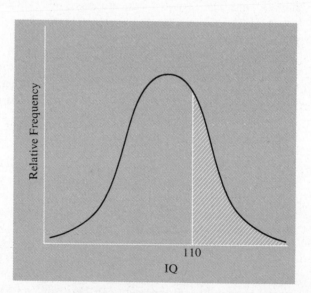

Fig. 8–1.   Theoretical relative frequency of the IQs of 10-year-old American children. The probability of randomly selecting a child who has an IQ higher than 110 is indicated by the proportion of the total area under the curve represented by the crosshatched area.

population would have an IQ greater than 110. According to the conception of probability discussed above, this probability value should be given by the theoretical relative frequency of scores that exceed 110. The approach to determining the probability here rests on equating the **area** between the curve and the abscissa with the concept of **theoretical relative frequency**, just as was done in Chapter 4. If the area under the curve in Figure 8–1 represents theoretical relative frequency, then the proportion of the total area that lies between 110 and $+\infty$ (indicated in the figure by crosshatching) represents the theoretical relative frequency of children having scores greater than 110. To formalize:

> The proportion of the total area under the curve of a theoretical relative frequency distribution that exists between any two points represents the probability of obtaining the events contained within the interval delimited by those two points.

Consider another example: What is the probability that a child randomly selected from the population would have an IQ between 85 and 115? This probability is given by the proportion of the total area that exists between IQs of 85 and 115 as illustrated in Figure 8–2.

The standard normal distribution described in Chapter 4 is a theoretical relative frequency distribution based upon unlimited samplings of normal distributions. It is presented in Table A of Appendix II and it has $\mu = 0$ and

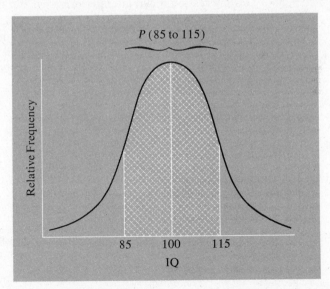

Fig. 8–2.   The proportion of the total shaded area that is crosshatched represents the probability of randomly selecting a child who has an IQ between 85 and 115.

$\sigma = 1$. The percentiles listed in Table A correspond not only to the proportion of cases existing between two points on the normal curve but also to the probability of obtaining a score value located between the two points. There are many other theoretical relative frequency distributions whose percentiles are also tabled. These distributions will form the basis for making probabilistic statements as outlined in the remainder of this book.[3]

Hypothesis testing: an informal example

The reason for discussing sampling distributions and probability (defined in terms of theoretical relative frequency) is that these two concepts are the basis of statistical inference. Although the specific techniques of statistical inference vary depending upon the research question being asked, the logic of the general approach is quite similar in diverse cases. In order to illustrate this logic and the terminology that accompanies it, a relatively simple example will be used.[4] In subsequent chapters, problems closer to those actually encountered in social science will be presented.

Suppose it is known that for a population of humans asked to learn and remember 15 nouns the mean number of correct nouns recalled after an 80-minute retention period is 7 and the standard deviation is 2. In symbols, $\mu = 7$ and $\sigma_x = 2$. Now suppose that a psychologist and a physician have a hunch that injecting the drug physostigmine between a learning session and a memory testing session might alter performance, but they don't know whether it will improve or retard memory. The scientists decide to make some relatively informal observations, first on one person and then on a small group to which they will administer the drug. Later, they plan to conduct a formal experiment comparing a large group of drugged with a large group of non-drugged subjects.

A single case   First the scientists randomly select one person from the population of available subjects. They teach that person the 15 nouns, administer the drug, wait 80 minutes and then test the person's recall of the nouns. If this one drugged person performs quite differently from the mean of the non-drugged population, it is possible that the drug has some effect on learning performance in this situation.

---

[3]The equation for a theoretical relative frequency distribution, such as the standard normal and several other distributions to be considered, is known as a **probability**, or **density, function**. To determine the probability of the occurrence of events located between any two points on the dimension, this function is integrated between these points by the methods of calculus. The values in Appendix II, Table A, and other tables in this book represent the results of such a process.

[4]Based upon, but not identical to, research reported by K. L. Davis et al., "Physostigmine: Improvement of Long-Term Memory Processes in Normal Humans," *Science* 201 (1978): 272–74.

Suppose the drugged subject remembers 11 nouns correctly. Although this is clearly more than the population mean of 7, is it enough more to warrant pursuing this line of research? How differently would the person have to perform to be sufficiently different from what would be expected of non-drugged subjects?

The strategy the scientists employ is quite simple. Suppose the drug has no effect. Then, the experimental subject should perform like a randomly selected member of the population of non-drugged subjects, which has a mean $\mu_x = 7$ and a standard deviation $\sigma_x = 2$. But if the experimental subject scores so well or so poorly that one would expect few (fewer than 5% by convention) of the non-drugged subjects to behave so extremely, then the scientist will be encouraged in the belief that the drug might have an effect.

Since 2.5% of a normal distribution falls below $P_{.025}$ and 2.5% falls above $P_{.975}$, determining whether a score of 11 falls below $P_{.025}$ or above $P_{.975}$ in the distribution of non-drugged people is a rather simple statistical problem. The scientists assume that the distribution of scores in this population is normal in form, and then they use the standard normal distribution to ascertain the percentile rank of a score of 11. Recall from Chapter 4 that

$$z_i = \frac{X_i - \mu_x}{\sigma_x}$$

translates any score in a normal distribution into a z score (also called a *standard normal deviate*) that can be related to the standard normal distribution. Assuming the drug had no effect, the experimental subject was thus a member of the normal population of non-drugged subjects and $X_i = 11$, $\mu_x = 7$, and $\sigma_x = 2$. Then

$$z = \frac{X_i - \mu_x}{\sigma_x} = \frac{11 - 7}{2} = \frac{4}{2} = 2.00$$

This means that if the experimental subject were a member of the non-drugged population, the subject would have performed 2.00 standard deviation units above the mean. Looking at Table A of Appendix II, we see that $z = 2.00$ falls at $P_{.9772}$ (i.e., 97.72% of the cases would fall below such a value) in a normal distribution. Since this is more extreme than the scientists' arbitrary cutoff of $P_{.975}$, the decision is that the drug might have an effect.

Recall that theoretical relative frequency may also be interpreted in terms of probability. Thus, the probability that the experimental person would score 11 or greater if the drug had no effect was less than .05. In fact, the probability was $1 - .9772 = .0228$ that a non-drugged subject would score 11 or higher. Since this probability is so low, perhaps the drug did have an effect.

A single group   Encouraged by this result, the scientists decide to randomly select 20 people and determine their performance under the presumed influence of the drug. The same logic as before is used: If the mean of the group of drugged subjects is sufficiently different from what would be expected of non-drugged subjects, the notion that the drug affects performance is supported. The scientists tentatively assume that the drug has no effect—that is, that the group of 20 subjects will have a mean comparable to the population mean of 7 for non-drugged subjects. If the performance of the drugged sample exceeds $P_{.025}$ or $P_{.975}$, the conclusion is that it is not very likely that they were comparable to non-drugged people: Their performance was too atypical, and perhaps the drug had some effect. If not, the experiment provides no support for the drug's effect.

The 20 subjects score an average of 8.4. Statistically, this situation looks just like the previous case, except that now there is a group of subjects with a mean score rather than a single subject with one raw score. The problem demands that an observed mean ($\bar{X} = 8.4$) be transformed into standard normal units (z). The translation formula derives from the fact that the expression

$$z = \frac{X_i - \mu_x}{\sigma_x}$$

is perfectly general. It can be read: Any observation minus the population mean of such observations divided by the population standard deviation of such observations will equal z. In the first example, the observation was a single score and the population of observations was the distribution of all non-drugged subjects. Now, simply permit a group mean ($\bar{X}$) to be the observation and the theoretical distribution of means (the sampling distribution of the mean) to be the population of observations. Hence, the observation will be $\bar{X} = 8.4$, the mean of such observations will be $\mu_{\bar{x}} = 7$ (since $\mu_{\bar{x}} = \mu_x = \mu$), and the standard deviation of such observations will be the standard deviation of the sampling distribution of means (i.e., the standard error of the mean) which will equal $\sigma_x/\sqrt{N} = 2/\sqrt{20}$. Thus, the observed mean of 20 subjects may be translated into standard deviation units by

$$z = \frac{\bar{X} - \mu_{\bar{x}}}{\sigma_{\bar{x}}} = \frac{\bar{X} - \mu}{\sigma_x/\sqrt{N}}$$

$$z = \frac{8.4 - 7}{2/\sqrt{20}} = 3.13$$

Looking again at Table A in Appendix II, we see that only .0009 of the cases in a normal distribution would be expected to fall above such a value. Hence, the probability is indeed less than .05 that a group of 20 non-drugged subjects would score so extremely, and one may conclude that their performance may have been caused by the drug.

Notice that although the formulas which were required to translate the observed value (X_i or $\overline{X}$) into standard normal units differ slightly in the two preceding examples, the logic behind the procedure is the same. Randomly select a subject or a group of subjects and tentatively assume that, despite the fact that they receive the drug, they are no different from non-drugged subjects. Arbitrarily select some percentiles (or probabilities) that define behavior that would provoke the interpretation that these experimental subjects might be different from non-drugged subjects. Translate a summary statistic of their performance (e.g., X_i or $\overline{X}$) into a form compatible with a theoretical relative frequency distribution and determine its percentile rank. If that rank exceeds the arbitrary limits set above, suspect that the tentative assumption that these people were just like non-drugged subjects may have been wrong, since their performance was very atypical or improbable. The next chapter examines the details of this logic.

FORMULAS

1. Standard score form

$$z = \frac{X - \mu}{\sigma_x} \qquad z = \frac{\overline{X} - \mu}{\sigma_{\overline{x}}}$$

2. Standard error of the mean

$$s_{\overline{x}} = \frac{s_x}{\sqrt{N}} \qquad \text{(sample or empirical)}$$

$$s_{\overline{x}} = \sqrt{\frac{N\Sigma X_i^2 - (\Sigma X_i)^2}{N^2(N - 1)}} \qquad \begin{array}{l}\text{(sample or empirical}\\\text{from raw scores)}\end{array}$$

$$\sigma_{\overline{x}} = \frac{\sigma_x}{\sqrt{N}} \qquad \text{(population or theoretical)}$$

EXERCISES

1. Discuss the appropriateness of the following samples for the stated populations.
   a. A researcher in education wants to test the effectiveness of two different teaching methods on college students in a university, using one method with an 8 A.M. class and the other with a 2 P.M. class.

   b. A social psychologist is investigating patterns of group dynamics in the context of a jury simulation experiment to find the factors involved in the change of attitudes that transpires during jury debate in young adults. Students taking Introduction to Psychology may volunteer to serve in an experiment. Eighty students select the jury simulation experiment from the 20 possible research projects available to them.

c. An advertiser wants to know if people like the new style of package that Wonderease Soap Powder comes in. The package has just been displayed on television, so the advertiser quickly sets up a telephone survey in a major city to inquire whether the people who saw the commercial liked the new packaging.

2. A population of 10 scores is listed below:

$$X_i = 2, 4, 4, 5, 5, 6, 7, 7, 8, 10$$

Write each score on a piece of paper and place it in a hat. Randomly select 10 samples of size 4 from this population and compute the mean. (Replace the selected numbers after each drawing of 4.) Create an *empirical* sampling distribution of the mean from these data. Estimate the *theoretical* standard error of the mean, first by using the standard deviation of the first sample you select in an appropriate formula and then by computing the standard deviation of the 10 samples you collected. How do you explain the differences in these values? Of what importance is the size of the sample in this process? What does the standard error of the mean tell you about the sample mean as an estimator of the population mean?

3. For the population standard deviations given below, what is the standard error of the sampling distribution of the mean for samples of size N?
   a. If $\sigma_x = 16$, $N = 16$
   b. If $\sigma_x = 15$, $N = 25$
   c. If $\sigma_x = 18$, $N = 36$

4. The sampling distribution of the mean is normal in form if either one of two conditions is met. What are they?

5. Under what circumstance are the mean and variance independent?

6. What is the relationship between an idealized experiment, probability, and theoretical relative frequency?

7. In a normal distribution with $\mu = 50$ and $\sigma_x = 10$, what is the probability of randomly sampling a subject who scores
   a. 62 or higher?
   b. 70 or higher?
   c. 40 or higher? 40 or lower?
   d. between 45 and 58?

8. Assuming a normal population distribution with $\mu = 80$, $\sigma_x = 20$, what is the probability of obtaining the following means for groups of subjects chosen by random sampling?
   a. $\bar{X} = 96$ or higher, $N = 4$?
   b. $\bar{X} = 92$ or higher, $N = 16$?
   c. $\bar{X} = 68$ or higher, $N = 25$? $\bar{X} = 68$ or lower, $N = 25$?
   d. $\bar{X}$ between 74 and 83, $N = 9$?

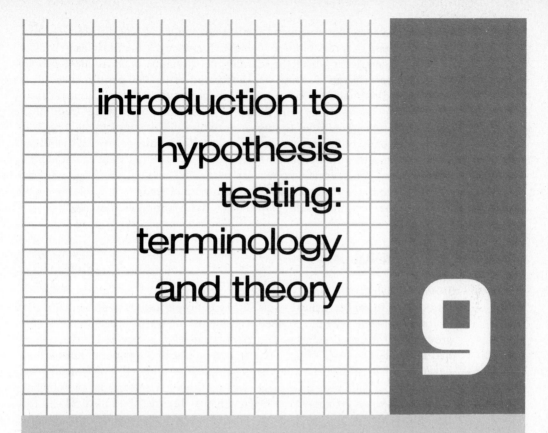

introduction to hypothesis testing: terminology and theory

9

N THE EXAMPLE that concluded the last chapter, sampling distributions and the concept of probability as relative frequency were illustrated with a simple example in which the learning performance of a group of people who had been given a drug was compared with the performance of a non-drugged population. The example was presented informally and phrased in common language. In this chapter, it is recast into more traditional statistical language in order to introduce the formal terminology, theory, and procedures of hypothesis testing. The chapter also introduces a distribution often used to determine probabilities in social science research: the Student's *t* distribution. When certain population parameters are unknown and must be estimated from quantities calculated on samples of small size, Student's *t* will provide more accurate results than the standard normal distribution.

STATISTICAL TERMINOLOGY

It will be helpful to summarize the example presented in Chapter 8 before introducing formal statistical terms for the concepts and processes involved in it. Recall that the scientists wanted to compare the learning performance of a group of 20 drugged subjects to a population of non-drugged subjects that had a population mean of $\mu = 7$ correct responses and a population standard deviation of $\sigma_x = 2$. The experimental procedure required two **assumptions**: first, that the sample of 20 subjects was randomly selected from the population of non-drugged people, and, second, that the distribution of scores for this population is normal in form. Then the scientists **hypothesized**, at least temporarily, that the drug actually had no effect on learning performance and that the group of 20 drugged subjects would perform just like a group of 20 non-drugged subjects (since they were random selections from the same population). If this hypothesis was true, then the mean performance of this small sample should not be unusually deviant from the non-drugged population mean, $\mu = 7$.

Before the experiment the scientist established some **decision rules**. Specifically, if the mean of the drugged group was a value between $P_{.025}$ and $P_{.975}$ in the sampling distribution of means based upon non-drugged subjects, there would be no cause to dispute the hypothesis that the drug has no effect. Such a result would be a reasonable outcome for non-drugged people. On the other hand, what if the observed mean for the drugged group was so different from the population mean of non-drugged subjects that such a value should be expected of non-drugged individuals *in only 5% or less of such samples*? That is, suppose the mean of the drugged group was lower than $P_{.025}$ or higher than $P_{.975}$? In that event the scientists would question or reject the hypothesis that the drug had no effect.

The experiment was then performed, and the sample mean of the 20 drugged subjects was observed to be $\overline{X} = 8.4$. This value was then translated by **computation** into a standard normal deviate (another term for z score when the population distribution is normal in form) by the formula

$$z = \frac{\overline{X} - \mu}{\sigma_x / \sqrt{N}} = \frac{8.4 - 7}{2 / \sqrt{20}} = 3.13$$

The value $z = 3.13$ corresponds to a percentile rank of $P_{.9991}$ in the standard normal distribution. Such an extreme value would not likely be found for a sample of 20 non-drugged subjects—indeed, the probability of observing a sample mean as extreme as $\overline{X} = 8.4$ is less than .05. Thus, the scientists **decide** that the drug might actually have an effect on learning performance. We will now apply more formal terminology to the concepts and processes described in the example.

Statistical assumptions and hypotheses

The kind of statistical inference we have described requires that certain **statistical assumptions** be made. In the example, the assumptions are that the 20 subjects were randomly and independently selected from a non-drugged population and that the sampling distribution of the mean is normal in form. The latter assumption is met if the population distribution of raw scores is normal or if N is large. Although the researchers attempted to sample randomly and did have some previously collected information about the normality of the population distribution, one can seldom be certain that these assumptions are, in fact, correct, and thus they must remain assumptions.

The investigators considered two **hypotheses** about what is true: Either the drug has an effect or it does not. Such hypotheses are mutually contradictory, and most often they exhaust the possibilities. Finally, one of the hypotheses is tentatively held to be true. In this case, the scientists temporarily assumed that the drug did not have any influence on memory.

> The hypothesis that is tentatively held to be true is called the **null hypothesis** and is customarily symbolized by H_0. The **alternative hypothesis** is represented by H_1.

In the present example the null and alternative hypotheses may be stated:

H_0: The observed mean is computed on a sample drawn from a population with $\mu = 7$ (i.e., the drug has no effect).

H_1: The observed mean is computed on a sample drawn from a population with $\mu \neq 7$ (i.e., the drug has some effect).

These assumptions and hypotheses are crucial to the statistical logic of the study. If the null hypothesis that the drug has no effect is tentatively held to be true, then the random sample of 20 people should behave like a typical sample of 20 non-drugged subjects. The assumption that the population distribution is normal in form means that the sampling distribution of the mean will also be normal, and that the percentiles of the standard normal distribution can therefore be used to determine the probability that a random sample of 20 should have a mean as deviant from $\mu = 7$ as the observed mean of 8.4 is. If this probability is small, then perhaps the hypothesis of no drug-effect, which was tentatively held true at the beginning, is actually false and should be rejected.

There are important differences between assumptions and hypotheses, even though both are statements about what is true in the population. **Hypotheses** represent a set of two or more contradictory and often exhaustive possibilities, only one of which can actually be the case. Thus, while all assumptions are presumed to be true, only one of the hypotheses *can* be true. Moreover, one of the hypotheses (the null hypothesis) is temporarily held to be true, and it is the reasonableness of this hypothesis that is being tested by these statistical procedures. The logic is as follows: Tentatively hold the null hypothesis to be true (i.e., assume the drug has no effect). If the results of the analysis could have reasonably occurred given that circumstance (i.e., if the mean of the drugged group could probably have been scored by a non-drugged group), there is no reason to question the validity of the hypothesis of no drug-effect. But if the analysis indicates that the likelihood is very small that a non-drugged group could have produced this mean, the null hypothesis is rejected in favor of the alternative hypothesis. Notice that the assumptions of random sampling and normality remain constant and are assumed true regardless of the outcome of the experiment; only the null hypothesis is tentatively held and then either rejected or not rejected.

Significance level

Recall part of the statistical logic described above: If the probability is very small that the observed sample mean could be produced by a non-drugged group, then one will reject the null hypothesis. How small does that probability need to be before the null hypothesis is rejected? That probability value is the significance level of the test of the null hypothesis:

The **significance level** (or **critical level**), symbolized by α (alpha), is the probability value that forms the boundary between rejecting and not rejecting the null hypothesis.

In social science research the precise value of α is customarily taken to be .05. In the current example, the significance level .05 indicates that unless the observed mean is so different from the population $\mu = 7$ that it would be expected of non-drugged subjects less than 5% of the time, the researcher will not challenge the validity of the null hypothesis, H_0. However, if the probability is less than .05 that the observed mean for the drugged group would be obtained by non-drugged subjects, the null hypothesis that the drug has no effect will be rejected and the conclusion will be that the drug probably did exert some influence. This result is then said to be significant at the .05 level, which may be written "$p < .05$," in which p stands for the probability of the null hypothesis. The significance level, in this case .05, represents the probability value that separates a decision to reject from a decision not to reject the null hypothesis.

Decision rules

Once the significance level has been established, it is used to formulate decision rules.

> **Decision rules** are statements, phrased in terms of the statistics to be calculated, that dictate precisely when the null hypothesis will be rejected and when it will not.

In the example above, the researchers selected the .05 level of significance, meaning that the null hypothesis would be rejected only if the observed sample mean deviated from the mean of the non-drugged population to an extent likely to occur in non-drugged samples less than 5% of the time. $P_{.025}$ and $P_{.975}$ of the sampling distribution of the mean mark the end points of a range of values that would include 95% of the means of non-drugged samples within their limits but exclude 5% of such means, $2\frac{1}{2}\%$ because they are extremely low and $2\frac{1}{2}\%$ because they are extremely high. Since the sampling distribution of the mean is normal in form, the standard normal distribution may be used to translate these percentile ranks into percentile points (i.e., z values). Looking at Table A in Appendix II, one can see that $P_{.975}$ corresponds to a z value such that .025 of the distribution falls to its right (above it). Scanning the rightmost column for .0250, one finds the corresponding z to be 1.96. Since the standard normal is a symmetrical distribution, we also know that .025 of the distribution falls to the left of (below) a z of -1.96. Thus, $P_{.025}$ and $P_{.975}$ correspond to the points $z = -1.96$ and $z = 1.96$.

This situation is pictured in Figure 9–1. The normal distribution shown in the graph represents the theoretical sampling distribution of means for samples of size $N = 20$ from the non-drugged population, but it is expressed in terms of standard normal deviates. Thus, the population mean ($\mu = 7$) is represented by $z = 0$. Since $P_{.025}$ and $P_{.975}$ correspond to $z = -1.96$ and $z = 1.96$, respectively, these points include 95% of the possible sample means,

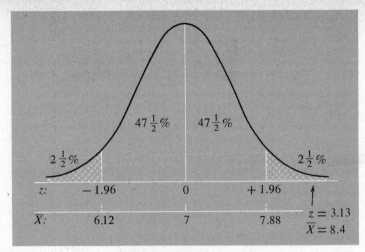

Fig. 9–1. Sampling distribution of the mean expressed in standard normal deviates (z) and sample means $(\overline{X})$

$47\frac{1}{2}\%$ on each side of $z = 0$. Thus, each of the crosshatched areas contains $2\frac{1}{2}\%$ of the cases, and these cases together constitute the 5% of extreme sample means that would result in rejecting the null hypothesis. Since the observed sample mean $(\overline{X} = 8.4)$ will be translated into a z-value equivalent, the formal statement of the decision rules is:

If z is between -1.96 and $+1.96$, do not reject H_0.
If z is less than or equal to -1.96 or greater than or equal to $+1.96$, reject H_0.

The same statements can also be expressed by using the symbols $<$ (less than), $>$ (greater than), $\leq$ (less than or equal to), and $\geq$ (greater than or equal to):

If $-1.96 < z < +1.96$, do not reject H_0.
If $z \leq -1.96$ or $z \geq +1.96$, reject H_0.

The z values of ± 1.96 in this example are called **critical values** because they separate z values that will result in a decision to reject or not to reject the null hypothesis. Thus, the crosshatched areas in Figure 9–1 define the **critical region** or **region of rejection** of the null hypothesis.

Computation

The statistical procedures take advantage of theoretical relative frequency distributions for which percentiles are already available. In the example we have been considering, the standard normal was used, but there will be other theoretical distributions employed in different contexts later in the text. The computational procedures essentially convert the observed statistic (e.g., the mean) into a standard normal deviate so that this value can be compared with the critical values previously established. In this case, $\overline{X} = 8.4$ was equivalent to $z = 3.13$.

In order to discriminate observed from critical values of the standardized variable, subscripts are often used. In our example,

$$z_{obs} = 3.13$$

represents the value corresponding to the observed data, whereas

$$z_{crit} = \pm 1.96$$

indicates the critical values of z.

Decision

In the example, $z = 3.13$ falls within the critical region (Figure 9–1), which dictates that the null hypothesis is to be rejected on the grounds that the obtained mean of 8.4 ($z = 3.13$) would be expected to occur in less than 5% of samples of 20 non-drugged subjects. Since it is unlikely that such a value would occur merely as a function of sampling error associated with selecting certain people as subjects and not others, the null hypothesis of no drug-effect is rejected.

It is important to understand the relationship between the logic of this statistical process and the concept of sampling error as introduced in the previous chapter. The sampling distribution of the mean reflects the extent to which the mean of one sample will differ from the mean of another sample, even if these samples are treated exactly alike. If the observed mean is not a likely value in this sampling distribution, we reject the null hypothesis. In that case, one can say that the observed mean was so discrepant from what would have been expected on the basis of random variation from sample to sample —i.e., on the basis of sampling error—that it is unlikely that a mean of this magnitude would be obtained merely because of chance variation from sample to sample. On the other hand, if the observed value is not sufficiently extreme to reject the null hypothesis, one can conclude that the mean of the drugged group could have occurred even if the drug had had no effect, since it could reasonably be expected to deviate from the population average on the basis of sampling error alone (random variation between samples, chance, etc.).

This interpretation can be understood by examining the formula for translating the observed sample mean into a standard normal deviate of the theoretical sampling distribution of the mean:

$$z = \frac{\overline{X} - \mu}{\sigma_{\bar{x}}}$$

The formula states that z is the ratio of the difference between the sample mean and the mean of the non-drugged population $(\overline{X} - \mu)$ relative to the amount of expected sampling error as expressed by its numerical index, the standard error of the mean $(\sigma_{\bar{x}})$. Because this formula yields the values that define the critical region, it is sometimes called a **critical ratio**.

In a sense, the statistical test simply asks how different the observed mean was from the population mean, relative to the sampling error one might expect. As this ratio increases in size, one begins to suspect that the difference between observed and population means is probably not simply a function of random variation between samples, chance, or—more precisely—of sampling error.

Note that the decision to reject the null hypothesis (when this is appropriate) is *not* equivalent to accepting the alternative hypothesis. One rejects H_0, one does *not* accept H_1. The reason for taking such care is that only one hypothesis, the null hypothesis, was tested. It was tentatively held to be true, and statistical procedures were carried out to determine if the data were reasonable under these conditions. Since they were not, H_0 was rejected as probably not a valid characterization of what really exists. Since it was H_0 that was tested, the decision made concerns H_0, not H_1.

Similarly, if the results indicate that the null hypothesis is not to be rejected, one does not accept it, since the results demand only that one not reject H_0. It may seem like quibbling to insist on a distinction between "not rejecting" and "accepting" the null hypothesis, but the failure to understand the difference often reflects a failure to understand the logic of this statistical procedure. Suppose the observed value of z is not in the critical region and thus is not large enough for the researchers to reject the null hypothesis. This does not necessarily say that the drug has no effect. It may influence learning performance, but not very much. Thus, the difference it produces is not large enough, relative to the amount of sampling error, to produce a z value sufficient to reject the null hypothesis in a sample of size N. The same difference might lead to rejecting H_0 in a larger sample. Therefore, one does not prove or accept the null hypothesis—one simply has no cause to reject it.

Decision errors

Generally, the purpose of statistical inference is to make an educated guess about what exists in a population when only a small sample of cases from that population has been studied. Since the decision is a guess, it may be wrong. In fact, if the significance level is .05, then in 5% of the cases in which the drug actually has no effect, the researchers will make the wrong decision, which is to reject the null hypothesis.

This is part of the logic of the procedure. You observe a sample mean and ask how likely such a mean is if the drug has no effect. You arbitrarily decide that if such a mean should occur less than 5% of the time, assuming H_0, you will suspect that the null hypothesis is incorrect. But suppose the null hypothesis is quite correct. Then, in 5% of the cases in which you observe a mean that deviates so much from the population value, the observed mean will be the result of sampling error and random variation. Following the decision rules, you will reject the null hypothesis in these cases, but you will be wrong: The drug *actually* has no effect, but you have *decided* that it does. This is called a type I error.

> A **type I error** occurs when the decision is to reject the null hypothesis when it is actually true. Given the validity of the null hypothesis, the probability that it is erroneously rejected by these procedures equals α, the significance level.

Another type of error is made when the null hypothesis is *actually* wrong—for example, as when the drug does indeed have an effect—but the *decision* is made not to reject it.

> A **type II error** occurs when the decision is not to reject the null hypothesis when it actually is false. The probability of this type of error is symbolized by β (beta).

The decision-making process is diagramed in Figure 9–2. The distribution labeled H_0 represents the sampling distribution of means (for samples with $N = 20$) for non-drugged subjects. The crosshatched areas in the tails of this distribution constitute the region of rejection. If the null hypothesis is actually true, then this area represents the probability of a type I error, since an observed mean in this region would lead to a decision to reject the null

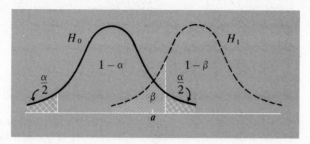

Fig. 9–2. Diagram of the relationship between type I (α) and type II (β) errors and power $(1 - \beta)$

9-1 The Four Possible Outcomes of a Simple Decision Process and Their Associated Probabilities

	Actual Situation	
Decision	H_0 is True	H_0 is False
Reject H_0	Type I error $p = \alpha$	Correct decision $p = 1 - \beta$ (power)
Do not reject H_0	Correct decision $p = 1 - \alpha$	Type II error $p = \beta$

hypothesis when it is actually true. Now shift your attention to the distribution labeled H_1, a theoretical distribution of possible means for drugged subjects. Assume that H_1 is, in fact, true and the drug has an effect. Because of the overlap in the distributions, it is possible that some sample means will fall close enough to the mean of the non-drugged population so that the decision will be not to reject the null hypothesis when it is actually false—a type II error. An observed mean corresponding to point a would be such a case. The diagonal-lined area labeled β in Figure 9-2 corresponds to means for drugged samples for which the statistical process would erroneously declare that the drug had no effect. It represents the probability of a type II error. Unfortunately, the distribution for drugged samples when the drug actually has an effect (labeled H_1) is rarely available, and hence it is usually impossible to estimate the precise size of β. However, as the likelihood of a type I error decreases, the probability of a type II error increases.

There are two other areas to notice in Figure 9-2. The portion of the H_0 distribution that is not in the region of rejection represents the probability of correctly refusing to reject H_0 ($p = 1 - \alpha$). The portion of H_1 that is not part of β represents the probability of correctly rejecting H_0 ($p = 1 - \beta$). The latter probability is called the **power** of a test. Table 9-1 summarizes these points.

Power of a statistical test

The power of a statistical test can be defined as follows:

The **power** of a statistical test is the probability that the test will lead to a decision to reject H_0 when H_0 is indeed false. *True*

A high power is a major requirement of a good statistical test. If H_0 is actually not true and should be rejected, we want to know that the statistical test we are using will lead to that conclusion. However, type II errors, in which the test concludes not to reject H_0 when H_0 is actually false, occur with a probability of β. Thus, the probability of correctly rejecting H_0 is $1 - \beta$. The actual value of this probability is difficult to obtain, since it requires one to know the H_1 distribution. That is, one must already know what actually exists in nature. We rarely have this information—indeed, that is why we are performing the statistical test in the first place. Obviously, researchers would like their statistical procedures to have as much power as possible. Despite the fact that we usually cannot precisely calculate the power of a statistical test, there are some ways of increasing this probability.

First, **the power of a statistical test increases as the significance level, α, increases.** Thus, a test performed at a significance level of .05 has more power than one at .01. Obviously, the less stringent the significance level, the more likely the null hypothesis will be rejected. But both correct and incorrect rejections will be more likely; simply having a larger value of α will increase the power but will also increase the probability of a type I error.

Second, **the power of a test increases with sample size.** Thus, the likelihood of correctly rejecting H_0 is better for a sample of 30 than a sample of 20.

Third, **some statistical tests are more powerful than others.** For example, the techniques described in Chapters 10–12 are called parametric tests; those discussed in Chapter 13 are nonparametric tests. In general, the more specific the assumptions that can be made about the nature of the data, the more powerful will be the statistical test. For example, most parametric tests require that the dependent variable be normally distributed, homogeneously variable across groups, and measured with an interval or ratio scale (see pages 22–26 for a description of types of scales of measurement). When these assumptions cannot be made, nonparametric tests (described in Chapter 13) may be used, but some power will probably be lost.

Selecting the significance level

Recall that the significance level is the probability of erroneously rejecting the null hypothesis when it is, in fact, true (type I error). How does one go about setting this level? The value of α reflects the investigator's feeling of how much error is tolerable in making a decision to reject H_0 on the one hand, balanced against how much power is wanted in the statistical test on the other. Ordinarily, social scientists accept the .05 level of significance as the value of α, but this is merely a convention that has grown up over the years, and there is nothing fixed about it. When testing some hypotheses, other levels of significance might be adopted depending upon how critical it is to avoid being wrong in rejecting the null hypothesis. For example, if you were a brain surgeon and were giving a test to determine whether a patient needed a very delicate type of surgery, it might be that you could not afford to risk an operation if the patient really didn't need one (i.e., if your diagnosis was wrong). Therefore, you might operate only if the probability was less than .01 or .001 that the test result was just a sampling error. However, the more you lower the probability of incorrectly rejecting the null hypothesis and performing the operation (type I error), the more you raise the probability that a patient who needs the operation will be incorrectly diagnosed as being able to do without it (incorrectly failing to reject H_0; a type II error). In any situation, it is important to remember that the level of significance is arbitrarily established on the basis of the researcher's tolerance for error in decision making and the desire for statistical power. However, depending upon the circumstances, the reader may feel differently from the researcher about the most appropriate relative sizes of error and power. Therefore, a reasonable procedure is for the researcher to adopt a significance level but also to report the actual probability found as a result of testing H_0, regardless of whether H_0 is rejected or not.

Directional tests

The test of the null hypothesis described in the drug example was a **nondirectional test**. It was nondirectional because the alternative hypothesis (H_1) did not specify the direction of the influence of the drug. Would it help or hinder learning? It only suggested the drug would influence performance, one way or the other. Consequently, the critical region was established so that extreme values in either direction would lead to a decision to reject H_0.

However, suppose the same drug had been found by other scientists to improve memory in rats, and the scientists then wondered if its administration would also improve memory in humans. Since it is very unlikely that the drug would actually hinder memory, the researchers would establish the alternative hypothesis as

H_1: The observed mean is computed on a sample drawn from a population with $\mu > 7$ (i.e., the drug facilitates memory).

This is a **directional** hypothesis, since it prescribes that the mean will be greater than, not just different from, 7.

In such an event, it would be unreasonable to divide the critical region into two parts so that extreme means in either direction lead to rejecting the null hypothesis. Suppose the previous research information had suggested that the probability was almost zero that drugged subjects would exhibit an extremely low mean. Therefore, in order to maintain the significance level (i.e., the probability of erroneously rejecting the null hypothesis) at .05, the region of rejection would have to include 5% of the scores in the right-hand tail of the distribution. Thus, when looking in Appendix II, Table A (the standard normal), one would search for the z value that has .0500 of the area lying to its right. The critical value is $z = 1.645$, so the decision rules would be:

$$\text{If } z_{obs} < 1.645, \text{ do not reject } H_0.$$

$$\text{If } z_{obs} \geq 1.645, \text{ reject } H_0.$$

A comparison of the regions of rejection for these nondirectional and directional tests is presented in Figure 9–3.

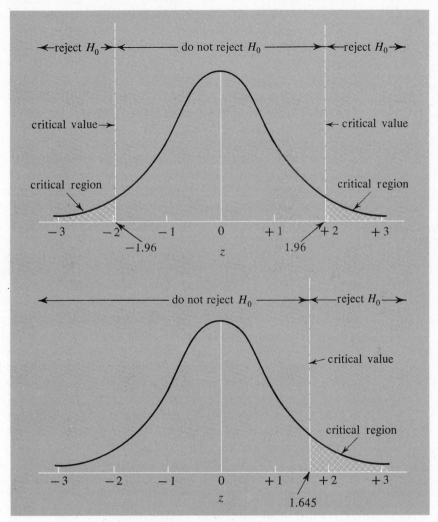

Fig. 9–3. Decision process for a nondirectional (two-tailed) test (*top*) and a directional (one-tailed) test (*bottom*)

Since a nondirectional test locates the critical region in both tails of the theoretical distribution, it is frequently called a **two-tailed test**. Conversely, since a directional test locates the critical region in only one end of the distribution, it is called a **one-tailed test**.

In determining whether a test should be directional or nondirectional, one considers whether there is other evidence or a theory that might predict the result. There needs to be some basis for presuming that a sample mean will deviate in only one direction from the population value. Then H_1 and the critical values are established to agree with that prediction. However, *this choice of a directional or nondirectional test must be established before one knows the experimental result*. Otherwise, if one waits to know the data and then decides to make a directional instead of a nondirectional test, the theoretical probabilities involved in the significance level (e.g., $\alpha = .05$) are no longer appropriate, because one will be tempted to apply the one-tailed test only in cases where z_{obs} is between the critical values for a directional and a nondirectional test. Therefore, the statistical hypotheses must be stated before the experimental observations are made.

Students frequently have difficulty understanding the rationale behind the size of the critical region in directional tests. One might look at Figure 9–3 and wonder, "If you have some reason for predicting that the sample mean will be greater than the mean of the population of non-drugged subjects, why do you select a *lower* critical value, which seems to make it easier to reject the null hypothesis?" It seems that it should be the other way around: If you know that the difference between means will likely be positive (to the right of $z = 0$), then the critical value of z should be more stringent (i.e., farther to the right) in order to balance out the advantage gained by being able to predict the direction of a difference. The explanation rests in the purpose of α, the significance level. Most researchers are interested in rejecting the null hypothesis. We want to have significant results. But, the significance level refers to the probability of rejecting the null hypothesis *by mistake*—that is, to the probability of deciding there is a difference when no difference actually exists. The problem, then, is to have a directional and a nondirectional test at the .05 level of significance comparable to each other in terms of the probability of falsely rejecting H_0. A nondirectional, or two-tailed, test must allow for error in rejecting the null hypothesis as a function of extreme differences in either direction due to sampling variation. However, if there is some reason to think that the sample mean will be larger than the population mean, because either empirical evidence or a theory suggests this will be the case, there is very little chance that an extremely large negative difference will be anything but sampling variation. To use a two-tailed, nondirectional, alternative in this instance would really constitute a test at the .025 level of significance, because it is known that extreme negative differences will be sampling error and not drug effects. Therefore, in order to ensure that the probability of a type I error is indeed the value of α, it is necessary to place all of the critical region in the right-hand tail.

HYPOTHESIS TESTING WHEN $\sigma_{\bar{x}}$ IS ESTIMATED BY $s_{\bar{x}}$

In the expression that translates a sample mean into a standard normal deviate,

$$z = \frac{\overline{X} - \mu}{\sigma_{\bar{x}}}$$

z is a standard normal deviate (if the sampling distribution of the mean is normal), μ is the population mean of the sampling distribution of $\overline{X}$'s determined for all possible samples of size N from the population, and $\sigma_{\bar{x}}$ is the theoretical standard deviation of such $\overline{X}$'s. Up to this point, the formulas presented have always involved a parameter, for example, $\sigma_{\bar{x}}$. Regrettably, the value of this parameter is usually not available, so it must be estimated.

The standard error of the sampling distribution of the mean, $s_{\bar{x}}$, which can be calculated on the basis of a single sample of scores ($s_{\bar{x}} = s_x / \sqrt{N}$), may be used for this estimation. However, when $s_{\bar{x}}$ is invoked to estimate $\sigma_{\bar{x}}$, the accuracy of $s_{\bar{x}}$ as an estimator, and thus the accuracy of the probability value derived from the standard normal distribution, depends upon the size of the sample, N. If the sample is quite large ($N > 50$ or 100), the probabilities derived by using the standard normal distribution as described above are fairly accurate. But most samples in psychology are not so large, and it happens that when $\sigma_{\bar{x}}$ is estimated by $s_{\bar{x}}$ calculated on a small sample, the standard normal does not provide sufficiently accurate probabilities. Fortunately, another theoretical sampling distribution exists that is appropriate for these situations.

Student's t distribution

The **Student's t distribution** is a theoretical sampling distribution developed by W. S. Gosset, who wrote under the name Student. When scores are transformed in the following way, the Student's t distribution, rather than the standard normal, is appropriate:

$$t = \frac{\overline{X} - \mu}{s_{\bar{x}}} = \frac{\overline{X} - \mu}{s_x / \sqrt{N}}$$

Notice that except for substituting t for z and $s_{\bar{x}}$ for $\sigma_{\bar{x}}$, the formula is the same as that used in the examples above. The t distribution, however, differs from the normal. In addition to permitting the parameter $\sigma_{\bar{x}}$ to be estimated by the sample statistic $s_{\bar{x}}$, there is a different t distribution for each size of sample upon which $s_{\bar{x}}$ is computed. The appropriate t distribution is determined not by the sample size, N, but by its **degrees of freedom**, $N - 1$ in this case. Figure 9–4 shows the t distribution for several degrees of freedom.

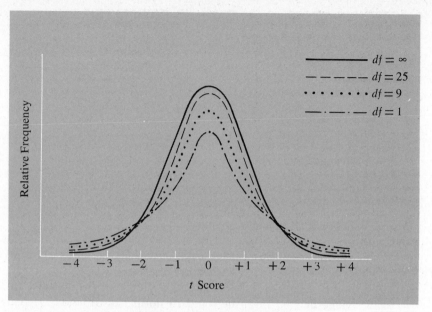

Fig. 9–4.   The t distribution for various degrees of freedom. With an
infinite number of degrees of freedom, the t and the standard
normal (z) distributions are identical.

It also depicts the standard normal ($df = \infty$). Notice that the t and z
distributions are quite different when the sample size (and thus the number of
degrees of freedom) is small, but as the number of degrees of freedom
increases the t distribution becomes more and more like the normal. In fact,
when the sample size is allowed to be theoretically infinite, the t and standard
normal distributions are identical.[1]

Degrees of freedom   Undoubtedly, one of the most difficult concepts
for students to grasp, but also one of the most necessary, is that of degrees of
freedom.

The number of **degrees of freedom** (**df**) for any statistic is the
number of components in its calculation that are free to vary.

The meaning of this definition is best explained by considering some examples. In a sense, every statistic in a specific context has a certain number of

---

[1]Some texts suggest that if the sample size is 30 or more, the normal and t distributions are
similar enough that the normal distribution (z) may be used even though $\sigma_{\bar{x}}$ has been estimated.
While an examination of the tables for z and t in Appendix II certainly indicates a great degree
of correspondence between the normal and t for samples of 30 or more, it will be the policy of
this book to suggest that the t distribution always be used when a parameter is being estimated,
regardless of sample size. This policy results in greater accuracy in determining probabilities and
in somewhat less confusion.

degrees of freedom associated with it. For example, the sample mean, $\bar{X}$, has N degrees of freedom. How does one know that? Begin by examining the formula for the mean:

$$\bar{X} = \frac{\Sigma X_i}{N}$$

The first question to ask is which components of this formula can vary at all and which cannot. When calculating a mean from any specified sample of scores, the N does not vary. It is fixed in value for that sample. In contrast, the scores, X_i, which compose the ΣX_i, do take on different values—they are free to vary. That is, each of the scores could assume any possible value, and knowing the values of all but one score does not tell us the value of the last score. If there are five scores, and we know four of them, (3, 7, 5, 2, ?), there is no way for us to determine the value of "?." Thus, we must know all N scores to calculate the mean. It is in this sense that all N scores are free to vary (i.e., to take on any values), and thus the number of degrees of freedom for the statistic $\bar{X}$ is simply N.

The task becomes more complicated when we consider the number of degrees of freedom for the variance. Again, begin with the formula:

$$s_x^{\,2} = \frac{\Sigma\left(X_i - \bar{X}\right)^2}{N - 1}$$

Neither the sample size, N, nor the mean, $\bar{X}$ is free to vary. For any specified sample, they are constants. However, the deviations between each score and the mean $(X_i - \bar{X})$ take on different values depending upon the scores in the distribution. How many of these N deviations are free to vary? It happens that all but one of these deviations are free to vary, due to the fact that the sum of the deviations of scores about their own mean is always zero, $\Sigma(X_i - \bar{X}) = 0$. Knowing $N - 1$ of these deviations, you can always determine the last one because its value is such that the sum of all the deviations is zero. For example, consider the distribution (5, 7, 9). The mean is 7. Note what happens if any two of the deviations are computed:

X_i	$\bar{X}$	$(X_i - \bar{X})$
5	7	-2
7	7	0
9	7	?
	$\Sigma(X_i - \bar{X}) = 0$	

Since the sum of the deviations from the mean is always zero, the third deviation must be $+2$. It is in this sense that the last deviation to be computed, regardless of which particular score in the distribution is remaining, is not free to vary but is determined. Knowing $N - 1$ of the deviations, you can determine the value of the last deviation. Therefore, only $N - 1$ deviations are free to vary—free to take on any value. Thus, the number of degrees of freedom for the sample variance (and sample standard deviation) is $df = N - 1$.

The degrees of freedom for t are also $N - 1$. The formula for t is

$$t = \frac{\bar{X} - \mu}{s_x / \sqrt{N}}$$

The population mean, the mean for this sample, and N are all constants. Thus, the degrees of freedom for t are the degrees of freedom for s_x, which we have seen is $N - 1$. When selecting the proper critical values for a statistical test involving the t distribution, one must know the degrees of freedom for t. The present application of the t distribution dictates that t have $N - 1$ degrees of freedom. Other applications of t discussed in the next chapter may have different numbers of df associated with them because they involve different formulas for t.

Application of the t distribution   Suppose the researchers decide to randomly select a sample of 20 subjects and give each of them the drug prior to the recall test; but suppose also that the standard deviation of the population of non-drugged people is not available. The population distribution is still regarded as being normal in form with $\mu = 7$ and the researchers wish to test the hypothesis that the observed mean is computed on a sample drawn from a population with $\mu = 7$. However, since the population standard deviation is not known, $s_x / \sqrt{N}$ is used to estimate $\sigma_{\bar{x}}$.

The expression

$$t = \frac{\bar{X} - \mu}{s_x / \sqrt{N}}, \quad df = N - 1$$

changes the observed sample mean into standardized t units. With certain assumptions similar to those invoked to use the standard normal, the t distribution may be employed in the same manner as the standard normal distribution.

The t distribution is found in Table B of Appendix II and deserves a brief examination. The table gives the critical values of t for various significance levels at each of several degrees of freedom. The degrees of freedom are listed in the leftmost column. To their right one finds various values of t for different significance levels. Since the t distribution (like the z) is symmetrical, the values shown in the table hold not only for the right-hand (positive) half of the distribution but also, with the addition of a minus sign, for the left-hand portion. For example, in a two-tailed test at the .05 level at $N - 1 = 20 - 1 = 19$ degrees of freedom, the table shows that $2\frac{1}{2}\%$ of the t's fall to the left of -2.093 and $2\frac{1}{2}\%$ fall to the right of 2.093. If a one-tailed test at 19 *df* at the .01 level of significance was appropriate, the entire 1% of the area under the curve would fall to the right of 2.539. Thus, the *t* table gives the critical values of *t* for various degrees of freedom and levels of significance for both directional and nondirectional tests.

One should observe two additional things about *t*. First, for any given level of significance the value of *t* that is required to reject H_0 (critical value of *t*) becomes smaller as the number of degrees of freedom increases. Consider the third column for a two-tailed test at the .05 level of significance. With only one degree of freedom, $t = 12.706$ is required to reject H_0, whereas if the $df = 2$, $t = 4.303$ is needed. If there are an infinite number of degrees of freedom, $t = 1.96$ is necessary, which is precisely the value of the standard normal distribution for this situation. Thus, as the number of degrees of freedom increases, the critical value of *t* decreases, until it reaches the same value that the standard normal would dictate.

Second, the smaller the standard error, the more likely the null hypothesis will be rejected. This cannot be seen from the *t* table, but it can be appreciated by examining the expression for *t*:

$$t = \frac{\bar{X} - \mu}{s_{\bar{x}}}$$

Since the ratio of two quantities becomes larger as the denominator becomes smaller, it can be seen that the value of *t* grows larger as the standard error of the mean ($s_{\bar{x}}$) becomes smaller. Thus, the smaller the standard error the more likely the null hypothesis will be rejected when it is false.

Hypothesis testing: a formal example

A formal presentation of the solution to this hypothesis-testing problem follows. Its format may be used as a model to follow when formal hypothesis-testing solutions are required.

Research question If the mean number of correct trials of the population of non-drugged subjects is 7, does the mean $\bar{X}$ of a random sample of 20 subjects that are given a drug differ sufficiently from that population to warrant the conclusion that the drug has an effect on memory performance?

Statistical hypotheses

H_0: $\bar{X}$ is computed on a sample from a population with $\mu = 7$.
H_1: $\bar{X}$ is computed on a sample from a population with $\mu \neq 7$.

Assumptions and conditions   It is assumed that (1) the subjects are randomly and independently sampled, and (2) the population of non-drugged subjects is normal with $\mu = 7$.

Significance level   Adopt the .05 level.

Critical values   Since the alternative hypothesis H_1 was nondirectional and $\alpha = .05$, t values must be selected so that $2\frac{1}{2}\%$ of the t distribution falls to the left of one critical value and $2\frac{1}{2}\%$ falls to the right of the other. Looking in Appendix II, Table B, with $N - 1 = 20 - 1 = 19$ df, these critical values are -2.093 and $+2.093$.

Decision rules   If t is between -2.093 and $+2.093$, do not reject H_0. If t is less than or equal to -2.093 or greater than or equal to $+2.093$, reject H_0. In symbols,

$$\text{If } -2.093 < t_{\text{obs}} < +2.093, \text{ do not reject } H_0.$$

$$\text{If } t_{\text{obs}} \leq -2.093 \text{ or } t_{\text{obs}} \geq +2.093, \text{ reject } H_0.$$

Computation   Suppose the experiment is conducted, the observed mean is 8.4 and the standard deviation is 2.3. Placing these values into the expression for t one obtains

$$t_{\text{obs}} = \frac{\overline{X} - \mu}{s_x/\sqrt{N}} = \frac{8.4 - 7}{2.3/\sqrt{20}} = 2.72$$

Decision   The observed t of 2.72 exceeds the critical values and falls within the province of the second decision rule to reject H_0. It is concluded that given the expected amount of variability in means of samples of size 20, the mean of 8.4 is too deviant from $\mu = 7$ to be attributed simply to sampling error. The experimenters conclude that the drug may have an effect on memory.

To summarize, the purpose of hypothesis testing is to help the researcher solve the problem of deciding whether the results obtained in an experiment represent chance or sampling variability, or whether the factors that are isolated or manipulated actually produce differences. On the basis of sample data, the scientist must make a decision about what exists in a population not available for study. First, certain assumptions about the sampling procedures and about the parameters of the populations are made. Second, usually two mutually exclusive hypotheses are offered about the population that reflect in statistical terms the researcher's scientific question. One of these hypotheses, the null hypothesis, is tentatively assumed to be true. Third, a significance level is selected; then, as a function of the nature of the alternative hypothesis, the significance level and the particular theoretical distribution to be used, critical values, and decision rules are established.

Fourth, the experiment is conducted and the salient aspects of the data converted into the chosen standardized theoretical distribution. Fifth, the observed transformed statistic is compared with the decision rules, and a decision to reject or not reject the null hypothesis is made. If the observed result deviates enough from what would be expected if the null hypothesis is true to fall in the area of rejection, the tenability of the null hypothesis is in doubt and it will be rejected. Conversely, if the observed value does not deviate markedly from what would be expected if the null hypothesis is true, the null hypothesis will not be rejected, and it will be concluded that the observed results apparently deviate from the hypothesized parametric values merely as a function of sampling error.

Confidence intervals for the mean

Suppose that a psychologist in a school system wants to know the average IQ of students in a given high school. It is rather expensive to give an IQ test to each student, so a random sample of 25 students is tested. Suppose the sample mean is 109 with a standard deviation of 15.

If the psychologist was required to estimate with one value the mean IQ in the population of all students in the high school, the estimate would be 109. After all, the sample mean is an unbiased estimator of the population mean, and if the sample was indeed random, one would feel somewhat confident that the sample mean of 109 was near the population mean. This is called **point estimation** because a single value is used as the estimator.

However, if you asked the psychologist whether the population mean was *exactly* 109, the answer would certainly be no. Well, how close is 109 to the population value? An approach to answering this question is to give a range of values such that you feel reasonably confident that the interval limited by these values includes the population mean. For example, the psychologist might say that the population mean would likely be within 6 points of the sample mean, and thus that the interval of 103–115 is likely to contain the population mean. This is called **interval estimation** because the estimator is an interval, not a single value.

It would be helpful to be more precise about this procedure: For example, one could make a probability statement concerning just how confident one was that the suggested interval actually contained the population mean. The interval to be constructed is called a **confidence interval** and the values describing the boundaries of such an interval are called **confidence limits**. The degree of confidence in the proposition that the stated interval actually contains the population mean is indicated by a probability value. Of course, one would expect that a very large interval would more likely contain the population value than a very small one (everything else being equal). Potentially there are any number of confidence intervals, each having a particular probability associated with it. The most commonly used confidence intervals are the "95% confidence interval" and the "99% confidence interval."

To understand how a 95% confidence interval for a sample mean is determined, consider the data from the above example. The sample mean was 109 and the standard deviation was 15 for a sample of $N = 25$. Now consider the sampling distribution of the mean. Given the data at hand, the sample mean of 109 is a good estimate of the mean of the sampling distribution of the mean (i.e., the population mean), and $s_x/\sqrt{N} = 15/\sqrt{25} = 3$ is an estimate of the standard deviation of this distribution of means (i.e., the standard error of the mean, $s_{\bar{x}}$). Now, 95% of the sample means will fall between $P_{.025}$ and $P_{.975}$ of the sampling distribution of the mean. Since the standard error of that distribution was estimated from the sample data, the t rather than z distribution will be used to determine these percentile points. Table B (Appendix II) may be examined under "level of significance for a nondirectional test" at .05 for $df = N - 1 = 25 - 1 = 24$. The value at this place in the table is 2.064. This is the t corresponding to $P_{.975}$. Because the t distribution, like the z, is symmetrical, $P_{.025}$ is -2.064. Therefore, 95% of the area in this particular t distribution falls between $t = -2.064$ and $t = 2.064$. These t values really represent the number of standard deviations—in this case, the number of standard errors—above and below the mean that the percentile points fall. Therefore, $P_{.025}$ is 2.064 times the standard error below the sample mean, while $P_{.975}$ is 2.064 times the standard error above the sample mean. Since the standard error in this example is 3 and the sample mean is 109, 95% confidence limits for the mean are

$$\bar{X} - 2.064s_{\bar{x}} = 109 - 2.064(3) = 102.808$$

and

$$\bar{X} + 2.064s_{\bar{x}} = 109 + 2.064(3) = 115.192$$

Thus, the probability is .95 that the interval from 102.808 to 115.192 contains the population mean. In other words, if a 95% confidence interval were computed on each of an unlimited number of samples drawn from this population, on the average 95% of such intervals would include the population mean value within their limits.

Note that the probability statement applies to the interval and not to the population mean. The population mean is a fixed value, whereas the sample mean and the confidence interval are different from sample to sample. Therefore, the statement "The probability is p that the population mean falls within the interval" is technically bad form, because it implies that the value of μ varies and might or might not happen to land in the stated interval. Actually, it is the interval which is variable, and thus a more correct statement is "The probability is p that the interval includes the population value."

Suppose one wants to be especially cautious and construct an interval that 99 times out of 100 would include the population mean. Following the same logic as above, one would go to the t table and find the t value for a two-tailed test at .01 with $df = 24$. The value listed is 2.797, and this indicates that 99% of the cases in a distribution of such sample means are likely to fall within $\pm 2.797 s_{\bar{x}}$ of the mean. Consequently, the 99% confidence limits would be

$$\bar{X} - 2.797 s_{\bar{x}} = 109 - 2.797(3) = 100.609$$

and

$$\bar{X} + 2.797 s_{\bar{x}} = 109 + 2.797(3) = 117.391$$

In general, confidence limits for the mean at the α percent level are defined by

$$\bar{X} \pm t_{\alpha/2}(s_{\bar{x}})$$

in which $\alpha/2$ instructs one to take the t corresponding to the percentile rank of one-half the desired confidence level. In the case of 95% confidence limits, one would find the t corresponding to the percentile rank of $\alpha/2 = .05/2 = .025$.

FORMULAS

1. **Standard form for a score**

$$z = \frac{X - \mu}{\sigma_x}$$

2. **Standard form for a mean**

$$z = \frac{\bar{X} - \mu}{\sigma_{\bar{x}}}, \quad \sigma_{\bar{x}} = \frac{\sigma}{\sqrt{N}}$$

N = Number of cases in the sample for $\bar{X}$

3. **Standard form for a mean, parameter estimated**

$$t = \frac{\bar{X} - \mu}{s_{\bar{x}}}, \quad s_{\bar{x}} = \frac{s_x}{\sqrt{N}}$$

N = Number of cases in the sample for $\bar{X}$

$$df = N - 1$$

4. **Confidence limits for the mean (α percent confidence)**

$$\bar{X} - t_{\alpha/2}(s_{\bar{x}})$$
$$\bar{X} + t_{\alpha/2}(s_{\bar{x}})$$

EXERCISES

1. What is the difference between statistical assumptions and hypotheses?
2. Some people say that a statistical test helps decide whether an observed experimental outcome was just a chance result. Explain what they mean.
3. Why is $(\bar{X} - \mu)/\sigma_{\bar{x}}$ called a critical ratio?
4. Why does a statistician never accept H_1? Accept H_0?
5. Distinguish between the two types of decision errors.
6. What is the relationship between α and β?
7. What is the power of a statistical test?
8. What is the basis of deciding between a directional and a nondirectional alternative?
9. What considerations go into selecting the value of α?
10. When do you use the standard normal (i.e., z) and when do you use the t distribution as the theoretical relative frequency distribution for making a statistical test?
11. List the assumptions, hypotheses, formulas, significance level (adopt the .05 level), critical values, decision rules, computation and decision for the following tests of hypotheses.

    a. Determine whether a mean of 84 based upon a sample of 25 is significantly different from a population mean of 81 with a population standard deviation of 10.

    b. What difference would it make in (a) above if you do not know the population standard deviation but you do know that the sample $s_x = 10$?

    c. Suppose males in a given area of the world average 62 inches in height. One tribe in the same area and of comparable genetic stock pierces the lips and molds the heads of its young children, while the other tribes do not. The natives believe this piercing and molding makes their males grow stronger. A psychological theory also suggests that a moderate amount of stress given in infancy produces skeletally larger adults. The average height of the 36 adult males in this special tribe is 64.5 inches and the standard deviation is 7. Is this information consistent with the theory and tribal beliefs? Why do these data not constitute proof that the piercing and molding causes this height advantage?[2]

    d. The average IQ of male students at State University who have at least one older brother or sister is 113. A random sample of 121 first-born males was similarly tested and found to have a mean IQ of 117 with a standard deviation of 16. Is it reasonable to conclude that first-born male students have a different IQ from that of later borns?

12. Determine 95% and 99% confidence intervals for the following cases:
    a. $\bar{X} = 30$, $s_{\bar{x}} = 6$, $N = 22$
    b. $\bar{X} = 46$, $s_x = 12$, $N = 16$

[2]Adapted from research reported by T. K. Landauer and J. W. M. Whiting, "Infantile Stimulation and Adult Stature of Human Males," *American Anthropologist* 66 (1964): 1007–28.

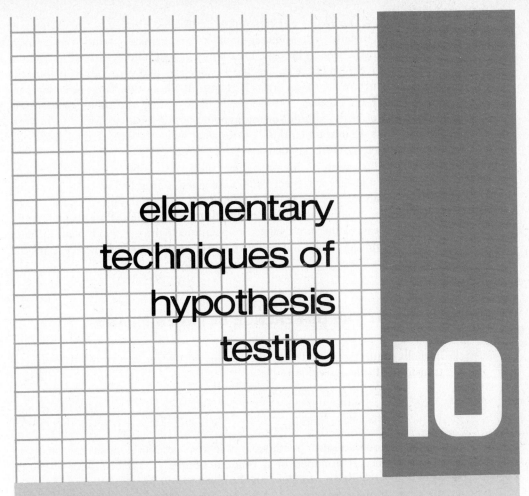

elementary techniques of hypothesis testing

10

THE GENERAL LOGIC for hypothesis testing was discussed in Chapters 8 and 9. Briefly, a hypothesis about a characteristic of the population is tentatively held to be true, data are collected, and statistical techniques are used to determine whether the results are likely to occur within the expected range of sampling error. If the results do not deviate markedly from what would be expected on the basis of sampling variations, there is no reason to doubt the validity of the hypothesis previously assumed to be true. If the results do deviate markedly from what would be expected on this basis, perhaps the hypothesis is not true.

The illustration of this procedure compared the mean number of nouns recalled by a group of subjects who were given a drug before testing with the known mean number recalled correctly by a population of adults. The question was whether the mean of the drugged sample differed significantly from the mean of the non-drugged population. If the observed mean differed by an amount that could not be attributed to sampling error, then perhaps the drug had an effect.

INFERENCES ABOUT THE DIFFERENCE BETWEEN MEANS

The illustration required that the sample mean for the drugged group be transformed into a t value by

$$t = \frac{\overline{X} - \mu}{s_{\overline{x}}}$$

However, good scientific methodology seldom permits a researcher simply to compare the performance of a random sample of subjects with a known population value, as was done in the illustration. Moreover, even if a very large group of non-drugged subjects had been tested and its mean determined, that group might have been assessed at a different time than the drugged group, or under different circumstances. For example, perhaps people remember better in the morning, after a good night's sleep. If the drugged group is not tested at the same time and under the same conditions as the non-drugged population, such procedural factors—not the drug—may produce differences in the performance of the two groups.

A better and more common procedure would be to select two random samples of 20 subjects each. Before the testing session, the people in one group would receive the drug, while those in the other group would be given an injection of saline solution, a neutral, harmless, and ineffectual substance. Thus, each group would be under whatever stress is associated with being given an injection. Further, two assistants would be used. One would administer the injections and the other, without knowing which subjects had received the drug and which the saline, would give the memory tests. Since neither the subjects nor the research assistant who administered the tests would know which injection a subject had received, this study would be a double-blind experiment (see Prologue), and differences between the groups could not be produced by the expectations of the subject or experimenter. The goal of these procedures is to ensure that the only difference between the groups is the presence of the drug versus saline. Then, if there is a difference in their performance that is not obviously attributable to sampling error, one can conclude more certainly that the drug *caused* the difference. (Observe that statistical evaluation of the data is only one part of the research enterprise—as pointed out in the Prologue, the proper statistics cannot make up for a poorly designed, carelessly executed experiment or set of observations.)

Given that the experiment has been well designed and rigorously executed, how does one determine the probability that H_0 is true when there are two samples to be compared and no population parameters available? Consider again the formula:

$$t = \frac{\overline{X} - \mu}{s_{\overline{x}}}$$

This expression is quite general, and it can be read, "A value minus the population mean of such values divided by an estimate of the standard error of such values is distributed as t, a theoretical sampling distribution." In the above expression, the "value" was a single sample mean, $\overline{X}$.

The point of the two-sample experiment just described is to ask whether the difference between the mean of the drugged group (call it $\overline{X}_1$) and the mean of the saline group (label it $\overline{X}_2$) could occur through sampling error alone. If the two groups performed exactly the same, then the difference between their means would be zero: $\overline{X}_1 - \overline{X}_2 = 0$. However, the difference is not likely to be precisely 0 but instead will vary with sampling error. Thus, one needs to know the probability that the difference between the two groups, $(\overline{X}_1 - \overline{X}_2)$, could be just sampling error.

To obtain a formula for this question, permit the "value" in the earlier expression for t to be the difference between the two means, substituting $(\overline{X}_1 - \overline{X}_2)$ for $\overline{X}$. In words, the difference between two sample means minus the difference between the means of the two populations from which the samples are drawn divided by an estimate of the standard error of such differences between sample means will be distributed as t. In symbols,

$$t = \frac{(\overline{X}_1 - \overline{X}_2) - (\mu_1 - \mu_2)}{s_{\overline{x}_1 - \overline{x}_2}}$$

where μ_1 is the mean of the population from which the first sample was drawn, μ_2 is the mean of the population from which the second sample was drawn, and $s_{\bar{x}_1 - \bar{x}_2}$ is an estimate of the standard error of the difference between two sample means.

More concretely, suppose we carry out the two-sample drug-versus-saline experiment described above. We calculate the mean learning perform-ance for each group, and then determine the difference between these sample means, $\bar{X}_1 - \bar{X}_2$. This is the first term in the numerator of the formula given above. The second term ($\mu_1 - \mu_2$) indicates that each of the two samples is drawn from a separate population. The drugged subjects come from a population with a mean of μ_1 and the saline subjects are representatives of another population, which has a mean of μ_2.

Now, stop to think what the purpose of the experiment is and how this statistical analysis will serve that purpose. We want to know whether the observed difference between sample means is too large to be attributable to sampling error. If it is sampling error, then the drug and saline solution will have equal effects on memory performance in the long run, that is, *in their respective populations*. Indeed, this is the null hypothesis—that the drug and the saline do not have different effects on the mean of the population. If this is true, there will be no difference between the means of those populations, or $\mu_1 - \mu_2 = 0$. Since the logic of these statistical procedures demands that we tentatively hold the null hypothesis to be true, the term $\mu_1 - \mu_2$ will be zero under H_0.

The denominator of the formula is an estimate of the standard error of the difference between the sample means drawn from the same population. Under the null hypothesis, the populations of drugged and saline-injected people are tentatively presumed to be identical. But even so, the mean of one sample will not precisely equal the mean of the other sample. They will differ somewhat from each other because of sampling error. One can imagine selecting two samples from a common population, computing the difference between their means, collecting another two samples, etc. These differences would form a distribution of differences between pairs of means. As a distribution of a statistic, $(\bar{X}_1 - \bar{X}_2)$, this would constitute an empirical sampling distribution, and its standard deviation would be the standard error of the difference between pairs of means drawn from a common population. This population standard error may be estimated by $s_{\bar{x}_1 - \bar{x}_2}$, which is calcu-lated on the basis of sample data. This value reflects the extent to which two sample means from identical populations will differ from one another on the basis of sampling error alone.

In the present case, what we would like to know is whether the drug produces a different effect than does neutral saline solution on the memory performance of adults. This question can be posed in the form of two mutually exclusive and exhaustive hypotheses stated in terms of the means of the populations of drugged and saline-injected subjects. The null hypothesis is that the drug and saline solution have equivalent effects on their respective population means. Thus, H_0: $\mu_1 = \mu_2$. Conversely, the drug might affect learning differently than does saline, in which case the population means would not be identical. thus, H_1: $\mu_1 \neq \mu_2$. But, the statistical logic demands that we tentatively presume the null hypothesis to be true: We will suppose that we have two samples drawn from identical populations and that whatever difference we observe in their mean values is simply sampling error. Given the null hypothesis, at least for the moment, the formula reduces to

$$t = \frac{(\overline{X}_1 - \overline{X}_2) - \overbrace{(\mu_1 - \mu_2)}^{\text{equals 0 by } H_0}}{s_{\bar{x}_1 - \bar{x}_2}} = \frac{\overline{X}_1 - \overline{X}_2}{s_{\bar{x}_1 - \bar{x}_2}}$$

The formula now expresses the ratio of the observed difference between sample means to a measure of the expected sampling variation in such differences. This ratio represents a value in the t distribution. Using Table B of Appendix II, which lists critical values for the t distribution, we can determine the probability that the observed difference between means is simply sampling error. If this probability is small, suggesting that such a difference could not be reasonably expected from sampling error alone, the null hypothesis that these samples come from identical populations will be rejected and we will conclude that the drug has some differential effect over saline. The logic is the same as in the previous two chapters; only the formula is different.

The task remaining is to determine an expression for $s_{\bar{x}_1 - \bar{x}_2}$. It happens that the standard error of the difference between means varies, depending upon whether the scores in the two groups are **independent** or **correlated**. If different subjects compose the two groups, the scores in these groups will be independent. However, if the same subjects are measured twice—such as before a treatment is administered and again afterward—the scores in the two groups will be correlated. This happens because some factors that determine an individual subject's performance will influence it on both assessments. If children are given a special mathematics training program, it is likely that individual pupils who score high relative to the group before training will also score high after it. Thus, the before-and-after scores will be correlated. In this event, the estimate of $s_{\bar{x}_1 - \bar{x}_2}$ based upon independent groups is not appropriate and special techniques are required. We will consider both cases.

The difference between means: independent groups

Research question   A developmental psychologist was interested in how divorce affects the nursery school behavior of young boys.[1] It was

[1]Based on, but not identical to, a study by E. M. Hetherington, M. Cox, and R. Cox, "Play and Social Interaction in Children Following Divorce," paper presented at National Institute of Mental Health Divorce Conference, 1978.

10–1 Summary of the Test of Difference between Means
for Independent Groups
(Divorced vs. Non-Divorced Example)

Hypotheses

H_0: $\mu_1 = \mu_2$

H_1: $\mu_1 \neq \mu_2$ (nondirectional)

Assumptions and Conditions

1. The subjects in each group are **randomly** and **independently** sampled.
2. The groups are **independent**.
3. The population variances are **homogeneous**.
4. The population distribution of $(\bar{X}_1 - \bar{X}_2)$ is **normal** in form.

Decision Rules (from Table B)

Given: .05 significance level, a nondirectional test, and
$N_1 + N_2 - 2 = 17 + 15 - 2 = 30$ *df*

If $-2.042 < t_{obs} < 2.042$, do not reject H_0.
If $t_{obs} \leq -2.042$ or if $t_{obs} \geq 2.042$, reject H_0.

Computation

Divorced (Group 1)	Non-Divorced (Group 2)
$\bar{X}_1 = 40$	$\bar{X}_2 = 35$
$s_1^2 = 16$	$s_2^2 = 18$
$N_1 = 17$	$N_2 = 15$

$$t_{obs} = \frac{\bar{X}_1 - \bar{X}_2}{\sqrt{\left[\dfrac{(N_1 - 1)s_1^2 + (N_2 - 1)s_2^2}{N_1 + N_2 - 2}\right] \cdot \left[\dfrac{1}{N_1} + \dfrac{1}{N_2}\right]}}$$

$$= \frac{40 - 35}{\sqrt{\left[\dfrac{(17 - 1)16 + (15 - 1)18}{17 + 15 - 2}\right] \cdot \left[\dfrac{1}{17} + \dfrac{1}{15}\right]}}$$

$t_{obs} = 3.43$

Decision

Reject H_0.

possible that a child four to six years old might become more aggressive as a response to the divorce of his parents. To test this, the psychologist had research assistants observe over a two-week period the number of aggressive acts of 17 boys whose parents had recently divorced (for our purposes, the divorced group) and for 15 boys whose parents had not divorced (the non-divorced group). The assistants did not know which group any of the boys belonged to (this is a single-blind study). A formal summary of the statistical treatment of this experiment is given in Table 10–1, and the details are discussed below.

Statistical hypotheses The hypotheses are as follows:

$$H_0: \mu_1 = \mu_2$$

$$H_1: \mu_1 \neq \mu_2$$

in which μ_1 is the population mean for the divorced group, and μ_2 is the mean for the non-divorced group.

Notice two things. First, the statistical hypotheses are stated in terms of population parameters, not statistics. This is because the decision to be made involves what is true about the populations, boys of divorced and non-divorced parents, not just about the two specific samples observed in this experiment. Second, the alternative hypothesis, H_1, does not specify whether μ_1 is greater or less than μ_2. It just states that the population means are different. Thus, the statistical test should be sensitive to differences in either direction, $\mu_1 > \mu_2$ and $\mu_1 < \mu_2$, and a nondirectional, or two-tailed, test is required.

Assumptions and conditions First, the subjects must be **randomly** and **independently** sampled. This means that each subject in the population has an equal opportunity of being selected and that the inclusion of one subject in the sample does not influence the probability of selecting any other member of the population.

Second, these procedures are for studies in which the two groups are **independent** of each other. The next section will describe procedures to be used with correlated groups.

Third, the population variances of the two groups are presumed to be equal, a characteristic called **homogeneity of variance**. Unfortunately, it is often difficult to decide whether this condition is met by the data to be analyzed. Since the variance of one sample from a population is not likely to equal the variance of another sample from that population, the question is: How different can the two sample variances be before one suspects that their population parameters are not equal? While precise statistical tests are available that can be used to make this decision, many statisticians feel their application to this problem is not very worthwhile. In addition, moderate violations of this assumption can occur—especially if the number of subjects is approximately the same in each group—without seriously biasing the results. If the number of observations in one group is the same as in another and there are at least five subjects in each group, one sample variance may be as much as twice the size of the other without markedly altering the results of the statistical analysis.[2] Thus, it is best to have equal numbers of subjects in each group and adequate-sized samples; then homogeneity of variance probably will not be a problem.

Fourth, the population sampling distribution of $(\overline{X}_1 - \overline{X}_2)$ must be **normal** in form in order to use the percentiles of the t distribution. This requirement is satisfied either if the two population distributions of raw scores are both normal or if the sample sizes are reasonably large regardless of the form of the population distributions.

Formula The formula relating the difference between sample means for independent groups to its standard error is

$$t = \frac{\overline{X}_1 - \overline{X}_2}{s_{\bar{x}_1 - \bar{x}_2}} = \frac{\overline{X}_1 - \overline{X}_2}{\sqrt{\left[\dfrac{(N_1 - 1)s_1^2 + (N_2 - 1)s_2^2}{N_1 + N_2 - 2}\right] \cdot \left[\dfrac{1}{N_1} + \dfrac{1}{N_2}\right]}}$$

The standard error of the difference between means is estimated on the basis of the variances of the two samples, s_1^2 and s_2^2. The degrees of freedom for t are the sum of the df for each sample variance:

$$df = (N_1 - 1) + (N_2 - 1) = N_1 + N_2 - 2$$

Significance level Employ the .05 level of significance.

Critical values The critical values will depend upon the degrees of freedom, the nature of the alternative hypothesis, and the level of significance. A glance at Table 10–1 shows that there are 17 and 15 subjects in the groups and thus the degrees of freedom are $(N_1 + N_2 - 2) = (17 + 15 - 2) = 30$. The alternative hypothesis is nondirectional and the .05 level has been assumed. From Table B in Appendix II, the critical values are found to be -2.042 and 2.042.

Decision rules If the observed (computed) value of t is between -2.042 and $+2.042$, the null hypothesis, that there is no difference, will not be rejected, and one will conclude that the difference between means is within the range of sampling error. In this event the results would not provide

[2]G. E. P. Box, "Some Theorems on Quadratic Forms Applied in the Study of Analysis of Variance Problems: II. Effect of Inequality of Variance and of Correlations of Errors in the Two-Way Classification," *The Annals of Mathematical Statistics*, 25 (1954): 484–98.

evidence supporting the effects of divorce on the aggressive behavior of young boys in nursery school. Conversely, if the observed value of t is greater than or equal to 2.042 or less than or equal to -2.042, the null hypothesis will be rejected. The reasoning is that the probability is too remote that such an observed difference in means reflects sampling error, and that it is likely that the divorce experience has influenced social behavior. In symbols,

$$\text{If } -2.042 < t_{\text{obs}} < 2.042, \text{ do not reject } H_0.$$
$$\text{If } t_{\text{obs}} \leq -2.042 \text{ or if } t_{\text{obs}} \geq 2.042, \text{ reject } H_0.$$

Computation   To find the value of the observed t, one proceeds by listing the relevant known information and the formula for t, calculating the components of the formula, and computing the result. This work is presented in Table 10–1. The $t_{\text{obs}} = 3.43$.

Decision   The observed value of t conforms to the second decision rule, reject H_0. The probability that such a difference between means would occur merely as a function of sampling error is so small that it is likely that the two samples have been drawn from different populations. If the experiment has been well conceived, the implication is that the divorce and its aftermath produced the difference in observed means.

The difference between means: correlated groups

Sometimes the two means to be compared are obtained from the same subjects or from matched pairs of subjects (such as identical twins; pairs of individuals matched for age, sex, social class, IQ; etc.). In either case, some of the factors that influence scores in one group will also influence them in the other group (especially the factors associated with individual subjects), and thus the scores in the two groups will be correlated to some extent. In that event, the assumption of independent groups cannot be made, and alternative procedures must be used.

Research question   Suppose it is important to ascertain whether stimulation of one area of the brain is more reinforcing to a rat than stimulation of another part. Electrodes are sunk into the two areas in each of 10 rats. After an initial training session, the rats are allowed to press two bars, one stimulating the first area of the brain and the other stimulating the second area. The number of presses on each bar is recorded; the mean number of presses for stimulation in area 1 will be compared with the mean for area 2. In addition, suppose a set of experiments by other investigators and a current theory strongly suggest that area 1 should be more effective in producing

10-2 Summary of the Test of Difference between Means for Correlated Groups (Brain Stimulation Example)

Hypotheses

H_0: $\mu_1 \leq \mu_2$
H_1: $\mu_1 > \mu_2$ (directional)

Assumptions and Conditions
1. The subjects are **randomly** and **independently** sampled.
2. The scores of the two groups are **correlated**.
3. The population distribution of the D_i is **normal** in form.

Decision Rules (from Table B)
Given: .01 significance level, a directional test, and
$N - 1 = 10 - 1 = 9 \ df$

If $t_{obs} < 2.821$, do not reject H_0.
If $t_{obs} \geq 2.821$, reject H_0.

Computation

Subject	X_1 (Area 1)	X_2 (Area 2)	$X_{1i} - X_{2i} = D_i$	D_i^2
a	58	42	16	256
b	45	50	-5	25
c	61	23	38	1444
d	55	50	5	25
e	58	45	13	169
f	90	85	5	25
g	26	30	-4	16
h	35	20	15	225
i	42	50	-8	64
j	48	60	-12	144

$N = 10 \quad \Sigma X_1 = 518 \quad \Sigma X_2 = 455 \qquad \Sigma D_i = 63 \quad \Sigma D_i^2 = 2393$

$\overline{X}_1 = 51.8 \quad \overline{X}_2 = 45.5 \qquad \overline{D} = 6.3$

$$t_{obs} = \frac{\Sigma D_i}{\sqrt{\dfrac{N\Sigma D_i^2 - (\Sigma D_i)^2}{N - 1}}} = \frac{63}{\sqrt{\dfrac{10(2393) - (63)^2}{10 - 1}}}$$

$t_{obs} = 1.34$

Decision
Do not reject H_0.

responses than area 2. Therefore, the research question is whether area 1 is a more reinforcing site for brain stimulation than is area 2. A summary of the statistical procedure testing this question is presented in Table 10–2.

Statistical hypotheses   Because theory and empirical evidence suggest that stimulation of area 1 may be more reinforcing than stimulation of area 2, the hypotheses are directional rather than nondirectional. That is, the null hypothesis states that μ_1 either equals or is less than μ_2, while the alternative hypothesis specifies that μ_1 is larger than μ_2. In symbols,

$$H_0: \mu_1 \leq \mu_2$$
$$H_1: \mu_1 > \mu_2$$

Formula   As can be seen from Table 10–2, the basic data are two scores per subject, one for area 1 and one for area 2. The computational routine for this case differs from the preceding example and rests upon the following fact:

The difference between two means equals the mean difference between pairs of scores.

This states that if the difference between each subject's two scores is computed, the mean of such differences will equal the mean over all subjects of scores for area 1 minus the mean of the scores for area 2. Suppose D_i represents the difference between the pair of scores for the ith subject $(X_{1i} - X_{2i})$. Then if there is absolutely no difference between the two sample means ($\overline{X}_1 - \overline{X}_2 = 0$), $\overline{D}$ will also equal zero. Because of sampling error, $\overline{D}$ rarely will be exactly zero even if the two populations are identical. However, if the means of the two samples are both estimates of the same population mean value, as the null hypothesis states, the positive and negative differences should cancel out and the sampling distribution of $\overline{D}$ will have a mean of zero. Consequently, the statistical question reduces to the probability that $\overline{D}$ should deviate from zero relative to the sampling error of $\overline{D}$:

$$t_{\text{obs}} = \frac{\overline{D} - 0}{s_{\overline{D}}} = \frac{\sum D_i / N}{\sqrt{\dfrac{N \sum D_i^2 - (\sum D_i)^2}{N^2(N-1)}}} = \frac{\sum D_i}{\sqrt{\dfrac{N \sum D_i^2 - (\sum D_i)^2}{N-1}}}$$

The N in the above formula refers to the number of subjects (i.e., the number of pairs of scores). In this example (Table 10–2), $N = 10$.

Assumptions and conditions   The data are pairs of scores and it is assumed that the two sets of scores are **correlated**. Often this results from having the same or closely matched subjects contribute both scores. It is assumed that the pairs are **randomly** and **independently** sampled and that in the population the D_i are **normally** distributed.

Significance level   Suppose that the implications of this study are such that a type I error—incorrectly rejecting the null hypothesis—would be exceptionally undesirable. Therefore, the .01 level of significance will be selected.

Critical values   The test to be performed will involve a directional alternative. This means that the test is one-tailed: The critical region lies in only one of the two tails of the t distribution. Since additional information dictates that area 1 should be more reinforcing than area 2, the converse possibility, that area 1 might be poorer in reinforcing bar pressing, is so unlikely that the probability that this should occur is nearly zero. Under these conditions it is no longer appropriate to expect that a negative value of t might occur, and so the left-hand tail of the theoretical distribution is not included in the critical region. As a result, the critical value appropriate for this test is such that 1% of the area under the curve falls to its right. Looking in the table of t values with $N - 1 = 10 - 1 = 9$ degrees of freedom, one finds the .01 level, one-tailed critical value is 2.821.

Decision rules   The decision rules are as follows:

$$\text{If } t_{\text{obs}} < 2.821, \text{ do not reject } H_0.$$
$$\text{If } t_{\text{obs}} \geq 2.821, \text{ reject } H_0.$$

Computation   The raw data and computation are illustrated in Table 10–2.

Decision   The observed value of t ($t_{\text{obs}} = 1.34$) is less than the critical value of 2.821, and according to the decision rules, one does not reject the null hypothesis. An observed $\overline{D}$ of 6.3 is within the realm of sampling error for correlated samples of size 10 drawn from populations having the same mean. That is, if the two areas of the brain were identical in their reinforcing potential, with the amount of variability involved it would be quite possible to obtain a sample having a $\overline{D} = 6.3$, and the researcher must conclude that the data provide no evidence that stimulation of area 1 has more reinforcing value than area 2.

What if area 1 actually had produced substantially *fewer* responses than area 2, despite the prediction to the contrary? If a directional test had been selected before the experiment was conducted (and this decision must be made beforehand), then a large negative t would conform to the first decision rule, which dictates that one not reject the null hypothesis. This would be the decision even if t were very large, such as -4.00.

INFERENCES ABOUT CORRELATION COEFFICIENTS

In the previous section inferences about differences between means have been considered. But tests of the values of other population parameters can also be made: For example, one might be interested in inferences about the degree of relationship between two variables. This section is concerned with whether a significant relationship exists—whether the observed r could come from a population in which the parameter ρ (rho, the population correlation coefficient) is actually zero—and whether two r's drawn from independent groups of subjects are significantly different from each other.

The significance of r

Suppose a correlation of .74 is found between two variables. One would want to know if an r of this magnitude is merely an imperfect reflection of a population in which the correlation is actually zero or whether an $r = .74$ faithfully mirrors a nonzero relationship in the population. What is the probability that the former is the case?

Research question Some tests of infant development contain items that are designed to assess the amount and nature of vocalizations made by the infant. For example, a bell is rung in front of the infant and the presence and extent of the youngster's vocalization in response to this stimulus is measured. Although many items on such infant scales tend to assess motor development and do not seem to relate to later tested IQ, it might be reasonable to inquire whether there is a relationship between vocalization in 12-month-old female infants and verbal intelligence at 26 years of age.[3] A formal summary of the following procedure appears in Table 10–3.

Statistical hypotheses The data reveal a correlation of .74 between infant vocalization and adult verbal intelligence for a group of 27 females. However, it is possible that no relationship actually exists in the population from which the people were sampled, that is, perhaps $\rho = .00$. In such an event, the observed relationship of .74 is merely a function of sampling error. Theoretically, it would be possible to continue to select sample after sample of size 27 and compute a correlation on each. The distribution of such sample r's forms the sampling distribution of the correlation coefficient, and its standard deviation is the standard error of r. The null hypothesis states that

[3]Inspired by J. Cameron, N. Livson, and Nancy Bayley, "Infant Vocalizations and Their Relationship to Mature Intelligence," *Science* 157 (1967): 331–33; R. B. McCall, P. S. Hogarty, and N. Hurlburt, "Transitions in Infant Sensorimotor Development and the Prediction of Childhood IQ," *American Psychologist* 27 (1972): 729–48; T. Moore, "Language and Intelligence: A Longitudinal Study of the First Eight Years," *Human Development* 10 (1967): 88–106.

10–3 Summary of the Test of Significance of a Correlation Coefficient (Infant Vocalization–IQ Example)

Hypotheses
H_0: ρ = .00
H_1: $\rho \neq$ .00 (nondirectional)

Assumptions and Conditions
1. The subjects are **randomly** and **independently** sampled.
2. The population distributions of both X and Y are **normal** in form.

Decision Rules (from Table C)
Given: .05 significance level, a nondirectional test, and $N - 2 = 27 - 2 = 25$ df

If $-.3809 < r_{obs} < .3809$, do not reject H_0.
If $r_{obs} \leq -.3809$ or $r_{obs} \geq .3809$, reject H_0.

Computation
r_{obs} = .74

Decision
Reject H_0.

the observed value of r is typical of such a distribution of sample r's drawn from a population in which this correlation is actually .00. The alternative hypothesis dictates that the observed r is too extreme to be considered a member of such a sampling distribution, and thus it probably reflects a relationship of some nonzero magnitude in the population. Since there is no theory or empirical basis for predicting whether such a relationship, if it exists, is positive or negative, the alternative hypothesis is nondirectional. In symbols,

$$H_0: \rho = .00$$

$$H_1: \rho \neq .00$$

Formula   The formula relating a sample correlation coefficient to a theoretical sampling distribution of r's follows the same general pattern as the previous formulas:

$$t = \frac{r_{obs} - \rho}{s_r}$$

This states that a sample correlation, r, minus the population correlation, ρ, divided by an estimate of the standard error of sample correlations, s_r, is distributed as t with $N - 2$ degrees of freedom. Since the null hypothesis dictates that $\rho = 0$, this formula reduces to $t = r/s_r$. Substituting the computational expression for s_r and then making a few algebraic simplifications, one has

$$t = \frac{r_{obs}}{s_r} = \frac{r_{obs}}{\sqrt{\dfrac{1 - r_{obs}^2}{N - 2}}} = \frac{r_{obs}}{\dfrac{\sqrt{1 - r_{obs}^2}}{\sqrt{N - 2}}}$$

$$t = \frac{r_{obs}\sqrt{N - 2}}{\sqrt{1 - r_{obs}^2}} \qquad \text{with } df = N - 2$$

It is possible to substitute the observed value of r (in this case $r_{obs} = .74$) into the above formula, compute t_{obs} (which equals 5.50), and determine its probability under H_0 by looking in the table of t values as before. However, since this conversion is required so often and since the value of t depends only upon r_{obs} and upon its degrees of freedom, critical values of r for various df have been computed and are presented in Table C of Appendix II. To use the table, decide whether the alternative hypothesis requires a directional or nondirectional test and select the significance level. Then locate the appropriate column and find the row corresponding to the degrees of freedom ($df = N - 2$). The critical value r_{crit} is given at the intersection. For the present data, a nondirectional test at the .05 level with $df = 27 - 2 = 25$ has a critical value of $r_{crit} = .3809$. Since the sampling distribution is symmetrical and this is a nondirectional test, there are two critical values, $-.3809$ and $.3809$.

Assumptions and conditions   It is assumed that the subjects are **randomly** and **independently** sampled and that the population distributions of both X and Y are **normal** in form.[4] Notice that one can *compute* a correlation on two variables regardless of the nature of their distributions, but normality is required to ensure the accuracy of this *statistical test* of the significance of r.

Significance level   Adopt the .05 level.

Critical values   In this case, the critical values can be expressed in terms of r rather than t by using Appendix II, Table C. For a nondirectional test at .05 with $df = N - 2 = 27 - 2 = 25$, the critical values of r are $-.3809$

---

[4]Technically, the joint distribution of X and Y is bivariate normal.

and .3809. Therefore,

$$\text{If } -.3809 < r_{\text{obs}} < .3809, \text{ do not reject } H_0.$$
$$\text{If } r_{\text{obs}} \leq -.3809 \text{ or } r_{\text{obs}} \geq .3809, \text{ reject } H_0.$$

Computation   Once r_{obs} is determined, no further computation is required.

Decision   Since the observed r ($r_{\text{obs}} = .74$) exceeds the critical value .3809, the second decision rule is used and H_0 is rejected. The probability is very low that an observed correlation of .74 could be drawn from a population in which the correlation between these two variables is actually zero. The implication of the rejection of the null hypothesis is that there is a relationship in the female population between vocalization at age 1 and verbal intelligence at age 26.

The difference between two correlations: independent samples

Research question   The preceding results apply to females. It might be the case that the relationship between vocalization at age 1 and verbal IQ at age 26 is stronger or weaker for females than for males. To examine this possibility, a sample of 22 males was measured at 12 months and at 26 years of age in precisely the same manner as described for females. Whereas the correlation for females was .74, the correlation for males was .09. What is the probability that such a difference in observed correlations would occur merely as a function of sampling error when, in fact, in the population there is no difference between males and females in the degree of this relationship? Note that the two correlations being considered, .74 and .09, have been computed on independent (i.e., different) groups of subjects. The following procedures apply only to such cases. Somewhat different techniques are required in order to compare two correlations computed on the same group of subjects.[5] A summary of the procedures for independent groups is given in Table 10–4.

Statistical hypotheses   If there is really no difference in the degree of correlation for the two groups (sexes), then the correlation coefficient in the population for the males should equal that parameter for females ($\rho_1 = \rho_2$). On the other hand, if there is a difference in the magnitude of the relationship for the two groups, their population values should not be equal. The alternative in this case is nondirectional (two-tailed) and the hypotheses are as

[5]See G. A. Ferguson, *Statistical Analysis in Psychology and Education* (New York: McGraw-Hill, 1966), pp. 188–89.

10–4 Summary of the Test of Difference between Two Independent Correlation Coefficients (Infant Vocalization–IQ Example)

Hypotheses

H_0: $\rho_1 = \rho_2$

H_1: $\rho_1 \neq \rho_2$ (nondirectional)

Assumptions and Conditions

1. The subjects in each group are **randomly** and **independently** sampled.
2. The groups are **independent**.
3. The population distributions of X and Y for both groups are **normal** in form.
4. Both N_1 and N_2 are **greater than 20**.

Decision Rules (from Table A)

Given: .05 significance level and a nondirectional test

If $-1.96 < z_{obs} < 1.96$, do not reject H_0.

If $z_{obs} \leq -1.96$ or $z_{obs} \geq 1.96$, reject H_0.

Computation

Group 1 (females): $N_1 = 27$, $r_1 = .74$, $z_{r_1} = .950$ (from Table D)

Group 2 (males): $N_2 = 22$, $r_2 = .09$, $z_{r_2} = .090$ (from Table D)

$$z_{obs} = \frac{z_{r_1} - z_{r_2}}{\sqrt{\dfrac{1}{N_1 - 3} + \dfrac{1}{N_2 - 3}}}$$

$$= \frac{.950 - .090}{\sqrt{\dfrac{1}{27 - 3} + \dfrac{1}{22 - 3}}}$$

$$z_{obs} = 2.80$$

Decision

Reject H_0.

follows:

$$H_0: \rho_1 = \rho_2$$
$$H_1: \rho_1 \neq \rho_2$$

Formula   The test of these hypotheses is accomplished by first trans-forming the obtained correlation coefficients according to the following expression:

$$z_r = \tfrac{1}{2}\log_e(1 + r) - \tfrac{1}{2}\log_e(1 - r)$$

Fortunately, this formula has been computed for many values of r and tabled. This r-to-z transformation is given in Table D of Appendix II. To use the table, simply locate the value of r in the left-hand column and, immediately to its right, read the corresponding transformed value, z_r. In the case of $r_1 = .74$, the transformed value $z_{r_1} = .950$; for $r_2 = .09$ the z_{r_2} is .090.

Once the r-to-z transformations have been made on each of the two values of r, the difference between the transformed correlation coefficients (z_r) relative to the standard error of such differences is given by

$$z_{\text{obs}} = \frac{z_{r_1} - z_{r_2}}{\sqrt{\dfrac{1}{N_1 - 3} + \dfrac{1}{N_2 - 3}}}$$

Note that the distribution employed as the theoretical distribution is the standard normal (observe that z refers to the standard normal deviate and z_{r_i} refers to the transformed correlations). Because the standard normal is being used and not the t, it is necessary to have an adequate number of cases in the two samples. It is probably best to have N_1 and N_2 each greater than 20.

Assumptions and conditions   There are several assumptions required for performing this test. First (as always), the sample r's are computed on **randomly** and **independently** selected subjects. Second, it is necessary for the two samples to be **independent**, involving different or unmatched subjects. If this condition is not met, the formula for the sampling error of the difference between correlation coefficients will not be appropriate, for reasons similar to those discussed under the testing of the difference in means for independent versus correlated groups. Third, the X and Y distributions for both groups of subjects must be **normal** in form. Fourth, the two **sample sizes must be greater than 20** in order for the statistical test to be accurate.

Significance level   Assume the .05 level.

Critical values   A nondirectional test at the .05 level using the standard normal distribution requires an observed z_{obs} in excess of ± 1.96 (Appendix II, Table A).

Decision rules

$$\text{If } -1.96 < z_{\text{obs}} < 1.96, \text{ do not reject } H_0.$$
$$\text{If } z_{\text{obs}} \leq -1.96 \text{ or } z_{\text{obs}} \geq 1.96, \text{ reject } H_0.$$

Computation   The computation is summarized in Table 10–4. Note that the information given in the problem is stated first. Then the r's are transformed to their corresponding z_r values with the use of Appendix II, Table D. The values are substituted into the formula and a $z_{\text{obs}} = 2.80$ obtained.

Decision   The observed value of z ($z_{\text{obs}} = 2.80$) exceeds the critical values of ± 1.96 and thus conforms to the second decision rule, reject H_0. This means that the probability is very remote that two such correlation coefficients could be drawn from a common population purely on the basis of sampling error. Therefore, it is likely that these r's represent two populations that have different magnitudes of relationship between infant vocalization and later verbal intelligence. Specifically, it would appear that such a relationship is stronger for females than for males.

Comparison of H_0: $\rho = .00$ and H_0: $\rho_1 = \rho_2$

It is important to distinguish between the null hypothesis that a given correlation is zero (H_0: $\rho = .00$) and the null hypothesis that two correlations are equal (H_0: $\rho_1 = \rho_2$). Suppose the correlation for females is .74 but the correlation for males is .35. As illustrated, the $r = .74$ is significant; that is, it is unlikely that such a value has been computed on a sample drawn from a population that possesses no correlation at all ($\rho = .00$). More informally stated, the significant correlation for females implies that there is a relationship between infant vocalization and later verbal IQ for them. However, if the correlation $r = .35$ calculated on a sample of 22 males is tested for the null hypothesis $\rho = .00$, H_0 will not be rejected, implying that the relationship is not significant for males and that there is no evidence of a relationship between infant vocalization and verbal IQ for them. Now, when the two correlations (.74 and .35) are compared to determine if they are significantly different from each other (H_0: $\rho_1 = \rho_2$), the test fails to reject the null hypothesis, suggesting that the relationship is not different for males and females. But this combination of results does not seem to make sense. How is it possible that there is a relationship for females but not for males, even though females and males do not differ significantly in the amount of the relationship?

This anomaly stems from the fact that tests of hypotheses often are interpreted in terms of a yes-no decision. Either there is or there is not a relationship. Although sometimes it is useful to think in this manner, attempting to make a dichotomous decision is a somewhat artificial procedure when the tool employed to make that decision is a probability value that may range from .00 to 1.00 and ordinarily does not fall neatly into one of two distinct classifications. To illustrate more clearly by citing an extreme example, suppose that with a sample of 32 girls and 32 boys, the correlations are .35 and .34 respectively. With a nondirectional test at the .05 level, the relationship would be judged significant for girls but not for boys, yet the two correlations certainly are not significantly different from each other. The lesson to be learned from this example is that one should refrain from making simple dichotomous decisions, especially when not rejecting the null hypothesis. Remember, not being able to reject H_0 does not entitle you to say that there is no relationship. Rather, the evidence at hand is not sufficient to conclude that there is a relationship. In this case, the subtle difference between not rejecting H_0, which is the appropriate conclusion, and accepting or (worse yet) "proving" H_0, which are not appropriate conclusions, becomes important.

Another point illustrated by this example is that finding a relationship for females but finding no evidence that supports a relationship for males does not necessarily mean that the two sexes are different in the characteristic being examined. Performing a test of H_0: $\rho = 0$ separately on both males and females does not address the same statistical question as making a test of H_0: $\rho_1 = \rho_2$.

How does one interpret a situation like the one described, in which the combination of results is ambiguous? The correlation of .74 for girls is substantial enough to warrant the conclusion that there is probably a relationship between vocalization and verbal IQ for females. However, the data may be interpreted as being inconclusive for males (assuming $r = .35$), in that while the correlation is not substantial enough to warrant a rejection of the null hypothesis, the observed correlation was not so different from the substantial correlation of .74 for females. Other interpretations are also possible.

A COMPARISON OF THE DIFFERENCE BETWEEN MEANS AND CORRELATION

It will be instructive to compare the implications of a test of the difference between means and a test of the significance of a correlation. Fortunately, a very interesting comparison between testing for mean differences and using correlational procedures exists.[6] The relative role of heredity and environment

[6]This example suggested by results reported in M. Skodak and H. M. Skeels, "A Final Follow-Up Study of One Hundred Adopted Children," *Journal of Genetic Psychology* 75 (1949): 85–125; M. P. Honzik, "Developmental Studies of Parent–Child Resemblance in Intelligence," *Child Development* 28 (1957): 215–28.

as factors in development has been a focus of interest for psychologists for some time. More specifically, people have been concerned with whether heredity or environment is responsible for intelligence. One way to approach this issue is to investigate the IQs of adopted children and their biological and foster mothers. In general, the results have shown a correlation of approximately .38 between a measure of intellectual performance for the biological mothers and one for their children, but a correlation of almost zero between an estimate of the IQ of the foster mothers based upon their years of education and the IQ of the children they reared. By itself, this information would appear to suggest that heredity may be more important than environment in determining intelligence. However, the mean IQ of the children was approximately 21 points higher than the mean IQ of their biological mothers but quite close to the estimated mean IQ of their rearing mothers. Thus the children had an average IQ that was much more like the mean IQ of their foster mothers than that of their biological mothers, evidence that seems to implicate environment in the determination of intelligence. How is this pattern of results possible, and what does it say about the procedures for testing a difference between means in contrast to a correlation?

Consider the hypothetical data presented in Figure 10–1. These data have been exaggerated somewhat to illustrate the statistical point more

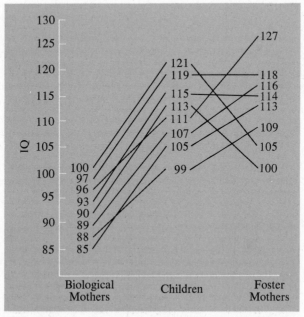

Fig. 10–1. Diagram contrasting correlation and difference between means

clearly. Notice first that the scores are plotted according to their value, with higher IQ scores at the top, and that they are clustered into scores for biological mothers, children, and foster mothers. The lines connecting pairs of scores designate which biological mother goes with which child and which child goes with which foster mother. A correlation is high if the lines linking corresponding scores do not cross excessively. Notice that the children tend to line up within their group in much the same order as the biological mothers within theirs, and thus the correlation between biological mother and child in this hypothetical example is fairly high. However, the extensive crossing of lines relating child's IQ and foster mother's IQ suggests that the ordering of children and foster mothers is quite unpredictable and thus the correlation is low. In contrast, the general height (vertical distance from the X-axis) of a group on this graph represents the mean of the group, and it is plain that in this respect the children are more like their foster mothers and less like their biological mothers. Therefore, in general the possibilities are that two groups of scores may have a correlation but no mean difference, a mean difference but no correlation, no correlation and no mean difference, or a correlation and a mean difference. Further, the correlations may be either positive or negative.

Correlations and mean differences often present two distinct types of information, both of which are valuable to the interpretation of the general results. Very often, researchers perform only one of these two types of data analyses when, in fact, both could be performed and would provide different types of information on the processes at work in the experiment or observation. For example, sometimes a test is given before a certain experimental treatment is introduced and then a posttest is given after the experimental treatment. Usually, the researcher looks for a difference in the means of the two groups of scores. Perhaps an experimental teaching method has been used and the investigator wants to know if the performance of the group taught by this method has improved significantly over its performance prior to the experimental program. Suppose that some improvement is noted and the test of the difference between means is significant. If there is also a correlation between the pretest and posttest, then one can also presume that the program affected each student to approximately the same extent: The better students were still at the top of the group on the posttest, and so on. However, if there is no significant correlation, such a result would suggest that although the treatment did raise the mean of the group, it seemed to influence some students more than others, changing their relative ranking within their group. As a consequence, a researcher might wish to investigate the characteristics of students who tended to show marked improvement and the characteristics of those who did not.

Therefore, the difference between the means of two sets of scores and the correlation between those sets are independent types of statistics, and they convey different messages about the events and processes being studied.

THE INTERPRETATION OF SIGNIFICANCE

It should be apparent that the interpretation of scientific results is a complex matter. Of paramount importance is the fact that statistical procedures provide the researcher with a set of tools that can help interpret the results of experiments, but statistical procedures cannot replace the adequate collection of data under well-controlled conditions.

This proposition implies two important points. First, if the data are sloppily collected or inappropriate to the research question being posed, all the statistical significance in the world will not make the experiment worthwhile. Statistics do not add to or change the meaning of the data; they merely help reveal whatever conclusions lie hidden within the collection of measurements.

The second point is that *statistical* significance and *scientific* significance are not the same. That is, an experimenter may find emphatic differences between groups, but this information may be totally useless and uninformative. What is important, or scientifically significant, is difficult to define because scientists differ in their opinions. Further, a given item of data may seem trivial today but become very important in 10 or 20 years. Generally speaking, significant research is a result that has potential implications for a wide variety of issues (applied or theoretical) or has the potential to influence the thought or actions of many people.

Probability and significance

There are several problems associated with declaring a difference simply as significant or nonsignificant. First, the critical value used to make such a decision is based upon the significance level, which the researcher arbitrarily selects. Thus, whether or not a statistical manipulation is significant depends upon the arbitrary selection of the probability of a type I error (see pages 217–18). Fortunately (or unfortunately) there is some agreement among behavioral scientists as to what level that probability should be. Usually it is set at .05, and sometimes (though rarely) at .01 or .001. In fact, many of the tables in Appendix II of this book list critical values corresponding to only these significance levels.

However, as we have seen on page 218, the significant–nonsignificant dichotomy is rather artificial. Moreover, a single pair of scores in a small sample can sometimes determine whether the relationship is significant or nonsignificant. But, despite the apparently artificial dichotomy between results labeled significant and nonsignificant, techniques of statistical inference were developed largely for just this purpose—to decide whether or not an event has been observed that deviates from the limits of sampling error. To decide after a statistical evaluation that the observation may be or may not be a chance result is nearly to leave us where we started—with subjective decisions. Therefore, the custom of dichotomous decisions remains.

Probability and N

The interpretation of statistical results also depends upon the number of subjects sampled because the critical value often varies with N. Suppose a correlation of .35 exists in the population, and suppose that such a value has been obtained in samples of various sizes. With a nondirectional test at .05, an r of .35 would be statistically significant only if 32 or more subjects had been sampled. An examination of Table C of Appendix II shows that the value of r required for significance at .05 decreases as the number of subjects in the sample (i.e., degrees of freedom) increases. Notice that in a sample of 102 cases, a correlation of .195 is significant at the .05 level (two-tailed), but an $r = .75$ is not quite significant if only 7 subjects are involved. In general, the more subjects, the more likely it is that a significant result will be obtained, in terms of either correlations or mean differences, if such a condition is actually present in the population.

On the one hand, this is as it should be. The probability value one obtains is intended to be a measure of the uncertainty in the situation. If a coin is tossed four times and it comes up heads on three of those tosses, one does not immediately conclude that the coin is biased. The coin might be biased, but how much money would you bet on such a proposition with only that much evidence? Not too much. On the other hand, if the coin is tossed 10,000 times, and 7500 of those tosses result in heads despite the fact that the same percentage of tosses was heads, one would be far more confident in concluding that the coin is biased. Therefore, it is not unreasonable that the probability of a given event should be a function of the number of cases involved.

FORMULAS

1. Difference between means

a. Independent groups

$$t_{obs} = \frac{\overline{X}_1 - \overline{X}_2}{\sqrt{\left[\frac{(N_1 - 1)s_1^2 + (N_2 - 1)s_2^2}{N_1 + N_2 - 2}\right] \cdot \left[\frac{1}{N_1} + \frac{1}{N_2}\right]}}$$

$$df = N_1 + N_2 - 2$$

b. Correlated groups

$$t_{obs} = \frac{\sum D_i}{\sqrt{\dfrac{N\sum D_i^2 - (\sum D_i)^2}{N-1}}}$$

$$df = N - 1$$

2. Correlations

a. Test that the correlation is zero

Look for the critical values of r in Appendix II, Table C.

b. Difference between two independent correlations

$$z_{obs} = \frac{z_{r_1} - z_{r_2}}{\sqrt{\dfrac{1}{N_1 - 3} + \dfrac{1}{N_2 - 3}}}$$

(Consult Appendix II, Table D, for r-to-z_r transformations)

EXERCISES

1. Why must it be assumed that the population distribution of the difference between means is normal, and under what conditions can such an assumption be made?
2. Why is the test of the difference between means for independent groups not appropriate when the same subjects contribute to both groups of scores?
3. A theory suggests that when people experience cognitive dissonance (e.g., conflicting motives or thoughts), they will attempt to reduce this dissonance by altering their perceptions of the circumstances. For example, college students were asked to perform a boring task for a long period of time. Then some were paid $20 and some were paid $1 to tell the next student that the task was really interesting and fun. Later, in private, they rated their own actual feelings about the task. The theory predicts that the group that was paid $20 will not experience as much dissonance when lying about the task because $20 is a fair wage for that kind of fib. They should rate the task as being quite boring. However, the group that was paid $1 should experience more dissonance about lying and as a consequence tend to view the task more favorably in order to justify their lie. The ratings (the higher the number, the more interesting the subject rated the task) for the two groups are below.

$1	$20
5	2
4	6
6	3
9	3
10	1
7	4
8	5
7	3
9	2
6	

In view of the theory, evaluate the data,[7] using the formal organization illustrated in this chapter.

4. Mothers frequently say that they can recognize the cries of their own baby, even in the first few days of their infant's life. Moreover, they apparently are able to distinguish between a pain and a hunger cry. A neonatologist set up the following experiment. Mothers listened to tape recorded sets of five cries from different babies, one of which was their own. They had to decide which of the five was their baby. Each mother heard 20 such comparisons, 10 in which the cries of the infants were associated with hunger and 10 in which the cries were produced by a slight pinprick to the foot. The researcher was confident, on the basis of a theory and some other data, that the mothers would be more successful in picking out their own infant when a hunger cry was involved, since they had more experience with that situation, than if the cries were expressions of pain. The data are below. Test the scientist's theory, using the data below (number of identifications correct out of 10):

Mother	Hunger Cry	Pain Cry
a	4	4
b	6	5
c	6	4
d	7	6
e	9	6
f	2	5
g	6	8
h	5	3
i	3	2

5. Suppose two types of psychotherapy were to be compared for their ability to alleviate anxiety. Two groups of 12 college students were selected from a large group of students in a freshman speech course. The 24 students were picked following a series of interviews, questionnaires, and physiological measurements, because they had the highest composite scores for anxiety about giving a speech. Each student then received ten sessions of therapy that followed one of two forms. One type, **behavior therapy**, attempted to teach the subject to relax and mentally associate this state of relaxation with the thought of giving a speech. The other group received **insight therapy**, in which the therapist attempted to discern the causes of each person's anxiety and give the person insight into the problem. Later, each student had to give another speech and the same type of composite score of anxiety was measured. Below are the differences between the first and second scores for each subject. High scores indicate a great deal of improvement. Evaluate these data in terms of the effectiveness of the two methods of therapy.[8]

Behavior Therapy	Insight Therapy
4	2
−2	3
3	−4
7	0
6	−1
5	5
9	2
2	1
4	4
5	0
7	1
1	3

6. Construct a numerical example in which two groups of eight scores are significantly different if the same eight subjects each contributed a score to both groups (i.e., for correlated groups) but obviously not significantly different if the same scores are con-

[7]Inspired by L. C. Festinger and J. M. Carlsmith, "Cognitive Consequences of Forced Compliance," *Journal of Abnormal and Social Psychology* 58 (1959): 203–10.

[8]Inspired by G. L. Paul, *Insight vs. Desensitization in Psychotherapy: An Experiment in Anxiety Reduction* (Stanford: Stanford University Press, 1966).

sidered to be from independent groups of subjects.

7. Compute the correlation for the data in exercise 4 (p. 256) and test to determine if the *r* is significant.

8. Suppose that the correlation between the IQs of 33 sets of identical twins is .87 and the correlation between 28 sets of fraternal twins is .53. Test the obvious theoretical prediction that the correlation is higher for identical than for fraternal twins.[9]

9. A new program is developed to enrich the kindergarten experience of children in preparation for first grade. Hillmont school system tries out the new curriculum in one classroom for a year. Pupils in the classroom using the new curriculum and those in another classroom using the old curriculum are each tested at the beginning of the school year (the pretest) and again at the end of the year (the posttest). The test gives a score of 10 if the pupil performs at grade level. The scores for the two classes are given below.

a. Determine separately for each group if performance relative to grade level improved from pretest to posttest. Assume $\alpha = .05$ and a nondirectional alternative hypothesis.

b. Notice that the pretest means for the two groups are nearly identical, 9.89 and 10.0. Consequently, the effects of the new and the old curriculums can be compared by assessing the difference between the two groups on their respective posttests. Assume $\alpha = .05$ and a nondirectional alternative hypothesis.

c. Calculate the correlation between pretest and posttest separately within each group and test its significance. Assume $\alpha = .05$ and directional hypotheses (i.e., positive *r*).

d. Determine whether the correlations in c above are different from one another. Assume $\alpha = .05$ and a nondirectional alternative hypothesis. What does this result imply for the effects of the new versus the old curriculum?

New Curriculum		
Pupil	**Pretest**	**Posttest**
a	9	16
b	7	12
c	14	13
d	12	10
e	8	14
f	7	11
g	12	15
h	8	10
i	12	14

Old Curriculum		
Pupil	**Pretest**	**Posttest**
j	12	13
k	8	9
l	7	6
m	10	11
n	11	10
o	14	15
p	6	8
q	9	9
r	11	11
s	12	12

[9]Inspired by L. Erlenmeyer-Kimling and Lissy F. Jarvik, "Genetics and Intelligence: A Review," *Science* 142 (1964): 1477–78.

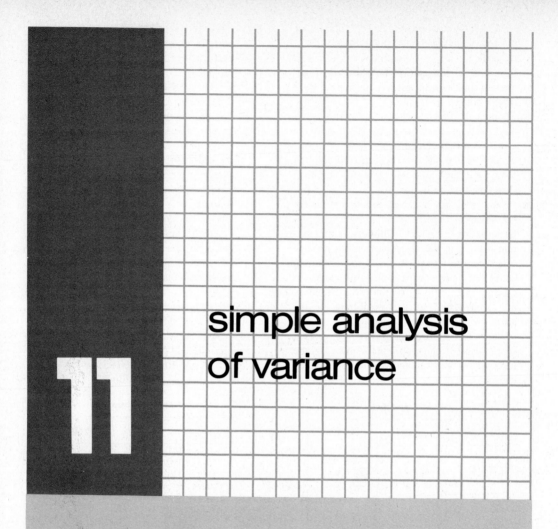

11

simple analysis of variance

TECHNIQUES THAT COMPARE the difference between two means were examined in the previous chapter. The null hypothesis for those tests is that the two samples are drawn from populations having the same mean (H_0: $\mu_1 = \mu_2$). However, very often a researcher wants to compare means from more than two samples and asks what the probability is that these several samples are drawn from populations having the same mean (H_0: $\mu_1 = \mu_2 = \ldots = \mu_p$). A statistical procedure that addresses this broader question is called the analysis of variance.

To illustrate, consider an experiment on the need of normal adults to dream. Some years ago it was discovered that an observer can tell whether a person is dreaming by monitoring his or her brain waves, EEG, heart rate, respiration rate, eye movements, and so on.[1] Among other things, this discovery made it possible to study the effects of disturbing people while they are dreaming, thus depriving them of the opportunity to dream. For example, the researcher might ask whether deprivation of dreaming has any effects on a person's temperament during the waking day. If normal adults need to dream, perhaps they become anxious and irritable if the opportunity to dream is curtailed.[2] Suppose one group of subjects has their sleep interrupted several times during the night, but never during or immediately before a dream. A second group is aroused the same number of times, but on two occasions during each night this wakening is at the onset of a dream. Last, a third group is awakened an equal number of times, but only after they have started to dream. Therefore, three groups of subjects can be characterized as undergoing either no, some, or much dream interruption. These procedures are carried out in a special sleep laboratory for six consecutive nights. During each day the subjects are interviewed and given tests to evaluate how anxious and irritable they are. Each subject is assigned a total score over the entire six-day period, with a high score indicating a very upset individual.

Suppose the means for the three groups (no, some, and much dream interruption) are 4.00, 8.00, and 17.75, respectively. This seems to indicate that the prevention of dreaming produces increased irritability in normal adults. But the observed differences between means could be a function of sampling error if there is no real difference among these groups in the population. The null hypothesis is that the population means are equal to one another and to a common value—H_0: $\mu_1 = \mu_2 = \mu_3 = \mu$. The statistical problem is to determine the probability that the null hypothesis of no difference in the population is true given the available data.

In the previous chapter techniques were presented for analyzing the difference between two means. Although it would be possible to use the t test on each pair of means (AB, AC, BC) that can be formed from the three samples (A, B, C), this approach is inappropriate for several reasons. First,

---

[1]See N. Kleitman, "Patterns of Dreaming," *Scientific American* 203 (1960): 82–88.
[2]Based on, but not identical to, W. Dement, "The Effect of Dream Deprivation," *Science* 131 (1960): 1705–07.

the amount of computational labor required increases rapidly as the number of groups increases. For example, though there are only 3 pairs of means for three groups, there are 10 pairs for five groups and 28 pairs for eight groups.

Second, when the .05 level of significance is adopted, one expects to reject the null hypothesis incorrectly (a type I error) in 5 of every 100 *independently* sampled pairs by chance or sampling error alone. But when several t tests are used, the pairs of means are not mutually independent: Out of k groups, each mean will be part of $k - 1$ pairs, and an extreme mean obtained by chance for a single group will lead to an incorrect decision in $k - 1$ cases. Therefore, one does not know just what the probability of a type I error would be if t tests are performed.

Third, frequently the researcher wants to ask a broader question than whether specific pairs of means are different. For example, the scientist might be less interested in whether no versus two interruptions or two versus many interruptions makes a difference than in the general question of whether the extent of dream interruption—from none to much deprivation—influences temperament. If pairs are tested, some pairs may be significantly different while others are not. Although this specific information might be important, it leaves somewhat ambiguous the answer to the general question of whether a relationship exists between dream deprivation and temperament, because the information is dependent on which pairs are tested. The analysis of variance gives a single answer to this general question.

LOGIC OF THE ANALYSIS OF VARIANCE

The purpose of simple analysis of variance is to determine the probability that the means of several groups of scores deviate from one another merely by sampling error.[3] The analysis of variance partitions the variability in the total sample in much the same way as was illustrated in the discussion of correlation and regression that opened Chapter 6. In that context, the total variability of Y_i is divided into a portion that is attributable to X and a portion that is not associated with X. In the case of the analysis of variance, the total variability in the scores is partitioned into a portion that reflects differences among the means of the groups and a portion that is not influenced by the differences among means.

The partitioning of variability is performed in such a way that two estimates of the variance of the scores in the population are computed. One of these estimates is based upon the deviation of the group means about the grand mean (the mean over all scores in the total analysis). The size of the estimate is influenced both by the variability of individual subjects (since

---

[3]The discussion of the analysis of variance in this chapter and the next assumes that the factors in the research design are fixed rather than random. This assumption is explained in Chapter 12.

their scores are involved in both the group and grand means) and by differences between group means. Because this variance estimate is based upon the deviation of group means about the grand mean, it is called the **between-groups estimate** of the population variance. The second estimate of the variance of the scores in the population is based upon the deviation of scores about their respective group means. This estimate is influenced not by differences in group means but only by the random variability of individual subjects. Since the estimate is determined by the deviation of scores within each group about the mean of that group, it is known as the **within-groups estimate** of the population variance.

Conceptually, the two variance estimates differ only in that the between-groups estimate is sensitive to differences between the group means whereas the within-groups estimate is not. As the group means become increasingly different from one another, the between-groups variance estimate grows larger. However, the null hypothesis being tested is that in the population all the group means are equal (H_0: $\mu_1 = \mu_2 = \mu_3$). Since the null hypothesis is tentatively held to be true while its validity is being tested, the between-groups estimate should not be influenced by population differences between means: Under H_0, the means are assumed to be equal to one another. Thus, given the null hypothesis, the between-groups estimate is influenced only by the same random variation in scores that determines the within-groups estimate.

Therefore, the analysis of variance provides two estimates of the population variance. Given the null hypothesis, the estimates should be identical except for sampling error. The probability that two independent variance estimates differ from one another only by sampling error can be determined by taking the ratio of the two sample variances, in this case,

$$\frac{s^2_{\text{between}}}{s^2_{\text{within}}}$$

Under the null hypothesis that these variances estimate the same population value, this ratio has a known theoretical distribution called F, the percentile points of which are listed in Table E of Appendix II. If the ratio is so large that the probability is exceedingly small that s^2_{between} and s^2_{within} estimate the same population variance, then one may assume that the additional influence of differences between group means has inflated the value of s^2_{between}, causing the F ratio to be unusually large. In that case, the null hypothesis should be rejected.

Notice that although this technique is called the analysis of *variance*, its purpose is really to assess differences in group *means*. It accomplishes this goal by comparing two variance estimates, one of which can be influenced by differences in group means and one of which cannot. The remainder of this section is devoted to expanding and clarifying the rationale behind the technique.

11-1 General Notation for Simple Analysis of Variance

	Group 1	Group 2	...	Group p
	X_{11}	X_{12}	...	X_{1p}
	X_{21}	X_{22}	...	X_{2p}
	X_{31}	X_{32}	...	X_{3p}
	X_{41}	X_{42}	...	X_{4p}
	$\vdots$	$\vdots$	$\vdots$	$\vdots$
	$X_{n_1 1}$	$X_{n_2 2}$	...	$X_{n_p p}$

	Group 1	Group 2	...	Group p	Grand Total
Group Totals	T_1	T_2	...	T_p	T
Group Means	$\overline{X}_1 = \dfrac{T_1}{n_1}$	$\overline{X}_2 = \dfrac{T_2}{n_2}$	...	$\overline{X}_p = \dfrac{T_p}{n_p}$	Grand Mean $\overline{X}$

X_{ij} = score for the ith subject in the jth group

p = number of groups

n_j = number of subjects in the jth group

$$N = n_1 + n_2 + \cdots + n_p = \sum_{j=1}^{p} n_j = \text{total number}$$

of subjects over all groups

T_j = total of all scores in group j

$\overline{X}_j$ = mean for group $j = T_j / n_j$

T = grand total over all subjects

$\overline{X}$ = grand mean = T/N

Notation and terminology

Notation   It will help to provide some notation that will serve as a convenient language with which to discuss the analysis of variance. Table 11-1 presents such notation for the general case. Any score, X_{ij}, is written with two subscripts. The first, or i, subscript denotes the subject's number within its group, and the second, or j, subscript indicates the group to which that subject belongs. Thus, X_{24} is the second subject in the fourth group. Table 11-1 is arranged to show several subjects in each of the groups. The i subscript runs up to n_1 for the first group, where n_1 indicates the total number of subjects in the first group. Similarly, n_2 indicates the number of subjects in the second group, and n_j denotes the number of subjects in the jth group, where j signifies some group in the set. There are p groups, so the number of subjects in the last group is n_p.

The total of all scores in group j is symbolized by T_j and the mean of the jth group is $\bar{X}_j = T_j/n_j$. Thus, $\bar{X}_1$ is the mean of the first group and $\bar{X}_p$ the mean of the last or pth group. In general, $\bar{X}_j$ is the mean of some group (the jth group). The grand mean, symbolized by $\bar{X}$ without any subscripts, is the mean of all the scores from all the groups. It can be expressed as

$$\bar{X} = \frac{T}{N}$$

in which T is the total of all scores and N is the total number of subjects in the design. One should note that the grand mean equals the average of the group means *only if all n_j are equal.*

Variance terminology   The logic of the analysis of variance as briefly sketched above requires two estimates of the population variance, the within-groups and the between-groups variance estimates. It will help to refine some terminology and notation about variances at this point. Earlier in the text the sample variance (variance estimate) was defined to be

$$s^2 = \frac{\sum\limits_{i=1}^{n} \left(X_i - \bar{X} \right)^2}{N - 1}$$

The numerator of this fraction is the sum of squared deviations about the mean, and thus it is sometimes called the **sum of squares** and is represented by **SS**. The denominator is the **degrees of freedom** associated with the sum of squares, and its symbol is **df**. In these terms the variance estimate is a type of average sum of squares, and it is sometimes known as a **mean square** or **MS**. Consequently, a variance estimate can by symbolized by

$$s^2 = \frac{\sum\limits_{i=1}^{n} \left(X_i - \bar{X} \right)^2}{N - 1} = \frac{SS}{df} = MS$$

Although this expression uses a different vocabulary from that used earlier, there is no important conceptual difference. The reason for the translation is that the new terms, sum of squares (SS) and mean square (MS), are customarily associated with the analysis of variance. Therefore, a variance estimate in the context of the analysis of variance is called a mean square, or MS.

Hypotheses   The purpose of the analysis of variance is to estimate the probability that the means of several groups differ from one another merely by sampling error. The null hypothesis is that in the population the group means are equal to one another and to the grand mean over all the groups. If μ_p represents the population mean for the last, or pth, group and μ is the population grand mean, then the null hypothesis is symbolized by

$$H_0: \mu_1 = \mu_2 = \cdots = \mu_p = \mu$$

The alternative hypothesis states that in the population the group means are not all equal and is best expressed as follows:

$$H_1: \text{not } H_0$$

Partitioning of Variability

Derivation of *MS* Determining the two estimates of the population variance involves partitioning the total variability into two portions in a manner quite similar to what was done in regression and correlation. Recall that (as shown in Figure 6–1, page 133) the total variability in scores was divided into a portion associated with the dependent variable's relationship to the predicting variable and a portion not associated with the relationship (i.e., error). The same strategy is used in the analysis of variance: The total variability in scores is divided into a portion associated with the group membership of the subjects and a portion not associated with that affiliation. If one thinks of group membership as the predicting variable, the partitioning in regression and correlation and in the analysis of variance is essentially the same.

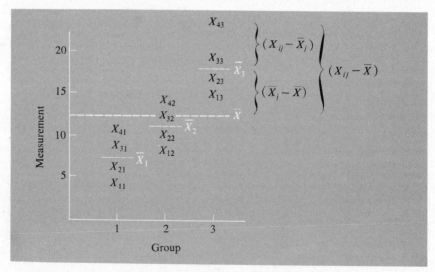

Fig. 11–1. Partitioning of deviations in the analysis of variance

The partitioning of deviations for a single score in the analysis of variance is illustrated in Figure 11–1. The analysis of variance deals with the relationship between two variables—group affiliation (e.g., dream deprivation), which is marked off along the abscissa, and some measurement or dependent variable (e.g., temperament), which is marked off along the ordinate. The scores for individual subjects are represented by X_{ij}, in which i signifies a specific person in the jth group. The height of the X_{ij} on the graph reflects the value of the dependent variable for that subject. The means for each of the three groups as well as the mean for all subjects (the grand mean) are indicated in white. The total variability of the scores is composed of their deviations from the grand mean of all scores. The total deviation for the particular score X_{43} in Figure 11–1 is the braced vertical distance between the score and the grand mean $(X_{ij} - \overline{X})$, which is composed of two parts: the deviation of the score from its own group mean $(X_{ij} - \overline{X}_j)$, plus the deviation of that group mean from the grand mean $(\overline{X}_j - \overline{X})$. In symbols,

$$(X_{ij} - \overline{X}) \quad = \quad (X_{ij} - \overline{X}_j) \quad + \quad (\overline{X}_j - \overline{X})$$

$$\downarrow \qquad\qquad\qquad \downarrow \qquad\qquad\qquad \downarrow$$

$$\begin{bmatrix} \text{Total deviation} \\ \text{from grand mean} \end{bmatrix} = \begin{bmatrix} \text{deviation of score from} \\ \text{group mean} \end{bmatrix} + \begin{bmatrix} \text{deviation of} \\ \text{group mean from} \\ \text{grand mean} \end{bmatrix}$$

It should now be clear that scores differ from one another for two reasons. First, scores are members of different groups, and if the group means differ from one another, the scores within a group will, on the average, differ by the same amount from the scores in other groups. Thus, the expression $(\overline{X}_j - \overline{X})$ indicates by how much the mean of the jth group deviates from the grand mean. This is the **between-group** source of variability. Second, the scores within a group differ from one another—they do not all equal their group mean. This fact is expressed in the quantity $(X_{ij} - \overline{X}_j)$, which is the difference between a score and its group mean. This is the **within-group** source of variability. Thus, scores differ from one another because they belong to different groups $(\overline{X}_j - \overline{X})$ and because of something unique to each subject that causes a subject's score to differ from the mean of its own group $(X_{ij} - \overline{X}_j)$. Another way to express the same concept is to transform the above formula slightly by shifting the grand mean to the right side of the expression:

$$\left(X_{ij} - \overline{X}\right) = \left(\overline{X}_j - \overline{X}\right) + \left(X_{ij} - \overline{X}_j\right)$$

$$X_{ij} = \overline{X} + \left(\overline{X}_j - \overline{X}\right) + \left(X_{ij} - \overline{X}_j\right)$$

Now, any score X_{ij} may be "built up" by starting with the grand mean of all the scores $(\overline{X})$, adding the amount of difference associated with being in the *j*th group $(\overline{X}_j - \overline{X})$, and then adding the amount of difference associated with that subject's unique deviation from the group mean $(X_{ij} - \overline{X}_j)$. If the grand mean is 10, the *j*th group mean is 14, and subject X_{ij} scores 2 units below the group mean, then $\overline{X}_j = 14$, $\overline{X} = 10$, $(\overline{X}_j - \overline{X}) = 4$, $(X_{ij} - \overline{X}_j) = -2$ and the score is given by

$$X_{ij} = \overline{X} + \left(\overline{X}_j - \overline{X}\right) + \left(X_{ij} - \overline{X}_j\right)$$
$$= 10 + 4 - 2$$
$$X_{ij} = 12$$

The above deviations are for a single score. Variability is usually measured in squared deviations, and it needs to be determined for all subjects in the study. Therefore, to obtain one expression for the total variability, begin with

$$\left(X_{ij} - \overline{X}\right) = \left(\overline{X}_j - \overline{X}\right) + \left(X_{ij} - \overline{X}_j\right)$$

Then square each side and sum over the n_j scores within each group and over the *p* groups.

$$\sum_{j=1}^{p} \sum_{i=1}^{n_j} \left(X_{ij} - \overline{X}\right)^2 = \sum_{j=1}^{p} \sum_{i=1}^{n_j} \left[\left(\overline{X}_j - \overline{X}\right) + \left(X_{ij} - \overline{X}_j\right)\right]^2$$

When these operations are carried out, the result is

$$\underbrace{\sum_{j=1}^{p} \sum_{i=1}^{n_j} (X_{ij} - \overline{X})^2}_{} = \underbrace{\sum_{j=1}^{p} n_j(\overline{X}_j - \overline{X})^2}_{} + \underbrace{\sum_{j=1}^{p} \sum_{i=1}^{n_j} (X_{ij} - \overline{X}_j)^2}_{}$$

$$\downarrow \qquad\qquad \downarrow \qquad\qquad\qquad \downarrow$$

$$\begin{bmatrix} \text{Total squared} \\ \text{deviations} \\ \text{in } X \end{bmatrix} = \begin{bmatrix} \text{squared deviations} \\ \text{between group} \\ \text{means} \end{bmatrix} + \begin{bmatrix} \text{squared deviations} \\ \text{within} \\ \text{groups} \end{bmatrix}$$

Notice that each component of the above expression is a sum of squared deviations (SS). Thus the expression can be written

$$SS_{\text{total}} = SS_{\text{between}} + SS_{\text{within}}$$

This statement indicates that the total sum of squares is composed of the sum of squared deviations of group means from their grand mean (SS_{between}) plus the sum of the squared deviations of scores from their respective group means (SS_{within}). Notice that the sum of squared deviations between a group mean and the grand mean is multiplied by n_j. The multiplication is done because this deviation is meant to apply to each subject in the group, and there are n_j subjects in the *j*th group.

Recall from the general rationale of the analysis of variance that rather than two **sums of squares**, one sensitive to group differences and the other not, two **mean squares** (MS) with these characteristics are required. A variance estimate is a sum of squares divided by its degrees of freedom. Therefore, the degrees of freedom for the sums of squares discussed above must be determined.

For a sample variance, the number of degrees of freedom (df) equals one less than the number of cases considered for the sum of squares. In the notational scheme described at the beginning of this section, there are N total cases, n_j cases in the jth group, and a total of p groups. The total degrees of freedom in the sample is one less than the total number of cases in the sample, or

$$df_{\text{total}} = N - 1$$

The degrees of freedom for SS_{between} should be one less than the number of means in the total sample, because SS_{between} is the sum of the squared deviations of each mean about the grand mean. Since there are p means,

$$df_{\text{between}} = p - 1$$

Last, the SS_{within} is composed of the sum of the squared deviations of scores about their respective group means. Within any single group j the sum of squares has $n_j - 1$ degrees of freedom. To obtain the within-group degrees of freedom, these df for single groups must be summed across all p groups:

$$\sum_{j=1}^{p} (n_j - 1)$$

Distributing the summation sign,

$$\sum_{j=1}^{p} n_j - \sum_{j=1}^{p} (1)$$

allows us to replace the first term by its equivalent, N, and the second term by its equivalent, p (recall that the sum from 1 to p of a constant is p times the constant). Thus, the number of degrees of freedom for SS_{within} is

$$df_{\text{within}} = N - p$$

It happens that the df_{total} equals the sum of the degrees of freedom for SS_{between} and SS_{within}:

$$df_{\text{total}} = df_{\text{between}} + df_{\text{within}}$$
$$N - 1 = (p - 1) + (N - p)$$

Therefore, the degrees of freedom for the total sample, like the sum of squares, is the sum of a between- and a within-group component.

To obtain a mean square, one divides a sum of squares by its degrees of freedom. Therefore, the mean square sensitive to between-group differences is

$$MS_{between} = \frac{SS_{between}}{df_{between}} = \frac{\sum_{j=1}^{p} n_j (\overline{X}_j - \overline{X})^2}{p-1}$$

and the mean square not sensitive to group differences is

$$MS_{within} = \frac{SS_{within}}{df_{within}} = \frac{\sum_{j=1}^{p} \sum_{i=1}^{n_j} (X_{ij} - \overline{X}_j)^2}{N-p}$$

One must be careful *not* to suppose that, because

$$SS_{total} = SS_{between} + SS_{within}$$

and

$$df_{total} = df_{between} + df_{within}$$

the total mean square equals the sum of the between and within mean squares. It does not:

$$MS_{total} \neq MS_{between} + MS_{within}$$

Comparison of $MS_{between}$ and MS_{within}   The logic of the analysis of variance rests on the fact that one of the variance estimates just discussed, $MS_{between}$, can be influenced by population differences between the means of the several groups, while the other, MS_{within}, cannot be so influenced. This section attempts to make that vital difference more obvious.

First, consider why two subjects from *different* groups do not have the same score. Recall the example discussed at the beginning of the chapter. Suppose the temperament score for the first subject in the some-deprivation group was 10 and that of the first subject in the much-deprivation group was 16. Why are these two scores different? There are two major possibilities. First, one subject had only some dream deprivation while the other had much. Perhaps the difference in temperament between them is related to their group affiliation—that is, to the amount of dream deprivation. This contribution to differences between scores stems from possible **treatment effects**. A **treatment effect** occurs when the population mean corresponding to a given group in the analysis differs from the average of the population means for all groups. The treatment effect is estimated by the deviation between a group mean and the grand mean, $(\overline{X}_j - \overline{X})$. However, if the null hypothesis is true, there are no treatment effects because the population means of all groups are identical. But if H_1 is the case, population group means will differ from one another and treatment effects are said to exist. Therefore, one possible reason two subjects do not have the same score is treatment effects.

In addition to potential treatment effects, two people assessed under identical circumstances usually do not respond the same way because of a variety of uncontrolled factors. For example, some subjects are happier than others because they tend to cope well with almost any circumstance. The entire collection of such potential causes of differences, presumably unrelated to differences in treatments, is called **error**. Therefore, differences or variability in scores between different groups may stem from treatment effects and/or error.

In contrast, consider why two scores from subjects in the *same* group are not identical. Both subjects have the same amount of dream deprivation, so differences in treatment condition cannot be a determinant of variability within a group. However, all the influences subsumed under the concept of error may still be operative. Therefore, in contrast to variability between groups, variability within groups is a function of error but not of treatment effects.

In brief,

$$MS_{\text{between}} \text{ reflects treatment effects } + \text{ error}$$
$$MS_{\text{within}} \text{ reflects error}$$

It is important to remember that under the null hypothesis all treatment effects are presumed to be zero, so MS_{between} and MS_{within} both reflect only error and are comparable estimates of the population variance. Since they are estimates, they will probably not precisely equal each other but will vary as a function of sampling error. But in the long run we would expect them to tend toward the same value.

The fact that MS_{between} is sensitive to treatment effects while MS_{within} is not may be illustrated with a numerical example. Suppose one consults a table of random numbers and selects three groups of five numbers each. Since these are random numbers, the three groups are analogous to three samples of size five drawn from a common population (sampling under the null hypothesis). These numbers are displayed in the top half of Table 11–2. The group means are 4, 6, and 5; the grand mean is 5. The between-groups variance estimate is computed below the listing of the groups—$MS_{\text{between}} = 5.0$. The SS_{within} for each group is presented and the within-groups estimate is at the right: $MS_{\text{within}} = 7.33$.

Now suppose that there are group differences or treatment effects. To implement this condition, 3 is added to each score in group 2 and 12 is added to each score in group 3. Consider the two variance estimates presented in the lower half of Table 11–2. Notice that the MS_{within} is precisely the same as it was before the effects were introduced. This is a reflection of the fact that the variability of scores within a group is not influenced by adding a constant to each score in the group. However, the value of MS_{between} has indeed changed as a function of the treatment effects. Now $MS_{\text{between}} = 215.0$. This is because a different constant was added to each of the three groups and thus to the three group means, and MS_{between} is sensitive to these differences. It should be

11–2 Numerical Illustration: $MS_{between}$ Is Influenced by Group Differences and MS_{within} Is Not

Group 1		Group 2		Group 3	
7	$SS_1 = \Sigma(X_{i1} - \bar{X}_1)^2$	2	$SS_2 = \Sigma(X_{i2} - \bar{X}_2)^2$	9	$SS_3 = \Sigma(X_{i3} - \bar{X}_3)^2$
5		6		3	
3	$SS_1 = 20$	9	$SS_2 = 38$	5	$SS_3 = 30$
4		9		6	
1	$df_1 = 4$	4	$df_2 = 4$	2	$df_3 = 4$
$\bar{X}_1 = 4.0$		$\bar{X}_2 = 6.0$		$\bar{X}_3 = 5.0$	

$$MS_{within} = \frac{SS_1 + SS_2 + SS_3}{df_1 + df_2 + df_3}$$

$$MS_{within} = \frac{20 + 38 + 30}{4 + 4 + 4} = 7.33$$

$$\bar{X} = 5.0$$

$$MS_{between} = \frac{\sum_{j=1}^{p} n_j(\bar{X}_j - \bar{X})^2}{p - 1}$$

$$MS_{between} = \frac{5(4 - 5)^2 + 5(6 - 5)^2 + 5(5 - 5)^2}{3 - 1} = 5.0$$

Treatment effect (added to each score)

0   +3   +12

Group 1	Group 2	Group 3
7	5	21
5	9	15
3	12	17
4	12	18
1	7	14

$SS_1 = 20$ $SS_2 = 38$ $SS_3 = 30$

$df_1 = 4$ $df_2 = 4$ $df_3 = 4$

$\bar{X}_1 = 4.0$ $\bar{X}_2 = 9.0$ $\bar{X}_3 = 17.0$

$\bar{X} = 10.0$

$$MS_{within} = \frac{SS_1 + SS_2 + SS_3}{df_1 + df_2 + df_3}$$

$$MS_{within} = \frac{20 + 38 + 30}{4 + 4 + 4} = 7.33$$

$$MS_{between} = \frac{\sum_{j=1}^{p} n_j(\bar{X}_j - \bar{X})^2}{p - 1}$$

$$MS_{between} = \frac{5(4 - 10)^2 + 5(9 - 10)^2 + 5(17 - 10)^2}{3 - 1} = 215.0$$

clear from this numerical example that the within-groups variance estimator is not influenced by differences in group means (i.e., treatment effects), whereas the between-groups estimate is influenced by treatment differences.

The F test

The final step in the rationale of the analysis of variance is to compare the two variance estimates under the assumptions of the null hypothesis. Recall that the null hypothesis states that there are no treatment effects, which implies that in the population the group means are all equal to one another and to the grand mean:

$$H_0: \mu_1 = \mu_2 = \cdots = \mu_p = \mu$$

The alternative hypothesis is that these conditions are not so, that there are treatment effects and thus differences between group means:

$$H_1: \text{not } H_0$$

If the null hypothesis is true, $MS_{between}$ and MS_{within} should both estimate the same parameter, just as they did in the numerical example above before treatment effects were added. Therefore, in the long run, the ratio of these two variances, or mean squares,

$$F = \frac{MS_{between}}{MS_{within}}$$

should vary only because of sampling error. The percentiles of the F distribution may be used in order to determine the probability of obtaining an F ratio of a specified size purely by sampling error. If the probability is very small that an observed F of this size is merely a function of sampling error, then perhaps the tentatively assumed null hypothesis of no treatment effects is wrong. Presumably, since $MS_{between}$ but not MS_{within} is influenced by treatment effects, a high F ratio means that $MS_{between}$ is large relative to MS_{within} because of treatment effects.

The percentile points for the theoretical sampling distribution of the ratio of two independent variance estimates are listed in Table E of Appendix II. However, just as there is a different t distribution for each number of degrees of freedom, there is a separate F distribution for each *combination* of degrees of freedom for the mean square in the numerator and in the denominator of the F ratio.

Independence The use of the F distribution in the analysis of variance requires that the sample variances in the ratio be independent. As discussed previously, estimates of the population mean and variance are independent if they are based upon samples from populations that are normally distributed.

Thus, if the population of scores X_{ij} is normally distributed, $\overline{X}_j$ and s_j^2 are independent. Where several groups of scores are sampled from a normal population, the collection of all the group means $(\overline{X}_j)$ is independent of the collection of all the group variances (s_j^2). MS_{between} is based upon the $\overline{X}_j$ while MS_{within} involves the deviations of scores about their group mean (s_j^2). Therefore, MS_{between} and MS_{within} are two independent estimates of the population variance if the population is normally distributed. In this case the F distribution may be used.

> The ratio of two independent variances is distributed as F with $N_1 - 1$ and $N_2 - 1$ degrees of freedom, where $N_1 - 1$ is the degrees of freedom of the variance in the numerator and $N_2 - 1$ is the degrees of freedom of the variance in the denominator.

Table E in Appendix II lists the .05 (roman type) and .01 (boldface type) critical values for F's of various degrees of freedom. To locate a critical value, find the column of the table corresponding to the number of degrees of freedom for the variance in the numerator. Then locate the row corresponding to the degrees of freedom for the variance in the denominator of the ratio. Suppose the degrees of freedom for the numerator and denominator are 2 and 30, respectively. Locate the intersection of the column labeled 2 and the row labeled 30 degrees of freedom. There are two values at that point. The first (roman type) is 3.32; that is the critical value for a test at the .05 level. Below this figure you will find the value **5.39** (boldface type), which represents the critical value for a test at the .01 level.

Logic of the F test The logic of the F test can now be summarized as follows:

1. MS_{within} is an estimate of the population variance based upon the deviations of scores about their respective group means. It is not influenced by mean differences between groups (treatment effects).

2. MS_{between} is also an estimate of the population variance if the null hypothesis is true. It is based upon the deviations of group means about the grand mean. Since it is influenced by any treatment effects that exist in the population, it estimates the same population variance as MS_{within} only if those treatment effects are zero, that is, if the null hypothesis is true.

3. Since the two variance estimates are independent if the population distribution is normal and since the logic of hypothesis testing demands that the null hypothesis be tentatively assumed to be true, the ratio of the two variance estimates is distributed as F:

$$F = \frac{MS_{\text{between}}}{MS_{\text{within}}}$$

4. Under conditions of the null hypothesis, which states that the two MS's estimate the same population value, the ratio should approach 1.0 in the long run.[4] The observed value of F is compared to the sampling distribution of such ratios to determine the probability that such an F value could be obtained merely by sampling error.

5. As the observed F ratio becomes larger, the probability becomes smaller that an F of this size should be obtained merely by chance. If there are treatment effects, $MS_{between}$ will be sensitive to them and MS_{within} will not. Therefore, an improbably large F value suggests that the null hypothesis (no treatment effects) was not appropriate and that treatment effects, in fact, do exist in the population. Thus, the null hypothesis should be rejected.

Assumptions and conditions underlying the simple analysis of variance

Four assumptions must be made in order to perform the statistical manipulations necessary for the analysis of variance. These assumptions are essentially the same as those required for the t test for means of two independent groups (Chapter 10), now applied to several groups.

First, it is assumed that the groups involved in the analysis are composed of **randomly** and **independently** sampled subjects. Second, the groups of scores being analyzed must be **independent**; this will be the case if the subjects in one group are not the same individuals as those in another group.

The procedures for the analysis of variance outlined in this book all require independent groups. However, occasionally it is desirable not to have independent groups but to measure the same subjects under each of several different conditions and to have the analysis determine if these correlated groups of scores have different means. The procedures for this type of analysis are called **repeated measures analyses of variance** because subjects are measured under more than one condition.[5] The difference between the analysis of variance for independent groups and repeated measures analyses of variance is analogous to the difference between the t test for means of independent groups and the t test for means of correlated groups.

The third assumption required for the analysis of variance is **homogeneity of within-group variances**. That is, it is assumed that the populations from which the groups are drawn have equal variances. In symbols,

$$\sigma_1^2 = \sigma_2^2 = \cdots = \sigma_p^2$$

One reason for this assumption is that it enables one to pool the variability

[4]Actually, the ratio approaches 1.00 in the long run only as the number of subjects becomes very large.
[5]B. J. Winer, *Statistical Principles in Experimental Design*, 2nd ed. (New York: McGraw-Hill, 1971). For a simplified discussion of the problems and assumptions in performing these analyses, see R. B. McCall and M. I. Appelbaum, "Bias in the Analysis of Repeated-Measures Designs: Some Alternative Approaches," *Child Development* 44 (1973): 401–15.

about each group mean into a single estimate of the population variance, MS_{within}. Without homogeneity of variance, one group with a very large variance might contribute disproportionately to this single estimate, and MS_{within} would not be representative of the variability within each group.

There are procedures for testing the homogeneity of variances. The F distribution can be employed to test the difference between two sample variances, and there are methods that extend such a test to include several variances rather than just two.[6] However, some statisticians have argued that these procedures do not provide an entirely appropriate test of the homogeneity of variance.[7] Further, if a sufficient number of cases are sampled and the number of subjects in each group is the same, moderate violations of this assumption do not alter the result of the analysis of variance very much.

Finally, it is assumed that each sample is drawn from a population of scores that is **normal** in form. This condition should be reflected in each of the groups sampled: Each should have a relatively normal distribution.

As discussed previously, the population distribution must be normal so that the two variance estimates will be independent, permitting the F test to be used. If the means are independent from the variability of scores about these group means (as they are in a normal distribution), then the between-groups and within-groups variance estimates will be independent. Moreover, the probability levels for the F statistic are accurate only if the two variance estimates are based on normal distributions.

Violations of the assumption of normality are not terribly damaging if a sufficient number of cases are sampled and if the departure from normality is not severe. However, when the distributions are decidedly not normal or there are not many cases in each group, a nonparametric analysis may be performed. Some nonparametric techniques are presented in Chapter 13.[8]

COMPUTATIONAL PROCEDURES

General format

As was demonstrated in previous chapters, the definitional formulas for statistical quantities are often not the most convenient for computational purposes. Table 11–3 gives a general computational scheme for the simple analysis of variance. Section A lists the scores in the p groups and several quantities needed for the computations. Under each group are the total of all

---

[6]Winer, *Statistical Principles*.

[7]W. L. Hays, *Statistics for the Social Sciences*, 2nd ed. (New York: Holt, Rinehart & Winston, 1973).

[8]See also: S. Siegel, *Nonparametric Statistics*, (New York: McGraw-Hill, 1956); J. V. Bradley, *Distribution-Free Statistical Tests*, (Englewood Cliffs: Prentice-Hall, 1968).

11-3 General Computational Procedures for the Simple Analysis of Variance

A. Data	Group 1	Group 2	...	Group p	Total Sample
	X_{11}	X_{12}	...	X_{1p}	
	X_{21}	X_{22}	...	X_{2p}	
	X_{31}	X_{32}	...	X_{3p}	
	$\ldots$	$\ldots$	...	$\ldots$	
	$X_{n_1 1}$	$X_{n_2 2}$	...	$X_{n_p p}$	
Totals	$T_1 = \sum_{i=1}^{n_1} X_{i1}$	$T_2 = \sum_{i=1}^{n_2} X_{i2}$	...	$T_p = \sum_{i=1}^{n_p} X_{ip}$	$T = \sum_{j=1}^{p} T_j$
n_j	n_1	n_2	...	n_p	$N = \sum_{j=1}^{p} n_j$
Means	$\bar{X}_1 = \dfrac{T_1}{n_1}$	$\bar{X}_2 = \dfrac{T_2}{n_2}$	...	$\bar{X}_p = \dfrac{T_p}{n_p}$	
Sum of squared scores	$\sum_{i=1}^{n_1} X_{i1}^2$	$\sum_{i=1}^{n_2} X_{i2}^2$	...	$\sum_{i=1}^{n_p} X_{ip}^2$	$\sum_{j=1}^{p} \left(\sum_{i=1}^{n_j} X_{ij}^2 \right)$
Squared sum of scores divided by n_j	$\dfrac{T_1^2}{n_1}$	$\dfrac{T_2^2}{n_2}$	...	$\dfrac{T_p^2}{n_p}$	$\sum_{j=1}^{p} \left(\dfrac{T_j^2}{n_j} \right)$

B. Intermediate Quantities

(I) $\dfrac{T^2}{N}$

(II) $\displaystyle\sum_{j=1}^{p}\left(\sum_{i=1}^{n_j} X_{ij}^{2}\right)$

(III) $\displaystyle\sum_{j=1}^{p}\left(\frac{T_j^{2}}{n_j}\right)$

C. Basic Formulas

$SS_{\text{between}} = (\text{III}) - (\text{I})$ $\qquad df_{\text{between}} = p - 1$ $\qquad MS_{\text{between}} = \dfrac{SS_{\text{between}}}{df_{\text{between}}}$

$SS_{\text{within}} = (\text{II}) - (\text{III})$ $\qquad df_{\text{within}} = N - p$ $\qquad MS_{\text{within}} = \dfrac{SS_{\text{within}}}{df_{\text{within}}}$

$SS_{\text{total}} = (\text{II}) - (\text{I})$ $\qquad df_{\text{total}} = N - 1$

D. Summary Table

Source	df	SS	MS	F
Between groups	$p - 1$	SS_{between}	MS_{between}	$\dfrac{MS_{\text{between}}}{MS_{\text{within}}}$
Within groups	$N - p$	SS_{within}	MS_{within}	
Total	$N - 1$	SS_{total}		

the scores for the group (T_j), the n_j, the mean ($\bar{X}_j$), the sum of squared scores ($\sum_{i=1}^{n_j} X_{ij}^2$), and the squared sum of the scores divided by n_j (T_j^2/n_j). (Remember that the sum of squared scores ($\sum X^2$) is determined by squaring each score and then summing. The squared sum of the scores ($\sum X$)^2 divided by n_j is calculated by first summing the scores, then squaring the sum and dividing by n_j.) To the right of these group quantities are the totals for the entire sample: the total of all scores (T, sum the T_j), the total number of cases (N, sum the n_j), the total sum of the squared scores

$$\sum_{j=1}^{p} \left(\sum_{i=1}^{n_j} X_{ij}^2 \right), \text{ sum the } \sum_{i=1}^{n_j} X_{ij}^2$$

and the squared sum of all the scores divided by n_j

$$\sum_{j=1}^{p} \left(T_j^2/n_j \right), \text{ sum all the } T_j^2/n_j$$

These are the basic quantities needed to calculate the sums of squares for the analysis of variance.

Section B lists three intermediate quantities that facilitate the computation of the SS. Quantity (I) is the squared total sum of scores divided by N (T_{total}^2/N, square T_{total}, and divide by N); quantity (II) is simply the total sum of the squared scores as found previously

$$\sum_{j=1}^{p} \left(\sum_{i=1}^{n_j} X_{ij}^2 \right)$$

and quantity (III) is the total of the squared sum of scores divided by n_j, also calculated previously

$$\sum_{j=1}^{p} \left(\frac{T_j^2}{n_j} \right)$$

Section C of the table lists the formulas for the three sums of squares, the three degrees of freedom, and the two mean squares required for the analysis of variance. Notice that the formulas for the SS are expressed in terms of the three intermediate quantities [(I), (II), (III)] previously computed. The degrees of freedom are determined by using p (the number of different groups) and N (the total number of subjects in the sample).

Section D presents the traditional summary table for the simple analysis of variance. The first column denotes the source of the variance estimate. This is followed by the degrees of freedom, sums of squares, mean squares, and F ratio.

Numerical example

Research question   We can illustrate the statistical procedure by continuing with the dream example in which the question was whether deprivation of dreaming produces differences in the temperaments of normal adults. The hypothetical results are presented in Tables 11–4 and 11–5. The means for the three groups (no, some, and much deprivation) are 4.00, 8.00, and 17.75, respectively. It would seem that the prevention of dreaming produces increased irritability and anxiety in normal adults. But the observed differences between means could be a function of sampling error, there being no real difference among the three groups in the population. Given the null hypothesis of no differences in the population means, what is the probability

---

11–4 Summary of the Simple
Analysis of Variance (Dream Example)

Hypotheses
  $H_0: \mu_1 = \mu_2 = \mu_3 = \mu$
  H_1: not H_0

Assumptions and Conditions
  1. The subjects in each group are **randomly** and **independently** sampled.
  2. The groups are **independent**.
  3. The population variances for the groups are **homogeneous** $(\sigma_1^2 = \sigma_2^2 = \sigma_3^2)$.
  4. The population distribution of scores is **normal** in form.

Decision Rules (from Table E)
  Given: .05 significance level, $df = 2, 12$

$$\text{If } F_{\text{obs}} < 3.88, \text{ do not reject } H_0.$$
$$\text{If } F_{\text{obs}} \geq 3.88, \text{ reject } H_0.$$

Computation (see Table 11–5)

$$F_{\text{obs}} = 24.50$$

Decision
  Reject H_0.

11-5 Computational Example of the Simple Analysis of Variance (Dream Example)

A.	Group 1 (No Deprivation)	Group 2 (Some Deprivation)	Group 3 (Much Deprivation)	Total Sample
	7	5	21	
	5	9	15	
	3	12	17	
	4	12	18	
	1	7		
		3		
Totals	$T_1 = (7 + 5 + \cdots + 1)$ $T_1 = 20$	$T_2 = (5 + 9 + \cdots + 3)$ $T_2 = 48$	$T_3 = (21 + 15 + \cdots + 18)$ $T_3 = 71$	$T = \sum_{j=1}^{p} T_j$ $= 139$
n_j	$n_1 = 5$	$n_2 = 6$	$n_3 = 4$	$N = \sum_{j=1}^{p} n_j$ $= 15$
Means	$\bar{X}_1 = \dfrac{20}{5} = 4.00$	$\bar{X}_2 = \dfrac{48}{6} = 8.00$	$\bar{X}_3 = \dfrac{71}{4} = 17.75$	
Sum of squared scores	$\sum X_{i1}^2 = 100$	$\sum X_{i2}^2 = 452$	$\sum X_{i3}^2 = 1279$	$\sum_{j=1}^{p}\left(\sum_{i=1}^{n_j} X_{ij}^2\right)$ $= 1831$
Squared sum of scores divided by n_j	$\dfrac{T_1^2}{n_1} = \dfrac{(20)^2}{5} = 80$	$\dfrac{T_2^2}{n_2} = \dfrac{(48)^2}{6} = 384$	$\dfrac{T_3^2}{n_3} = \dfrac{(71)^2}{4} = 1260.25$	$\sum_{j=1}^{p}\left(\dfrac{T_j^2}{n_j}\right)$ $= 1724.25$

B. (I) $\dfrac{T^2}{N} = \dfrac{(139)^2}{15} = 1288.07$

(II) $\displaystyle\sum_{j=1}^{p}\left(\sum_{i=1}^{n_j} X_{ij}^2\right) = 1831$

(III) $\displaystyle\sum_{j=1}^{p}\left(\dfrac{T_j^2}{n_j}\right) = 1724.25$

C. $SS_{\text{between}} = \textbf{(III)} - \textbf{(I)} = 1724.25 - 1288.07 = 436.18$ $\qquad df = p - 1 = 3 - 1 = 2$

$SS_{\text{within}} = \textbf{(II)} - \textbf{(III)} = 1831 - 1724.25 = 106.75$ $\qquad df = N - p = 15 - 3 = 12$

$SS_{\text{total}} = \textbf{(II)} - \textbf{(I)} = 1831 - 1288.07 = 542.93$ $\qquad df = N - 1 = 15 - 1 = 14$

$$MS_{\text{between}} = \dfrac{SS_{\text{between}}}{df_{\text{between}}} = \dfrac{436.18}{2} = 218.09$$

$$MS_{\text{within}} = \dfrac{SS_{\text{within}}}{df_{\text{within}}} = \dfrac{106.75}{12} = 8.90$$

D. Summary Table

Source	df	SS	MS	F
Between groups	2	436.18	218.09	$\dfrac{MS_{\text{between}}}{MS_{\text{within}}} = \dfrac{218.09}{8.90}$
				$= 24.50**$
Within groups	12	106.75	8.90	
Total	14	542.93		

Critical values $(df = 2, 12)$ $*F_{.05} = 3.88, p < .05$

$**F_{.01} = 6.93, p < .01$

that the observed difference between these three sample means is merely a function of sampling error? The details of the analysis are described below and summarized in Tables 11–4 and 11–5.

Statistical hypotheses   To answer the question above requires determining the probability that the population means for the three groups are in fact equal and thus equal to the grand mean. Symbolically stated, the null hypothesis is

$$H_0: \mu_1 = \mu_2 = \mu_3 = \mu$$

The alternative is

$$H_1: \text{not } H_0$$

Assumptions and conditions   The assumptions underlying this test have been explained on pages 274–75.

Significance level   Adopt $\alpha = .05$.

Critical values   The computational work presented in Table 11–5 shows that in this case the numerator of the F ratio has 2 degrees of freedom and the denominator has 12. At $\alpha = .05$ with $df = 2/12$, the critical value for F is 3.88 according to Table E of Appendix II.

Decision rules   The decision rules are:

If $F_{obs} < 3.88$, do not reject H_0.
If $F_{obs} \geq 3.88$, reject H_0.

The statement of the decision rules makes it look as if a directional test is being performed. In one sense, it is; in another sense, it is not. Since the mean square sensitive to treatment effects is always placed in the numerator of the F ratio, the F test in the analysis of variance rejects H_0 only when the value of F_{obs} is very large. In this sense, it is a one-tailed test because the critical region is in only one tail of the theoretical sampling distribution. However, large F values can occur regardless of which particular group means are highest. In fact, the analysis of variance detects only whether the means differ by more than sampling error—which particular means are larger or smaller than other means is not considered. Therefore, the analysis of variance is making a nondirectional test because the direction of differences between means is not specified. It is possible to state alternative hypotheses that do specify the ordering of group means or the form of the relationship between the levels of the factor and the dependent variable, but these techniques are beyond the scope of this text.[9] Consequently, students should not be concerned about one- versus two-tailed, directional versus nondirectional, tests in the analysis of variance in this text.

Computation   The calculations are presented in Table 11–5 and follow the general procedures set forth in Table 11–3.

Decision   The observed $F = 24.50$. Clearly this is greater than the required value of 3.88 at $\alpha = .05$. In fact, as presented at the bottom of Table 11–5, the observed value exceeds the critical value for the degrees of freedom when the significance level is .01. Customarily, the .05 level is used as the *minimum* significance level for stating that the data suggest the rejection of the null hypothesis. However, the researcher may indicate when the F exceeds the critical value at a higher level of significance. Frequently, the value of F is followed by * or** or *** if it exceeds the critical value for $\alpha = .05$, .01, or .001, respectively. Thus, since 24.50 is larger than the critical value for $\alpha = .01$, this fact is indicated by writing the F value in the summary table as 24.50**. (Actually, $F_{obs} = 24.50$ exceeds the .001 level, but our table gives only the .05 and .01 critical values.)

The obtained F value results in a decision to reject the null hypothesis. This means that the probability is very small that the three means (4, 8, and 17.75) differ merely by sampling error. Therefore, it is likely that the between-groups variance estimate has been influenced by treatment effects and that the population group means probably do differ from one another. The conclusion can be made that it is likely that the interruption and prevention of dreaming in adults leads to increased anxiety and irritability.

THE RELATIONSHIP BETWEEN *F* AND *t*

Since the analysis of variance is an extension of the *t* test to more than just two groups, there should be some relationship between *t* and *F*. More specifically, there should be some relationship between the results of an independent-groups *t* test and of the analysis of variance when only two groups are compared.

The general nature of the relationship between *F* and *t* is

$$F_{1, v} = t_v^2$$

in which the subscripts indicate the degrees of freedom.

[9]Winer, *Statistical Principles*.

This relationship applies both to the observed values and to the critical values of F and t when just two groups are being compared.

Suppose one considers only the first two groups of scores in the experiment on the deprivation of dreaming as presented in Table 11–5. One could perform a t test or an analysis of variance to test the difference between these two groups. It happens that because $F_{1, v} = t_v^2$, the two analyses would yield the same conclusion. For example, at $\alpha = .05$ (nondirectional) with 17 df, the critical value of t is $t_{\text{crit}} = 2.11$. At the same significance level with degrees of freedom equal to 1 and 17, the required F is $F_{\text{crit}} = 4.45$. If $F_{1, v} = t_v^2$, then the following should hold:

$$4.45 \overset{?}{=} (2.11)^2$$

It does. In addition, the observed values of F and t hold the same relationship:

$$F_{\text{obs}} = (t_{\text{obs}})^2$$

Therefore, F and t are directly related, and the analysis of variance is an extension of the t test between means to more than two groups.

EXERCISES

1. The following sets of numbers were obtained from a random-number table.

Group 1	Group 2	Group 3
3	4	8
6	7	6
1	3	5
6	0	7
9	6	2
		5
		9

   a. Compute the analysis of variance, following the format outlined in Table 11–5.
   b. Add 3 to each score in group 2 and 6 to each score in group 3. This process is analogous to the existence of what in the population? Now, compute the analysis of variance again and explain any dif-

   ferences between this result and that found in your answer to 1a.
   c. Return to the original numbers as given in the table above and add 20 to the last score in each group. Recompute the analysis of variance and explain any differences you observe.
2. Explain how the partition of variability in the analysis of variance is analogous to the partition of variability in regression and correlation.
3. Suppose the error component of your individual score is 4, the grand mean is 10, and the treatment effect associated with your group is -2. What is your score?
4. Suppose that an experiment was designed to determine what kind of discipline in the schools is most effective at preventing disruptive behavior. Third graders who were disruptive were either not responded to, verbally reprimanded in front of the class, or

privately told that they should not repeat that behavior. During the next ten days of class a record was kept of the number of days on which no disruptive behavior occurred. The results are given in the table to the right. Determine if the three approaches differed in their effectiveness.[10]

[10]Based on, but not identical to, K. D. O'Leary, K. F. Kaufman, R. E. Kass, and R. S. Drabman, "The Effects of Loud and Soft Reprimands on the Behavior of Disruptive Students," *Exceptional Children* 37 (1970): 145–55.

No Response	Public Reprimand	Private Reprimand
4	7	5
2	4	9
5	6	7
1	3	7
	5	8
		6

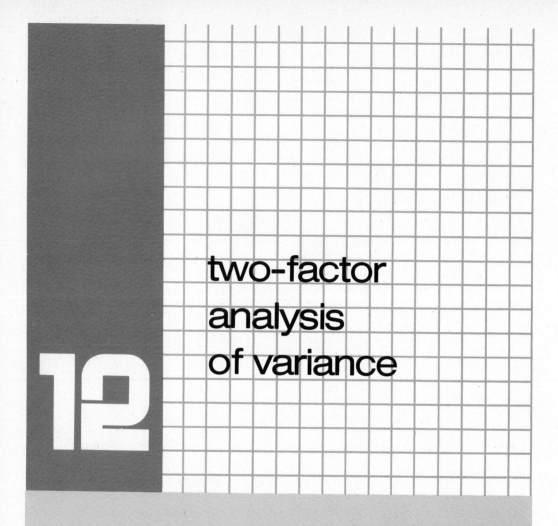

two-factor analysis of variance

12

THE SIMPLE ANALYSIS of variance as described in Chapter 11 tests the likelihood that the means from several sample groups actually estimate the same population mean and differ only because of sampling error. It was assumed that the groups were part of a classification scheme: different amounts of employment or dream deprivation, different teaching programs, etc. These groups represent different levels or types of a *single* category of events, treatments, stimuli, and so on. But scientists more frequently design their experiments to include groups belonging to more than one classification, and it is the purpose of this chapter to consider the extension of simple analysis of variance to the case in which the groups fall into a two-way classification scheme.

TWO–FACTOR CLASSIFICATION

Children watch the behavior of others, and they often imitate it, especially if they see that good things have happened to the other person as a consequence of such actions. Consider the following experiment to test this notion.[1] Forty boys and forty girls are randomly selected from a given school. All the children see one of two movies. In both films, an adult hits, pounds, pushes, and otherwise assaults a large plastic balloon doll with a weighted base (called a Bobo doll). The adult is called a model, presumably for the child to imitate. However, in a film seen by half the children of each sex the model is rewarded by another adult with praise and congratulations for the conquest of the doll, whereas the film seen by the other half of the children shows the model being punished with verbal reprimands for violent behavior. After the movie all the children are brought into a room with several age-appropriate toys, including the Bobo doll. In a 10-minute play session, the number of imitated aggressive advances to the Bobo doll are counted. There are four groups in this experiment: (1) boys who see the model rewarded, (2) boys who see the model punished, (3) girls who see the model rewarded, and (4) girls who see the model punished. These four groups represent different levels of two classification schemes—one based upon the sex of the subjects (male vs. female) and one based upon the type of reinforcement the model received (reward vs. punishment). Psychologists call reinforcement that a person sees given to another person, as in the films, *vicarious reinforcement*.

Classification terminology

A single classification scheme is called a **factor**, and each group within it represents a **level** of that factor. Notice that the levels do not necessarily

[1] Based on, but not identical to, A. Bandura, D. Ross, and S. A. Ross, "Imitation of Film-Mediated Aggressive Models," *Journal of Abnormal and Social Psychology* 66 (1963): 3–11.

12–1 Example of a Two-Factor Research Design
(Vicarious Reinforcement Study)

		Factor *B* (Vicarious Reinforcement)	
		*b*₁ (reward)	*b*₂ (punishment)
Factor *A* (Sex of Subject)	*a*₁ (males)	Males who see the model rewarded	Males who see the model punished
	*a*₂ (females)	Females who see the model rewarded	Females who see the model punished

represent different *amounts* of the factor, as is demonstrated by the levels of the sex factor, male and female. The above study has a **two-factor design** with two levels of each factor. Designating a factor by a capital letter and each level by a subscripted lower-case letter of the same character, this design can be expressed in a tabular fashion as shown in Table 12–1.

Table 12–2 shows the average number of imitative aggressive responses for each of the four groups. In addition to the group means, the numbers at

12–2 Means for the Vicarious Reinforcement Study

		Factor *B* (Vicarious Reinforcement)		
		*b*₁ (reward)	*b*₂ (punishment)	
Factor *A* (Sex of Subject)	*a*₁ (males)	25	5	15
	*a*₂ (females)	19	3	11
		22	4	13

the ends of the rows and columns, called **marginals**, express the mean number of imitative responses for males (15) and for females (11) *collapsed over*, or *ignoring*, reinforcement conditions, and the mean number for subjects watching a rewarded (22) and a punished (4) model collapsed over sex of subject. The grand mean number of responses over all subjects is 13.

Main effects and interactions

Three types of questions could be asked about the data from such an experiment. First, is there a significant difference between the levels of **factor A**? In this case, on the average, do boys imitate aggressive behavior more often than girls do? Specifically, does the difference between the means of 15 and 11 exist in the population, or is this observed difference merely a function of sampling error? Second, is there a significant difference between the levels of **factor B**? In this case, does the vicarious experience of seeing the model rewarded or punished influence the extent to which children will imitate? Is the difference between the means of 22 and 4 due to sampling error? Third, is there an **interaction** between the sex of the child and the type of vicarious reinforcement? Specifically, is the effect of seeing the model rewarded or punished different for boys than for girls? The same question may be turned around—does the difference between boys and girls in their tendency to imitate depend on whether the model is rewarded or punished?

> An **effect** is a difference among population means. A **main effect** is a difference among population means for levels of a factor collapsed over the other factors. An **interaction** occurs when the nature of the effect for one factor is not the same under all conditions or under all levels of another factor.

The meaning of these terms can be appreciated best by examining graphs of possible results of the two-factor example just described. In Figure 12–1, graph A depicts the case in which all the means are approximately the same. There are no main effects and no interaction. Graph B describes one main effect: Males imitate more often than females do. Notice that the reward and punishment conditions do not appear to have influenced the amount of imitation. Graph C illustrates a main effect for vicarious reinforcement with no effects for sex and no interaction. Children of both sexes imitated more if the model was rewarded than if the model was punished, and the amount of imitation was the same for both sexes.

In graph D, there are two main effects. Males imitated more than females, and the reward condition generated more imitation than the punish-

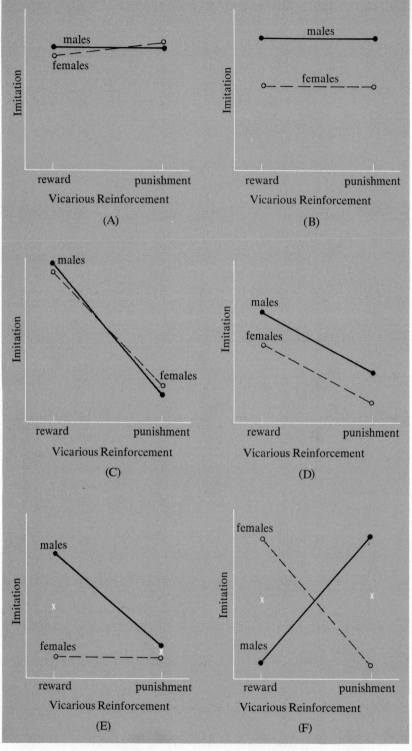

Fig. 12–1. Examples of possible main effects and interactions

ment condition. There is no interaction in graph D because the difference between males and females under both reinforcement conditions is about the same. Another way to say this is that the lines for the two sexes are approximately parallel, indicating that the reinforcement conditions had a comparable effect on both sexes.

Graphs E and F depict interactions between sex and vicarious reinforcement because the nature of the effects for one factor is not the same within both levels of the other factor. In graph E, whether the model was rewarded or punished made a difference for males but not for females. In graph F, the vicarious reinforcement had opposite effects for the two sexes. A rewarded model produced more imitation in females than a punished model, but the reverse was true for males. Notice that in graph E there is no main effect for vicarious reinforcement or for sex. That is, if males and females are averaged separately within the reward and then within the punishment conditions, these means (indicated by white x's) are not different from one another. Similarly, if the reward and punishment means are averaged separately within each sex, they are not different. Thus, graph F depicts an interaction with no main effects.

As can be seen in the above examples, any possible combination of results can occur. There may be one or two main effects with or without an interaction, or there may be an interaction with no main effects. However, when an interaction is found to be significant, generally one ignores the significance or insignificance of the main effects. That is, if there is an interaction, that fact means that the influence of one factor differed depending upon the level of the other factor. Since a main effect represents a difference in means for one factor collapsed over the levels of the other factor, the occurrence of an interaction indicates that collapsing over the other factor is not meaningful: The effects for factor A are not the same within each level of factor B. For example, consider graph E in Figure 12–1. The small x's indicate the means for the reward and punishment conditions collapsed over the sex of the subject. It might be the case that the difference between the two means (the x's) is significant, and one would conclude that a rewarded model produces more imitation than a punished model. However, if the interaction depicted in that graph is also significant, it means that vicarious reinforcement makes a difference only for males and not for females, thus qualifying the general interpretation of the main effect for reinforcement. The reverse of this situation occurs in graph F. Here the white x's are not different, and it is likely there would not be a significant main effect for reinforcement. But looking at the graph, it would be wrong to conclude that reinforcement has had no influence on imitation—its influence depends on the sex of the subject, and these differences are masked when the two sexes are averaged within reward and within punishment conditions. Therefore, if an interaction is significant, considerable care must be taken in interpreting the nature of main effects for factors that are involved in that interaction.[2]

LOGIC OF TWO-FACTOR ANALYSIS OF VARIANCE

The logic of two-factor analysis of variance is a direct extension of the rationale underlying simple analysis of variance.[3] To review, in simple analysis of variance the total sum of squares is partitioned into two components, each of which provides an estimate of the variability in the population if the null hypothesis—that the population mean of each group is the same value—is true. However, one of these variance estimates is based upon the deviations of the group means about the grand mean (between-groups estimate), while the other is based upon the deviations of the scores about their respective group means (within-groups estimate). The between-groups estimate is influenced by any treatment effects that may exist, whereas the within-groups estimate is not. Consequently, if the between-groups estimate is very large relative to the within-groups estimate, then the tentative presumption that these groups are all samples from the same population (i.e., the null hypothesis) may not be true. The ratio of the between-groups estimate to the within-groups estimate is distributed as F; by determining the percentile rank of the observed ratio in this theoretical relative frequency distribution, one can find the probability of obtaining such a ratio by sampling error alone. If the probability is sufficiently small, the null hypothesis of no difference between the population means is rejected.

In two-factor analysis of variance, the total sum of squares is partitioned into four components. The four variance estimates are called **mean squares** (MS), and they all estimate the same population variance if the null hypothesis is true.[4]

One estimate of the population variance (written MS_A) is based upon the deviations about the grand mean of the marginal means for the levels of

---

[2]This general "rule" does have an exception in the following event: An interaction exists, but all groups within a level of a factor are significantly higher than the corresponding groups within another level of that factor. The analysis would yield a main effect and an interaction. For example, suppose in graph E of Figure 12–1 the line for males was substantially higher than for females under each of the reinforcement conditions. If the males displayed more imitation than females under reward and under punishment conditions in addition to having the pattern difference displayed in graph E, then the main effect for sex as well as the interaction would be meaningful. Determining if a main effect should be interpreted when an interaction is also present is a complex issue and beyond the scope of this text.

[3]This discussion will be restricted to fixed factors, independent groups, and equal cell sizes as discussed later.

[4]Technically, there are several null hypotheses, as discussed in the material that follows.

factor A. This estimate is analogous to the between-groups estimate in a simple analysis of variance, involving only the groups of factor A (i.e., ignoring factor B). In the example, this estimate would be based upon the deviations of the means for males (15) and females (11) about the grand mean (13). MS_A is sensitive to differences between the means of the levels of factor A.

A second estimate of the variability in the population (written MS_B) is based upon the deviations about the grand mean of the marginal means for the levels of factor B. As such, it is analogous to MS_A, except the means for the levels of factor B (collapsed over factor A) are used. In the example, this estimate would be based upon the deviations of the reward (22) and punishment (4) means about the grand mean (13). MS_B is sensitive to differences between the means of the levels of factor B.

A third estimate of the variability in the population (symbolized by MS_{AB}) is based upon the deviations of each group mean from what would be predicted on the basis of knowledge of the two main effects. This mean square is sensitive to the possible interaction between factors A and B.

Finally, the fourth estimate of the population variability is derived in the same manner as the within-groups estimate in simple analysis of variance (MS_{within}). It is based upon the deviations of each score from its respective group mean. In contrast to the other mean squares, it is not sensitive to differences between individual groups or between levels of factors; thus, it can be used as a standard against which the other estimates may be evaluated.

The ratio of the mean square for factor A to the mean square within groups provides a test of the null hypothesis that the means of the levels of A differ from one another merely as a function of sampling error. Under the null hypothesis, the size of this ratio should not be very large because both variances presumably estimate the same value. However, since MS_A is sensitive to differences between the means for factor A but MS_{within} is not, the ratio will be large to the extent that those means deviate from one another. If they are very different, the ratio will be so large that the probability that the deviations derive merely from sampling error is exceedingly small, and the null hypothesis will be rejected. The conclusion will be that the means for the levels of factor A are significantly different from one another.

Similarly, the ratio of the mean square for factor B to MS_{within} reflects the extent to which the means for the levels of factor B differ from one another. The general rationale for testing the null hypothesis that such differences are merely a function of sampling error is similar to that just described for factor A.

Last, the ratio of MS_{AB} to MS_{within} tests the proposition that an interaction between factors A and B exists in the population.

In short, the logic of two-factor (or double classification) analysis of variance is a direct extension of that employed for simple analysis of

variance. The total sum of squares is partitioned into components, each
sensitive to a different aspect of the classification scheme. Under the null
hypothesis, all the groups in the design are randomly selected and have the
same population mean. If this is true, the sample means of the groups and the
means for the levels of the two factors should deviate only by sampling error.
Since MS_A, MS_B, and MS_{AB} are sensitive to different possible effects within
the design while MS_{within} is not, the ratio of each of these three mean squares
to MS_{within} provides F ratios that test the null hypothesis.

Notation

The raw score notation follows the general plan described for the simple
one-factor analysis of variance except that an additional subscript is necessary. This notation is summarized in Table 12–3. A single score is indicated
by X_{ijk}, which represents the ith individual in the jth level of factor A and the
kth level of factor B. There are a total of p levels of factor A and q levels of
factor B. It will be assumed that there are an equal number of subjects in
each of the jk groups and that this number is n.[5]

Any group in the experiment is called a **cell** of the design, and the
location of any cell is determined by the intersection of a specific row and
column as indicated in Table 12–3. Thus, the intersection of the jth row and
the kth column defines the jkth cell. The total of the scores for all the subjects
in that cell is represented by T_{jk} and the mean of that group is written $\bar{X}_{jk}$.
Therefore, the total of all the scores in cell 2, 3 (read "cell two-three") is T_{23}
and the mean of that cell is $\bar{X}_{23} = T_{23}/n$. The total of all the scores in a given
row is $T_{j.}$ in which j indicates the row and the dot (located where the column
designation should be) signifies that the scores have been summed over all
columns. Thus, the mean of the third row is written as $\bar{X}_{3.} = T_{3.}/nq$ since
there are q groups, each containing n subjects, in each row. Similarly, the
mean of the second column is defined by $\bar{X}_{.2} = T_{.2}/np$ or in the kth column,
$\bar{X}_{.k} = T_{.k}/np$. The dot now indicates scores have been summed over all
rows. The grand total and mean (i.e., over all rows and columns) are written
$T_{..}$ and $\bar{X}_{..}$.

Partitioning of variability

In simple analysis of variance, the total variability of scores is partitioned into
two parts, a between- and a within-groups component. In symbols,

$$SS_{total} = SS_{between} + SS_{within}$$

[5]The two-factor techniques described in this text demand that there be an equal number of
subjects per cell and that n be greater than 1. Special procedures do exist for the case in which
the groups do not have equal numbers of subjects and/or $n = 1$. See, for example, B. J. Winer,
Statistical Principles in Experimental Design, 2nd ed. (New York: McGraw-Hill, 1971).

12–3 Summary of Raw Score Notation for the Two-Factor
Analysis of Variance

		Factor B			Row
		b_1	b_2	$\cdots$	b_q	Means

		b_1	b_2	$\cdots$	b_q	Row Means
	a_1	X_{111} X_{211} X_{311} $\vdots$ X_{n11} $\bar{X}_{11} = \dfrac{T_{11}}{n}$	X_{112} X_{212} X_{312} $\vdots$ X_{n12} $\bar{X}_{12} = \dfrac{T_{12}}{n}$	$\cdots$ $\cdots$ $\cdots$ $\cdots$	X_{11q} X_{21q} X_{31q} $\vdots$ X_{n1q} $\bar{X}_{1q} = \dfrac{T_{1q}}{n}$	$\bar{X}_1 . = \dfrac{T_1 .}{nq}$
Factor A	a_2	X_{121} X_{221} X_{321} $\vdots$ X_{n21} $\bar{X}_{21} = \dfrac{T_{21}}{n}$	X_{122} X_{222} X_{322} $\vdots$ X_{n22} $\bar{X}_{22} = \dfrac{T_{22}}{n}$	$\cdots$ $\cdots$ $\cdots$ $\cdots$	X_{12q} X_{22q} X_{32q} $\vdots$ X_{n2q} $\bar{X}_{2q} = \dfrac{T_{2q}}{n}$	$\bar{X}_2 . = \dfrac{T_2 .}{nq}$
$\vdots$		$\vdots$	$\vdots$	$\cdots$	$\vdots$	$\vdots$
	a_p	X_{1p1} X_{2p1} X_{3p1} $\vdots$ X_{np1} $\bar{X}_{p1} = \dfrac{T_{p1}}{n}$	X_{1p2} X_{2p2} X_{3p2} $\vdots$ X_{np2} $\bar{X}_{p2} = \dfrac{T_{p2}}{n}$	$\cdots$ $\cdots$ $\cdots$ $\cdots$	X_{1pq} X_{2pq} X_{3pq} $\vdots$ X_{npq} $\bar{X}_{pq} = \dfrac{T_{pq}}{n}$	$\bar{X}_p . = \dfrac{T_p .}{nq}$
Column Means		$\bar{X}_{.1} = \dfrac{T_{.1}}{np}$	$\bar{X}_{.2} = \dfrac{T_{.2}}{np}$	$\cdots$	$\bar{X}_{.q} = \dfrac{T_{.q}}{np}$	**Grand Mean** $\bar{X}_{..} = \dfrac{T_{..}}{npq}$

n = the number of subjects in each group	T_{jk} = the total of all scores in the jkth group
a_j = the jth level of factor A	$\overline{X}_{jk}$ = the mean of the jkth group
b_k = the kth level of factor B	$T_{..}$ = the total of all scores
p = the number of levels of factor A	$\overline{X}_{i.}$ = the marginal mean for the ith row
q = the number of levels of factor B	$\overline{X}_{.j}$ = the marginal mean for the jth column
X_{ijk} = the ith score in the jkth group	$\overline{X}_{..}$ = the grand mean

In the case of a two-factor design, subjects also differ in score value because of between-group differences, but the group differences are now more complicated because there are two classification schemes. A subject belongs to a level in factor A that may be associated with a higher or lower average score, and the subject belongs to a level of factor B that may also be associated with a certain average score. In addition, there might be something special about being in a specific cell that would not have been predicted solely on the basis of factors A and B. Finally, subjects vary within groups as a function of individual differences and measurement error. Therefore, in two-factor analysis of variance, subjects' scores differ from one another because of group differences due to factor A (SS_A), factor B (SS_B), the interaction of factors A and B (SS_{AB}), and individual differences and measurement error (SS_{within}). In symbols,

$$SS_{total} = SS_A + SS_B + SS_{AB} + SS_{within}$$

Consider a numerical example as an illustration of these sources of variability in score value. Suppose Bill is a male in the experiment described above and has been shown the movie in which the adult's aggressive behavior was rewarded. Referring to Table 12–2, which shows the means for the groups in this study, one can see that Bill is a member of the male-reward group, located in the top left corner. Suppose Bill's score is 28. Why did Bill score 28 and not the grand mean, 13? To what shall we attribute the difference between any score X_{ijk} and the grand mean $\overline{X}_{..}$? In Bill's case,

$$\overline{X}_{ijk} - \overline{X}_{..} = 28 - 13 = 15$$

First, Bill is male and the males in the study tended to score a little higher than the females. Specifically, according to Table 12–2, males averaged 15 whereas the grand mean over all subjects was 13. Therefore, one source of variability resides in the difference between the mean of level a_1 of factor A ($\overline{X}_{1.}$) and the grand mean ($\overline{X}_{..}$). In Bill's case,

$$\left(\overline{X}_{1.} - \overline{X}_{..}\right) = 15 - 13 = 2$$

which suggests that 2 points of the 15-point difference between Bill and the grand mean is because of factor A—Bill is male.

A second source of variability is the fact that Bill saw the model rewarded while some other children saw the same behavior punished. Notice in Table 12–2 that the reward condition averaged 22 relative to the grand mean of 13. Therefore, another source of variability is the difference between the mean of level b_1 of factor B ($\overline{X}_{.1}$) and the grand mean ($\overline{X}_{..}$). For Bill,

$$\left(\overline{X}_{.1} - \overline{X}_{..}\right) = 22 - 13 = 9$$

which suggests that 9 points of the 15-point difference between Bill and the grand mean comes from factor B—Bill is in the reward group.

Thus, being male and seeing the model rewarded would explain $2 + 9 = 11$ points of the 15-point difference between Bill's score and the grand mean. But perhaps there is something special about the particular combination of being male and seeing the model rewarded that contributes to the scores of individuals in this group. According to the calculations just performed, the male-reward condition should have a mean that is 11 points above the grand mean, that is, a mean of $13 + 11 = 24$. But in fact, Table 12–2 indicates that the male-reward group has an average score of 25, 1 point higher than what would be predicted on the basis of differences associated with factor A and factor B. This 1-point difference is the variability associated with the AB interaction. Generally, the interaction variability comes from the difference between the particular group mean and the grand mean ($\overline{X}_{jk} - \overline{X}_{..}$) minus the effects for factor A ($\overline{X}_{j.} - \overline{X}_{..}$) and factor B ($\overline{X}_{.k} - \overline{X}_{..}$). Algebraically, this reduces to

$$\begin{aligned}
&(\overline{X}_{jk} - \overline{X}_{..}) - (\overline{X}_{j.} - \overline{X}_{..}) - (\overline{X}_{.k} - \overline{X}_{..}) \\
&\overline{X}_{jk} - \overline{X}_{..} - \overline{X}_{j.} + \overline{X}_{..} - \overline{X}_{.k} + \overline{X}_{..} \\
&\overline{X}_{jk} - \overline{X}_{j.} - \overline{X}_{.k} + \overline{X}_{..}
\end{aligned}$$

Applied to Bill,

$$25 - 15 - 22 + 13 = 1$$

which suggests that 1 point of the 15-point difference between Bill's score and the grand mean is associated with the specific combination of being a male who saw aggression rewarded—namely, the AB interaction.

The combination of the differences associated with factor A ($+2$), with factor B ($+9$), and with the AB interaction ($+1$) in conjunction with the grand mean (13) now accounts for the mean of 25 in Bill's male-reward group:

$$2 + 9 + 1 + 13 = 25$$

But not everyone in that group had a score of 25. Bill actually scored 28.

Therefore, the final reason subjects score differently from one another resides in individual differences and measurement error. This source of variability is reflected in the difference between an individual subject's score and the group mean ($X_{ijk} - \overline{X}_{jk}$), which in Bill's case was $28 - 25 = 3$.

Applying this discussion to all subjects, individuals vary in score value ($X_{ijk} - \overline{X}_{..}$) as a function of which level of factor A a subject is in ($\overline{X}_{j.} - \overline{X}_{..}$), which level of factor B ($\overline{X}_{.k} - \overline{X}_{..}$), which combination of factors A and B ($\overline{X}_{jk} - \overline{X}_{j.} - \overline{X}_{.k} + \overline{X}_{..}$), and individual differences and measurement error that produce differences between individual subjects and their respective group means ($X_{ijk} - \overline{X}_{jk}$). Algebraically, the sentence can be expressed as follows:

$$(X_{ijk}-\overline{X}_{..})=(\overline{X}_{j.}-\overline{X}_{..})+(\overline{X}_{.k}-\overline{X}_{..})+(\overline{X}_{jk}-\overline{X}_{j.}-\overline{X}_{.k}+\overline{X}_{..})+(X_{ijk}-\overline{X}_{jk})$$

$$\text{Total} = A + B + AB + \text{within}$$

In Bill's case,

$$28 - 13 = (15-13) + (22 - 13) + (25 - 15 - 22 + 13) + (28 - 25)$$
$$15 = 2 + 9 + 1 + 3$$
$$15 = 15$$

Sums of squares   Variability is expressed in terms of *squared* deviations summed over all subjects—that is, in terms of sums of squares. When the algebraic statement above is squared and summed over all subjects, the left side becomes $\sum_{i=1}^{n} \sum_{j=1}^{p} \sum_{k=1}^{q} (X_{ijk} - \overline{X}_{..})^2$ which is the total sum of squares, SS_{total}. When the right side is squared and summed, all the cross product terms cancel out and the expression reduces to

$$SS_{total} = SS_A + SS_B + SS_{AB} + SS_{within}$$

This is what is meant by partitioning. The total sum of squares is divided into separate components that sum to SS_{total} and represent different sources of variability (factor A, factor B, AB interaction, within groups).

Degrees of freedom   As before, the mean squares are obtained by dividing each sum of squares by its degrees of freedom. The total sum of squares involves the squared deviation of each score from the grand mean. Thus, the number of degrees of freedom for SS_{total} will be one less than the total number of observations in the entire design. If there are n subjects per group and pq groups, the total number of subjects is npq, which may be abbreviated by N. Therefore,

$$df_{total} = npq - 1 = N - 1$$

The sum of squares for factor A involves the squared deviation from the grand mean of each of the means for the levels of factor A. There are p such means, so the degrees of freedom for the Factor A variance estimate is

$$df_A = p - 1$$

Similarly, factor B has q levels, so the degrees of freedom is

$$df_B = q - 1$$

Determining the degrees of freedom for interaction is less intuitively clear. This sum of squares involves cell mean differences from a value determined by the grand mean and by the row and column means. Therefore, the task is to determine how many cell means in the table are free to vary given the grand mean and all marginal means. Consider, for example, a 3×2 table:

?	?		10
			20
10	15	20	15

A little thought indicates that once the values for any two cells in a row are established—for example, the ones with question marks—then the means of the remaining cells can be determined. The open cells must have a value that would make the average within that row (or column) equal to the given marginal average. Therefore, in the above example, there are two degrees of freedom for interaction. It happens that the number of cells free to vary is always the product of (the number of rows minus one) and (the number of columns minus one), because one vacant cell in any row or column can be determined. Consequently, the degrees of freedom for interaction is

$$df_{AB} = (p - 1)(q - 1)$$

The sum of squares within groups is composed of the squared deviations of the scores about their own mean. Within any one group, $n - 1$ of those scores are free to vary. Since there are pq groups in the design, the degrees of freedom for SS_{within} is pq times $(n - 1)$:

$$df_{within} = pq(n - 1)$$
$$= pqn - pq$$
$$df_{within} = N - pq$$

Notice that the degrees of freedom for the total equals the sum of the degrees of freedom for the component sources of variability:

$$df_{total} = df_A + df_B + df_{AB} + df_{within}$$

Mean squares   The mean squares for the several sources of variability are given by the sums of squares divided by their respective degrees of freedom:

$$MS_A = \frac{SS_A}{df_A}$$

$$MS_B = \frac{SS_B}{df_B}$$

$$MS_{AB} = \frac{SS_{AB}}{df_{AB}}$$

$$MS_{within} = \frac{SS_{within}}{df_{within}}$$

F ratios   Once the mean squares have been determined, F ratios can be constructed to test the significance of the several possible treatment effects:

$$F_A = \frac{MS_A}{MS_{within}}$$

$$F_B = \frac{MS_B}{MS_{within}}$$

$$F_{AB} = \frac{MS_{AB}}{MS_{within}}$$

In each case, under the null hypothesis the numerator and denominator are both estimates of the same population value. However, the numerators (in contrast to MS_{within}) are sensitive to differences between particular sets of group means. To the extent that these means differ from one another, the numerator of the F ratio will become large relative to its denominator. If the resulting ratio is so large that the probability is very small that an observed value of its size could be reasonably expected under the null hypothesis, then H_0 is rejected and one concludes that the differences between the means of the effect being considered are so great that they are not likely to be merely a result of sampling error.

Numerical illustration of treatment effects

It will be helpful to consider a numerical example showing how the presence of treatment effects, that is, large differences between group and grand means, is reflected in the results of an analysis of variance.

Table 12–4 presents a 2×2 analysis of variance using randomly selected numbers between 0 and 10 as the scores. Since the numbers are random, it can be assumed that the null hypothesis is true. The several mean squares should all estimate the same population variance. These different estimates will probably not equal one another; but if H_0 is true, they should deviate only by sampling error. The lower portion of the table presents the traditional summary of the analysis of variance. According to the F table given in Table E of Appendix II, the critical value for F with 1 and 16 degrees of freedom is 4.49 at the .05 significance level. As shown by the F_{obs} values at the bottom right of the table, none of the effects in this example is significant.

Now consider the analysis presented in Table 12–5. The same numbers are used, except that 10 has been added to each score in both the b_2 groups. This is analogous to introducing a treatment effect for factor B. Observe how the several sums of squares respond to this specific manipulation by comparing them to those found before the treatment effect was introduced (shown in Table 12–4). First, as one might expect, the sum of squares and mean square for factor B increase considerably as a function of this change (and so does the total sum of squares). However, notice that the sums of squares and mean squares for A, AB, and within groups do not change. This example illustrates how a selective introduction of a specific main effect is reflected only in the appropriate sum of squares and mean square.

Next, turn to the analysis presented in Table 12–6. Here the same random numbers used in Table 12–4 (without treatment effects) have been modified by adding 10 to only the scores in the upper right-hand cell, ab_{12}. This treatment should produce an interaction between factors A and B, since the pattern of results for the levels of B depends upon or interacts with the level of A. As anticipated, the analysis shows an increase in the sum of squares for interaction, SS_{AB}. But observe that the sums of squares for A, B, and the total also increase. Only the SS_{within} remains uninfluenced by this change. The inflation of SS_A and SS_B in response to the increment of a single cell results from the fact that such a manipulation does indeed differentially change the means of the respective levels of A and B. The SS_{within} is not altered because adding a constant to any group of scores does not change the variability of those particular scores about their mean. This example also illustrates that significant main effects may not be very meaningful when the interaction is significant.

These comparisons demonstrate how the analysis of variance is sensitive to different types of treatment effects. The tables show how SS_A and SS_B

12–4 Two-Factor Analysis of Variance with No Effects

		Factor B		
		b_1	b_2	
Factor A	a_1	5 4 3 4 2 $\overline{X}_{11} = 3.6$	8 9 4 2 5 $\overline{X}_{12} = 5.6$	$\overline{X}_{1.} = 4.6$
	a_2	6 7 5 8 4 $\overline{X}_{21} = 6.0$	6 9 5 9 3 $\overline{X}_{22} = 6.4$	$\overline{X}_{2.} = 6.2$
		$\overline{X}_{.1} = 4.8$	$\overline{X}_{.2} = 6.0$	$\overline{X}_{..} = 5.4$

$$SS_A = nq \sum_{j=1}^{p} (\bar{X}_{j.} - \bar{X}_{..})^2$$
$$= (5)(2)[(4.6 - 5.4)^2 + (6.2 - 5.4)^2]$$
$$SS_A = 12.80$$

$$SS_B = np \sum_{k=1}^{q} (\bar{X}_{.k} - \bar{X}_{..})^2$$
$$= (5)(2)[(4.8 - 5.4)^2 + (6.0 - 5.4)^2]$$
$$SS_B = 7.20$$

Interaction

$$SS_{AB} = n \sum_{j=1}^{p} \sum_{k=1}^{q} (\bar{X}_{jk} - \bar{X}_{j.} - \bar{X}_{.k} + \bar{X}_{..})^2$$

$$= 5[(3.6 - 4.6 - 4.8 + 5.4)^2 + (5.6 - 4.6 - 6.0 + 5.4)^2$$
$$+ (6.0 - 6.2 - 4.8 + 5.4)^2 + (6.4 - 6.2 - 6.0 + 5.4)^2]$$

$$SS_{AB} = 3.20$$

Within Groups

$$SS_{within} = \sum_{i=1}^{n} \sum_{j=1}^{p} \sum_{k=1}^{q} (X_{ijk} - \bar{X}_{jk})^2$$

$$SS_{within} = 75.60$$

Source	df		SS	$MS = \frac{SS}{df}$	$F = \frac{MS}{MS_{within}}$
A	$p - 1 =$	1	12.80	12.80	2.71
B	$q - 1 =$	1	7.20	7.20	1.52
AB	$(p-1)(q-1) =$	1	3.20	3.20	.68
Within	$N - pq =$	16	75.60	4.72	
Total	$N - 1 =$	19	98.80		

12–5 Two-Factor Design with a Treatment Effect Added to Level b_2

	Factor B		
	b_1	b_2 (+ 10)	
a_1	5	18	
	4	19	
	3	14	$\bar{X}_{1.} = 9.6$
	4	12	
	2	15	
	$\bar{X}_{11} = 3.6$	$\bar{X}_{12} = 15.6$	
a_2	6	16	
	7	19	
	5	15	$\bar{X}_{2.} = 11.2$
	8	19	
	4	13	
	$\bar{X}_{21} = 6.0$	$\bar{X}_{22} = 16.4$	
	$\bar{X}_{.1} = 4.8$	$\bar{X}_{.2} = 16.0$	$\bar{X}_{..} = 10.4$

$$SS_A = nq \sum_{j=1}^{p} (\bar{X}_{j.} - \bar{X}_{..})^2$$

$$= (5)(2)\left[(9.6 - 10.4)^2 + (11.2 - 10.4)^2\right]$$

$$SS_A = 12.80$$

$$SS_B = np \sum_{k=1}^{q} (\bar{X}_{.k} - \bar{X}_{..})^2$$

$$= (5)(2)\left[(4.8 - 10.4)^2 + (16.0 - 10.4)^2\right]$$

$$SS_B = 627.20$$

Interaction

$$SS_{AB} = n \sum_{j=1}^{p} \sum_{k=1}^{q} (\bar{X}_{jk} - \bar{X}_{j.} - \bar{X}_{.k} + \bar{X}_{..})^2$$

$$= 5[(3.6 - 9.6 - 4.8 + 10.4)^2 + (15.6 - 9.6 - 16.0 + 10.4)^2$$
$$+ (6.0 - 11.2 - 4.8 + 10.4)^2 + (16.4 - 11.2 - 16.0 + 10.4)^2]$$

$$SS_{AB} = 3.20$$

Within Groups

$$SS_{within} = \sum_{i=1}^{n} \sum_{j=1}^{p} \sum_{k=1}^{q} (X_{ijk} - \bar{X}_{jk})^2$$

$$SS_{within} = 75.60$$

Source	df		SS	$MS = \dfrac{SS}{df}$	$F = \dfrac{MS}{MS_{within}}$
A	$p - 1 =$	1	12.80	12.80	2.71
B	$q - 1 =$	1	627.20	627.20	132.88
AB	$(p - 1)(q - 1) =$	1	3.20	3.20	.68
Within	$N - pq =$	16	75.60	4.72	
Total	$N - 1 =$	19	718.80		

12-6 Two-Factor Design with a Treatment Effect Added to Cell ab_{12}

Factor B

	b_1	b_2	
a_1	5 4 3 4 2 $\bar{X}_{11}=3.6$	(+10) 18 19 14 12 15 $\bar{X}_{12}=15.6$	$\bar{X}_{1.}=9.6$
a_2	6 7 5 8 4 $\bar{X}_{21}=6.0$	6 9 5 9 3 $\bar{X}_{22}=6.4$	$\bar{X}_{2.}=6.2$
	$\bar{X}_{.1}=4.8$	$\bar{X}_{.2}=11.0$	$\bar{X}_{..}=7.9$

Factor A

$$SS_A = nq \sum_{j=1}^{p} (\bar{X}_{j.} - \bar{X}_{..})^2$$

$$= (5)(2)\left[(9.6 - 7.9)^2 + (6.2 - 7.9)^2\right]$$

$$SS_A = 57.80$$

$$SS_B = np \sum_{k=1}^{q} (\bar{X}_{.k} - \bar{X}_{..})^2$$

$$= (5)(2)[(4.8 - 7.9)^2 + (11.0 - 7.9)^2]$$

$$SS_B = 192.20$$

Interaction

$$SS_{AB} = n \sum_{j=1}^{p} \sum_{k=1}^{q} (\bar{X}_{jk} - \bar{X}_{j.} - \bar{X}_{.k} + \bar{X}_{..})^2$$

$$= 5[(3.6 - 9.6 - 4.8 + 7.9)^2 + (15.6 - 9.6 - 11.0 + 7.9)^2$$
$$+ (6.0 - 6.2 - 4.8 + 7.9)^2 + (6.4 - 6.2 - 11.0 + 7.9)^2]$$

$$SS_{AB} = 168.20$$

Within Groups

$$SS_{within} = \sum_{i=1}^{n} \sum_{j=1}^{p} \sum_{k=1}^{q} (X_{ijk} - \bar{X}_{jk})^2$$

$$SS_{within} = 75.60$$

Source	df		SS	$MS = \frac{SS}{df}$	$F = \frac{MS}{MS_{within}}$
A	$p - 1 =$	1	57.80	57.80	12.25
B	$q - 1 =$	1	192.20	192.20	40.72
AB	$(p-1)(q-1) =$	1	168.20	168.20	35.64
Within	$N - pq =$	16	75.60	4.72	
Total	$N - 1 =$	19	493.80		

reflect differences in the marginal means, how SS_{AB} is changed when some but not all cells within a single level of one factor are altered, and how SS_{within} remains uninfluenced by changes in cell means or level means. Consequently, if the null hypothesis is true and the groups are all random samples having the same population mean, these sums of squares all estimate the same value. However, to the extent that treatment effects exist for A, B, and/or AB, their respective mean squares will be inflated. But the mean square within groups will not be altered because it is not sensitive to differences between cell or level means. Consequently, the three F ratios, $\dfrac{MS_A}{MS_{within}}$, $\dfrac{MS_B}{MS_{within}}$, and $\dfrac{MS_{AB}}{MS_{within}}$, will test the existence of treatment effects in the population.

Assumptions and conditions underlying two-factor analysis of variance

Two-factor analysis of variance makes the same four assumptions and conditions that are required in a single-factor analysis (and for the same reasons), plus several additional ones.

First, the subjects must be **randomly** and **independently** sampled. Second, the groups in the design must be **independent** from one another. These assumptions are necessary so that the variability within one group is not correlated with the variability within another group. While all the analysis-of-variance procedures described in this text assume independent groups, techniques do exist for handling situations in which the groups are not independent, such as a before-after experiment in which the same subjects are measured before and then after a special treatment is administered.[6]

Third, the population distributions of scores for each group in the design are **normal** in form. The assumption of normality is made so that the variance estimates in each F ratio will be independent, and so that the probabilities of the F distribution will be appropriate. Fourth, it is assumed that the populations from which the groups are drawn have equal variances. **Homogeneity of variance** must be assumed so that MS_{within} will represent an appropriate pooling of the variability within each group in the design.

Fifth, the procedures require that the factors involved in the design be **fixed**. There are two types of factors, **random** and **fixed**. A random factor is one for which the levels are randomly selected. For example, suppose a researcher is interested in the effects of a new teaching program versus an old teaching program at schools in a particular school system. If the school

---

[6]See: "Repeated Measures Designs," in Winer, *Statistical Principles*; R. B. McCall and M. I. Appelbaum, "Bias in the Analysis of Repeated-Measures Designs: Some Alternative Approaches," *Child Development* 44 (1973): 401–15.

system is very large, perhaps not all the schools will be analyzed. Instead, three of them will be *randomly* selected, and they will compose the three levels of factor A (schools) in the analysis. In this case, factor A (schools) would be a random factor. However, if the researcher designates which particular schools are to be entered into the analysis, perhaps because of their size, socioeconomic makeup, or geographical location, then these schools would not be random selections from the population of schools but would represent a fixed set of schools. Factor A (schools) would then be a fixed factor. It is possible to have a two-factor analysis of variance in which one factor is fixed and the other is random. This is called a **mixed model**. The procedures outlined in this text are applicable only to the **fixed model** (both factors are fixed), since this is the most common of the three designs. Techniques for analyzing the random and mixed models can be found elsewhere.[7]

Finally, the techniques outlined here have assumed that there are an **equal number of cases** in each group and that there is **more than one observation per cell**. Again, procedures do exist for designs having unequal n.[8]

COMPUTATIONAL PROCEDURES

General format

Just as with the techniques discussed earlier, the definitional formulas for the two-factor analysis of variance are not convenient for computational purposes. A more satisfactory set of computational procedures is outlined in Table 12–7. Recall that n = the number of subjects in each group, N = the total number of subjects in the design, p = the number of levels of factor A, q = the number of levels of factor B, and X_{ijk} = the score for the ith subject in the jth level of A and kth level of B.

Section A of Table 12–7 displays the various cell, row, and column totals (sums). T_{jk} is the total for the jkth cell, $T_{j.}$ is the total for the jth level of A summed over all levels of B, $T_{.k}$ is the total for the kth level of B summed over all levels of A, and $T_{..}$ is the grand total accumulated over all observations in all conditions. If the data are laid out in rows and columns, all that has been done so far is to obtain the sum of scores within each group (T_{jk}), for rows ($T_{j.}$), for columns ($T_{.k}$) and for the total design ($T_{..}$).

Section B presents five intermediate quantities that facilitate the computation of the analysis of variance. Quantity (I) is the grand total squared then divided by the total N in the design. Quantity (II) is simply the sum of all squared scores. Quantity (III) is the sum of the squared totals for each

[7]Winer, *Statistical Principles*.
[8]Ibid.

12–7 Computational Formulas for Two-Factor Analysis of Variance

A. Table of Totals

| | | Factor B | | | | Row |
		b_1	b_2	$\cdots$	b_q	Totals
	a_1	T_{11}	T_{12}	$\cdots$	T_{1q}	$T_{1.}$
	a_2	T_{21}	T_{22}	$\cdots$	T_{2q}	$T_{2.}$
Factor A	$\vdots$	$\vdots$	$\vdots$	$\vdots$	$\vdots$	$\vdots$
	a_p	T_{p1}	T_{p2}	$\cdots$	T_{pq}	$T_{p.}$
Column Totals		$T_{.1}$	$T_{.2}$	$\cdots$	$T_{.q}$	$T_{..}$

B. Intermediate Quantities

$$\text{(I)} \ \frac{T_{..}^{2}}{N} \qquad \text{(II)} \ \sum_{i=1}^{n}\sum_{j=1}^{p}\sum_{k=1}^{q} X_{ijk}^{2} \qquad \text{(III)} \ \frac{\sum_{j=1}^{p} T_{j.}^{2}}{nq}$$

$$\text{(IV)} \ \frac{\sum_{k=1}^{q} T_{.k}^{2}}{np} \qquad \text{(V)} \ \frac{\sum_{j=1}^{p}\sum_{k=1}^{q} T_{jk}^{2}}{n}$$

C. Formulas and Summary Table

Source	df	SS	MS	F
A	$p-1$	(III) − (I)	$\dfrac{SS_A}{df_A}$	$\dfrac{MS_A}{MS_{\text{within}}}$
B	$q-1$	(IV) − (I)	$\dfrac{SS_B}{df_B}$	$\dfrac{MS_B}{MS_{\text{within}}}$
AB	$(p-1)(q-1)$	(V) + (I) − (III) − (IV)	$\dfrac{SS_{AB}}{df_{AB}}$	$\dfrac{MS_{AB}}{MS_{\text{within}}}$
Within	$N-pq$	(II) − (V)	$\dfrac{SS_{\text{within}}}{df_{\text{within}}}$	
Total	$N-1$	(II) − (I)		

level of factor A divided by nq, and quantity (IV) is the sum of the squared totals for each level of factor B divided by np. Quantity (V) is the sum of the squared totals for all the cells divided by n.

Section C presents the formulas for obtaining all the necessary entries in the analysis of variance summary table. Notice that the sums of squares are computed quite easily by adding and subtracting the intermediate quantities. The sum of the degrees of freedom and the sum of the SS should equal df_{total} and SS_{total}, respectively.

Formal example

The relative contribution of heredity and environment to behavior has been a controversial topic for many years. The most common position is called interactionism. This theory states that heredity and environment work in concert. One cannot talk about the role of heredity without considering the environment and vice versa; heredity and environment always interact to produce the behavior we observe.

One illustration of interactionism is that the same environmental experience affects different species in different ways. For example, consider an experiment in which 12 dogs from each of four different breeds (basenji, shetland sheepdog, wire-haired fox terrier, beagle) were either indulged or disciplined between the third and eighth week of their lives.[9] The indulged animals were encouraged in play, aggression, and climbing on their caretaker. In contrast, the disciplined dogs were restrained on their handler's lap, taught to sit, stay, come on command, etc. The indulged-disciplined treatment was inspired by reports that overindulged children often cannot inhibit their impulses in structured situations. Consequently, the test of the effects of these treatments was to take each animal into a room with the handler and a bowl of meat. The dog was hungry, but the handler prevented it from eating for three minutes by hitting the animal on the rump with a rolled newspaper and shouting no. After this period of restraint, the handler left the room, and the length of time it took the dog to begin to eat the meat (latency) was recorded. Presumably, if the observations on children translate to dogs, the indulged animals should go to the food more quickly (record shorter latencies) than the disciplined dogs.

Analysis   A simplified set of hypothetical data is presented in section A of Table 12–8. Factor A (rearing condition) has two levels, indulged and disciplined. Factor B is breed and has four levels. Note that the levels of factor B constitute separate categories, not a continuum. Consider some

[9]Inspired by, but not identical to, D. G. Freedman, "Constitutional and Environmental Interactions in Rearing of Four Breeds of Dogs," *Science* 127 (1958): 585–86.

12–8 Computational Example of a Two-Factor Analysis of Variance (Rearing × Breed Example)

A. Raw Data

		Factor B (Breed)			
		b_1 (Basenjis)	b_2 (Shetland)	b_3 (Terriers)	b_4 (Beagles)
Factor A (Rearing)	a_1 (Indulged)	1 4 3 1 2 2	7 10 10 9 6 8	6 9 7 8 5 10	9 7 10 10 8 9
	a_2 (Disciplined)	5 1 4 1 2 3	9 9 8 10 5 8	1 0 3 1 2 4	2 6 3 4 5 3

B. Table of Totals

	b_1	b_2	b_3	b_4	
a_1	$T_{11}=(1+\cdots+2)$ $=13$	$T_{12}=(7+\cdots+8)$ $=50$	$T_{13}=(6+\cdots+10)$ $=45$	$T_{14}=(9+\cdots+9)$ $=53$	$T_{1.}=161$
a_2	$T_{21}=(5+\cdots+3)$ $=16$	$T_{22}=(9+\cdots+8)$ $=49$	$T_{23}=(1+\cdots+4)$ $=11$	$T_{24}=(2+\cdots+3)$ $=23$	$T_{2.}=99$
	$T_{.1}=29$	$T_{.2}=99$	$T_{.3}=56$	$T_{.4}=76$	$T_{..}=260$

C. Intermediate Quantities

(I) $\dfrac{T_{..}^{2}}{N} = \dfrac{(260)^{2}}{48} = 1408.3333$

(II) $\displaystyle\sum_{i=1}^{n}\sum_{j=1}^{p}\sum_{k=1}^{q} X_{ijk}^{2} = (1^{2} + 4^{2} + 3^{2} + \cdots + 5^{2} + 3^{2}) = 1896.0000$

(III) $\dfrac{\displaystyle\sum_{j=1}^{p} T_{j.}^{2}}{nq} = \dfrac{(161^{2} + 99^{2})}{(6)(4)} = 1488.4167$

(IV) $\dfrac{\displaystyle\sum_{k=1}^{q} T_{.k}^{2}}{np} = \dfrac{(29^{2} + \cdots + 76^{2})}{(6)(2)} = 1629.5000$

(V) $\dfrac{\displaystyle\sum_{j=1}^{p}\sum_{k=1}^{q} T_{jk}^{2}}{n} = \dfrac{(13^{2} + 16^{2} + \cdots + 53^{2} + 23^{2})}{6} = 1801.6667$

D. Sums of Squares and Degrees of Freedom

$SS_A = \textbf{(III)} - \textbf{(I)} = 1488.4167 - 1408.3333 = 80.0834$
$df_A = p - 1 = 2 - 1 = 1$

$SS_B = \textbf{(IV)} - \textbf{(I)} = 1629.5000 - 1408.3333 = 221.1667$
$df_B = q - 1 = 4 - 1 = 3$

$SS_{AB} = \textbf{(V)} + \textbf{(I)} - \textbf{(III)} - \textbf{(IV)} = 1801.6667 + 1408.3333$
$\qquad\qquad - 1488.4167 - 1629.5000 = 92.0833$
$df_{AB} = (p - 1)(q - 1) = (1)(3) = 3$

$SS_{\text{within}} = \textbf{(II)} - \textbf{(V)} = 1896.0000 - 1801.6667 = 94.3333$
$df_{\text{within}} = N - pq = 48 - (2)(4) = 40$

$SS_{\text{total}} = \textbf{(II)} - \textbf{(I)} = 1896.0000 - 1408.3333 = 487.6667$
$df_{\text{total}} = N - 1 = 48 - 1 = 47$

E. Summary Table

Source	df	SS	MS	F
A (Rearing)	1	80.0834	80.0834	33.96**
B (Breed)	3	221.1667	73.7222	31.26**
AB (Rearing × Breed)	3	92.0833	30.6944	13.02**
Within	40	94.3333	2.3583	
Total	47	487.6667		

** $p < .01$

possible results. It may be that the disciplined dogs will refrain from eating the food while the indulged will feast quickly. In this case, one would expect that the analysis of variance would yield a significant effect for factor A. Another result might be that regardless of rearing condition, some breeds generally go to the food more rapidly than others, in which case there would be a significant effect for factor B. Last, the effect of indulged or disciplined rearing may depend on the breed. For one breed the rearing treatment might make a difference, whereas for another it might not (or it might even have an opposite effect). In that event, the analysis would yield a significant interaction between rearing and breed. If an interaction of this form turned out to be significant, then any main effects for rearing and breed would be reevaluated and perhaps disregarded. Section B of Table 12–8 gives the totals for each cell of the design and the marginal totals. Section C presents the computation of the intermediate quantities, Section D displays the calculation of the sums of squares and degrees of freedom, and Section E gives the summary of the analysis.

In the present example, the critical values of F for 3 and 40 degrees of freedom are 2.84 and 4.31 for tests at the .05 and .01 levels, respectively. The observed F for interaction is 13.02, a value that exceeds the critical level at $p < .01$. The F's for the main effects have been calculated for completeness of presentation. A formal summary of this analysis is presented in Table 12–9. Since there are three sources of variability to be tested, there are three sets of hypotheses, critical values, decision rules, and so on. Moreover, notice that the hypotheses are now stated in terms of **treatment effects** as a notational convenience. Observe that the symbol α_1 represents the difference in the population between the mean for a_1 and the grand mean: $\alpha_1 = \mu_{a_1} - \mu$. Under the null hypothesis, the population means for a_1 and a_2 are presumed to be equal to each other and thus to the population grand mean. Therefore, since $\mu_{a_1} = \mu$ and $\mu_{a_2} = \mu$, each treatment effect for factor A will be zero, and consequently the null hypothesis can be written H_0: $\alpha_1 = \alpha_2 = 0$ with the alternative being H_1: not H_0. The same logic applies to the hypotheses for factor B and the AB interaction.

A plot of the data, presented in Figure 12–2, will help to interpret the significant interaction. Note that the graph is not a polygon but a **bar graph** consisting of bars for the indulged and disciplined groups within each breed. The vertical height of the bars represents the mean length of latency to eat the meat in the test situation. A tall bar indicates a long hesitation. The bar graph, rather than connected points, is used in this context because the abscissa (breed) represents discrete categories (i.e., a nominal scale) and not points along a continuum.

An interaction implies that differences between levels of one factor are not the same within each level of the other factor. In terms of the example, the difference between indulged and disciplined dogs is not the same within each breed of animal. Notice in the graph that the basenjis ate the food

12–9 Summary of the Two-Factor Analysis
of Variance (Rearing × Breed Example)

Hypotheses

Factor A	Factor B
H_0: $\alpha_1 = \alpha_2 = 0$	H_0: $\beta_1 = \beta_2 = \beta_3 = \beta_4 = 0$
H_1: not H_0	H_1: not H_0

AB Interaction

H_0: $\alpha\beta_{11} = \alpha\beta_{12} = \cdots = \alpha\beta_{24} = 0$
H_1: not H_0

Assumptions and Conditions
1. The subjects are **randomly** and **independently** sampled.
2. The groups are **independent**.
3. The population variances for the groups are **homogeneous**.
4. The population distributions are **normal** in form.
5. The factors in the study are **fixed**.
6. The number of observations in each group is **equal** and **greater than 1**.

Decision Rules (.05 level; from Table E)

	Factor A ($df = 1/40$)	Factor B ($df = 3/40$)	AB Interaction ($df = 3/40$)
Do not reject H_0 if:	$F_{obs} < 4.08$	$F_{obs} < 2.84$	$F_{obs} < 2.84$
Reject H_0 if:	$F_{obs} \geq 4.08$	$F_{obs} \geq 2.84$	$F_{obs} \geq 2.84$

Computation

See Table 12–8

$F_{obs} = 33.96$	$F_{obs} = 31.26$	$F_{obs} = 13.02$

Decision

Since the F for interaction exceeds the critical level, reject H_0. It is concluded that the effects of indulgent and disciplined rearing are not the same for each breed of dog.

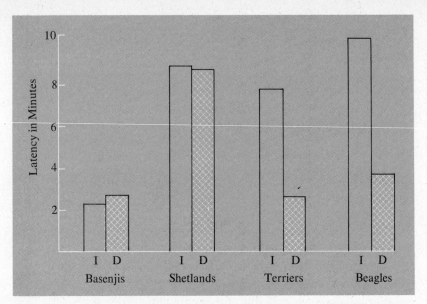

Fig. 12–2. Results in which a significant interaction occurred for the
rearing (indulged, disciplined) × breed experiment

quickly, regardless of whether they were indulged or disciplined. The shetlands also showed little difference between rearing conditions but seemed to
begin eating much later. In contrast, the rearing conditions did seem to affect
the terriers and beagles: The indulged dogs of these two breeds took longer to
go to the food than their disciplined companions—behavior opposite to that
widely suspected of indulged children.

Individual comparisons The significant interaction means that the
indulged-disciplined manipulation affected some breeds differently from
others. It is tempting to conclude more specifically that the rearing variable
affected terriers and beagles but not basenjis or shetlands. Technically,
however, the significant interaction does not apply to particular pairs of
comparisons within the total design. Rather, it indicates that the rearing
treatment had different effects on the set of breeds observed, all four breeds
considered at once.

   If the researcher had reason to be particularly concerned with specific
comparisons within a breed prior to the time of the experiment, the difference
between means could have been tested with a *t* test or other procedure for
making **a priori** (before the fact) comparisons within the total design. However, a theory or other justification for planning to examine specific pairs of
means within a total design is often not available prior to the experiment, and
the researcher decides to compare specific pairs only after the results are
available. As in the example described, the scientist performing the study may

have no reason before the experiment for testing the difference between indulged and disciplined animals within any one particular breed. However, after the data are collected, it is noted that there has been a significant interaction and that the graph suggests the rearing variable has affected the terriers and beagles but not the others. At this point, the researcher might wonder if there would be significant differences between the means for each of these two breeds. However, if a simple *t* test is performed now between these pairs of means, the resulting probability levels will not be appropriate. They are not correct because the researcher would probably not have tested them if the interaction had not been significant or if the graph had not suggested a difference for some breeds and not for others. In the group of such **a posteriori** (after the fact) comparisons that this researcher might make, there would be a high number of significant differences even if the null hypothesis of no difference is actually true, since the tests would be made only after the data suggested a significant difference in a pair. It is in this sense that a posteriori tests capitalize on chance, and thus some correction must be made. The theory behind choosing which correction to use in such after-the-fact comparisons is beyond the scope of this text, but the interested reader is referred to Winer's more advanced book on the analysis of variance.[10]

Interpretation   The preceding example revealed a significant effect for the interaction between rearing treatment and breed of dog. Judging from the graph of this interaction, one might conclude that the difference in rearing treatment affects some breeds but not others. The fact that a significant difference was found indicates that the observed differences in group means are sufficiently large that it is likely that they did not occur simply because of sampling error. Nevertheless, the statistical procedures themselves do not guarantee that these factors *caused* the observed differences. The analysis only reports that the difference observed probably was not random. The design and execution of the experiment, not the statistics, comment on the possible causal effects of the variables studied.

For example, quite sizable and significant effects might have been produced not by breed or rearing but by the behavior of the experimenter when preventing the dog from eating during the first three minutes of the test situation. Though it was not actually the case, suppose for the sake of the discussion that the experimenter tested all the shetlands first and punished the dogs very severely for going near the food. As a result, the dogs of both rearing groups tended to avoid the food, some for the entire ten-minute test situation. Feeling that the punishment was too severe, the experimenter could have tested the basenjis next and unconsciously reduced the level of punishment. However, since all the basenjis ate very quickly and since the experimenter sensed that none of the hypothesized rearing differences seemed to be

[10]Winer, *Statistical Principles*.

appearing, it is possible that the experimenter might have subsequently punished the remaining disciplined dogs harder than the indulged animals and thus produced the rearing differences reported. Although this example is somewhat contrived, experimenter bias has been found to influence the outcomes of experiments, and it is entirely possible that the effects suggested by the example can be produced by this kind of extraneous variable if appropriate experimental procedures are not followed.

As discussed in the Prologue, statistical procedures cannot discriminate between effects caused by the factors investigated and those caused by other influences that have been allowed to operate. Causal inferences can be made only when the design of the experiment is such that if a difference is found, there exists no possible cause for this difference other than the variables manipulated directly by the researcher. Statistics is a set of methods that serve to describe and quantify data in such a way that inferences may be made by an orderly process; the causes of the observed effects must be clarified through experimental design, not statistics.

EXERCISES

1. Distinguish (in your own words) between a main effect and an interaction. Give examples in terms of common events, and draw graphs to illustrate various combinations of effects.
2. List the assumptions for two-factor analysis of variance and explain why they are necessary.
3. The following is a fairly random set of data:

		Factor B	
		b_1	b_2
		3	5
		6	7
	a_1	5	8
		4	2
Factor A			
		6	8
		9	4
	a_2	4	6
		5	3

Compute the analysis of variance on these data. Now add 5 to each score in the a_2 level and recompute. Explain the differences and

similarities in results between the first and second computations. Now take the original data and add 10 to each score in cell ab_{11} and recompute. Compare all three analyses and explain the similarities and differences.

4. An important social topic is the effects of marijuana. Twenty college men are selected, ten who are regular users of marijuana, ordinary cigarettes, and alcohol and ten who are regular users of cigarettes and alcohol but have never tried marijuana.[11] Within each group, user and nonuser, five subjects smoked two marijuana cigarettes in the laboratory, while five subjects smoked placebo cigarettes that taste and smell like marijuana but have no effect. A short time after smoking, all subjects took a set of perceptual-motor tests of skills required for driving. The researcher wanted to know if

[11]Data and results are hypothetical but represent an extrapolation from a study cautiously reported by A. T. Weil, N. E. Zinberg, and J. M. Nelson, "Clinical and Psychological Effects of Marihuana in Man," *Science* 162 (1968): 1234–42.

marijuana had more of an effect on driving skills than a placebo, if users responded differently from nonusers, and if there was an interaction. The number of errors the subjects made are as follows:

Regular users receiving marijuana
4, 3, 7, 6, 5
Regular users receiving placebo
7, 3, 1, 8, 4
Naive users receiving marijuana
5, 2, 2, 7, 4
Naive users receiving placebo
6, 0, 3, 5, 5

Evaluate these data with the analysis of variance. Draw up a formal summary of the analysis, display all computation, graph any results, and interpret the experiment.

5. Suppose the same researchers were interested in comparing the high achieved by subjects who are regular users versus those who are naive to marijuana, relative to subjects who are given placebos. Twenty-four subjects participated, twelve who were regular users and twelve who were not (naive). Half of each group smoked two marijuana cigarettes in the laboratory and half smoked two placebos. An elaborate system of questions and tasks was given a short time after the smoking took place, and the resultant score reflected the degree of high as reported by the subject and as reflected in the clinical impressions of the attending physician. The greater the score, the greater the high. Following the procedures of the previous question, evaluate and interpret the following data:

Regular users receiving marijuana
7, 8, 8, 9, 7, 10
Regular users receiving placebo
3, 5, 6, 4, 5, 6
Naive users receiving marijuana
3, 5, 4, 4, 6, 2
Naive users receiving placebo
5, 4, 6, 3, 2, 3

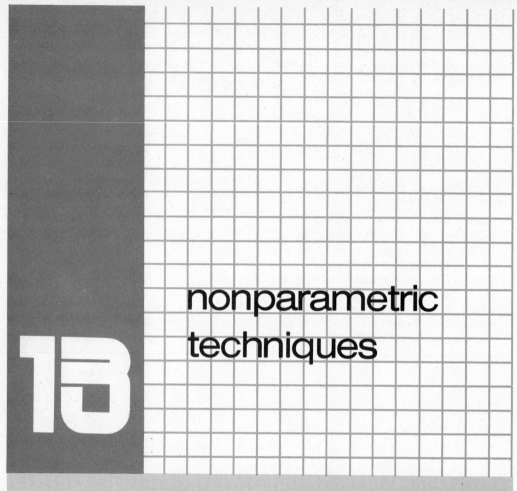

13

nonparametric techniques

MANY HYPOTHESIS-TESTING TECHNIQUES have been presented in the preceding chapters. In order to use these tests appropriately, several assumptions must be met concerning the population distributions. For example, the t test of the difference between two independent means requires the assumption that the two population distributions are normal and have the same variance. Since one almost never has precise knowledge about the population, one can only guess about the tenability of some of these assumptions. Furthermore, there are times when certain assumptions simply cannot be met. For example, if a very easy examination is given to a class, there may be many scores of 100% and only a few scores as low as 80%. This distribution would be markedly skewed to the left and decidedly not normal. Therefore, there is a need for statistical techniques that may be used when some of the assumptions required by the previously described procedures cannot be met.

PARAMETRIC AND NONPARAMETRIC TESTS

The statistical techniques described in the previous chapters represent tests on the values of certain parameters (e.g., μ or ρ) and have made certain assumptions about other parameters (e.g., $\sigma_1^2 = \sigma_2^2 = \cdots = \sigma_p^2$). Consequently, they are known collectively as **parametric** statistical tests. Methods that do not test hypotheses about specific parameters and require different (and sometimes fewer) assumptions are known as **nonparametric** tests. They are also called **distribution-free** tests, since they do not require the assumption that scores are from a normal distribution.

Although nonparametric tests are frequently used because certain assumptions cannot be made about the populations involved, a researcher may select a nonparametric technique to analyze data for other reasons as well. Sometimes the variables in question are measured with ordinal or even nominal scales (see Chapter 1). Parametric tests are not usually appropriate in such cases. For example, suppose an educator wants to correlate the scores from a new test of reading achievement with a teacher's rank ordering of pupils on reading proficiency. The teacher's rankings represent an ordinal scale, and a nonparametric index of the degree of relationship would be more appropriate than the Pearson r. Thus, when measurement is nominal or ordinal in character, a nonparametric test may be the only choice.

If nonparametric techniques do not require all the assumptions that parametric techniques do, why are they not always used? In fact, parametric procedures are used more often than nonparametric ones and are traditionally preferred when everything else is equal. One reason is that parametric tests are robust with respect to violations of some of their assumptions. That is, failure to have a perfectly normal distribution is not very damaging to the accuracy of the probability values obtained with the t test or the analysis of variance. Even rather substantial departures from normality have a relatively minor impact on the result of the test, especially if the sample size is large. Similar statements can be made about the robustness of these techniques when group variances are not equal. Consequently, *moderate* violations of the assumptions of normality and homogeneity of variance are often not strong reasons for choosing a nonparametric over a parametric test.

In addition, parametric methods usually have a greater **power-efficiency** than do nonparametric methods. As explained in Chapter 9, the power of a statistical test is the probability that the test will correctly reject the null hypothesis when that hypothesis is, in fact, false (correctly reject H_0).[1] Power-efficiency refers to the power of a test relative to the sample size, and permits one to compare the power of two different statistical tests. If the difference between the central tendency of two groups is being considered, a t test is more likely to detect a population difference than is an appropriate nonparametric test for the given N's.

Another reason parametric methods are often preferred is that they provide information nonparametric methods do not. For example, the two-factor analysis of variance includes a test for interaction. It is more difficult to assess this type of effect with nonparametric methods.

A final consideration is more subtle. Frequently, parametric and non-parametric tests do not address themselves to precisely the same question. They are sensitive to different aspects of the data. For example, if one wants to know whether the mean of one group differs from the mean of another in the population, a parametric test makes a rather direct assessment of this question, given the assumptions. A nonparametric test designed to make a similar evaluation may actually ask whether one distribution is different from another *in any way*, and distributions may differ not only in central tendency but also in variability, skewness, etc. Thus, although parametric and nonpara-metric tests are often intimately related, situations can occur in which a significant difference will be obtained with one technique and not with another because parametric and nonparametric tests often evaluate slightly different aspects of the data.

To summarize, parametric tests possess the advantages of being fairly robust with respect to violations of assumptions, having more power-efficiency (everything else being equal), and sometimes providing more infor-mation about a phenomenon (e.g., interactions in the analysis of variance). However, when departures from normality or homogeneity of variance are severe, or when the data are nominal or ordinal, a nonparametric method may be more appropriate.

---

[1]For a more detailed discussion of power-efficiency, see S. Siegel, *Nonparametric Statistics for the Behavioral Sciences* (New York: McGraw-Hill, 1956).

13–1 Summary of Statistical Tests and Their Distinguishing Characteristics

Tests of the Differences between Groups

Type of Sample	Distribution, Variance, and Minimum Scale Requirements
Two Independent Samples	
t test	Normality, Homogeneity, Interval
Mann-Whitney U test	Ordinal
Two Correlated Samples	
t test	Normality, Interval
Wilcoxon test	Ordinal
Several Independent Samples	
Simple analysis of variance	Normality, Homogeneity, Interval
Two-factor analysis of variance	Normality, Homogeneity, Interval
Kruskal-Wallis test	Ordinal

Tests of Association

Type of Sample	Distribution, Variance, and Minimum Scale Requirements
One Sample	
Pearson product-moment correlation	Normality, Interval
Spearman rank-order correlation	Ordinal
Chi-square test for $r \times c$ tables	Nominal
Two Independent Samples	
Test of the difference between two Pearson correlations	Normality, Interval

The techniques presented in this chapter are samples of some of the most common nonparametric methods. Many others exist and the interested reader is referred to books by Bradley and by Siegel[2] for more complete coverage of this area. Table 13–1 presents a summary of the statistical tests of hypotheses described in this text, organizing them around differences between groups and association; two or more independent or correlated samples; and major distribution (normality), variance (homogeneity of variance), and minimum measurement scale requirements.

TESTS ON INDEPENDENT SAMPLES

Pearson chi-square test

On occasion, the social scientist collects independent and random samples of observations and wants to compare them in terms of the similarity with which those observations are distributed among several discrete and mutually exclusive categories. Suppose that two or more groups of subjects are randomly selected. The groups may be boys and girls, students and nonstudents, athletes and nonathletes, etc. Now, suppose the social scientist has a set of several discrete categories into which any particular subject may be classified. For example, subjects may be judged to have warm, neutral, or cold personalities or to have several levels of agreement or disagreement with a given statement or political figure. No matter what the classification system, the categories must be exhaustive and mutually exclusive: Each subject must belong to one and only one category. A **chi-square test** can be used to analyze data from a study with this design.

Consider two randomly selected samples, one composed of college males and the other of college females. All subjects are given a statement to read describing woman's role in the family. They are then asked if they approve, disapprove, or are neutral in their feeling about the statement. The data from this set of observations might be as follows:

	Men	Women	Total
Approve	58	35	93
Neutral	11	25	36
Disapprove	10	23	33
Total	79	83	162

[2]J. V. Bradley, *Distribution-Free Statistical Tests* (Englewood Cliffs: Prentice-Hall, 1968); Siegel, *Nonparametric Statistics*.

Notice that the two samples, men and women, make up the columns, and the various categories of responses constitute the rows. The samples are independent, and a person's score can be tallied in only one of the table's six cells.

Sometimes this format is called a **contingency table**. The research question is whether men and women differ in their relative distributions of responses to the statement. Another way to phrase this same question is to ask whether the distribution of responses is dependent, or *contingent*, upon the sex of the respondent.

In this and each of the following presentations of a statistical technique, a summary of the procedures is presented in tabular form (e.g., Table 13–2) and its details discussed in the text.

Hypotheses   The statistical question is whether the groups differ in the relative distribution of observations among the different categories. The null hypothesis is that in the population the two groups do not differ in their relative frequency distribution of people among the categories. The alternative hypothesis is that the groups do differ. Notice that H_1 does not specify how they differ, but only that the two relative distributions are not the same in some way.

The hypotheses apply only to a population that possesses characteristics in the proportions specified by the marginal distributions for the sample. That is, the test of an association between sex and attitude pertains only to a population that is approximately half men and half women (in the sample, 79 and 83) and that is 2.5 to 3 times more approving (93) than neutral (36) or disapproving (33) in attitude toward the statement. The marginal distributions define the nature of the population; the question is whether an association between sex and attitude exists within that population. The hypotheses begin with the statement "Given the observed marginals" to remind us of this fact.

Assumptions and conditions   First, it must be assumed that the two samples are **independent** of each other. That usually implies that different and unrelated sets of subjects have been selected. Second, the subjects within each group must be **randomly** and **independently** sampled. Third, each observation must qualify for **one and only one category** in the classification scheme. Fourth, the **sample size must be relatively large**. The last assumption will be discussed again below.

Rationale and computation   The formula used to relate a statistic to a theoretical relative frequency distribution is based upon a comparison of the frequencies actually observed for men and women and the frequencies that would be expected if men and women had the same population distribution of response to the statement. Once again, the null hypothesis is tentatively presumed to be true—men and women have the same population distribution. As usual, the observed data will not conform precisely with this hypothesized state of affairs. Is the difference between the observed and expected frequencies a reasonable outcome of sampling error? If not, reject the null hypothesis.

A major requirement of this technique is that one be able to determine the set of frequencies that exists in the samples of men and women if their population distributions are identical, that is, if the null hypothesis is true. Since one does not know what proportion of college students in the population, men and women combined, approve, disapprove, or are neutral toward the statement, these proportions must be estimated from the observed data. For the data presented above, $\frac{93}{162}$ = 57.41% of the college students approved of the statement. Since there were 79 men, 57.41% of 79 or 45.35 should have approved the statement if there were no differences in the distribution of response for the two sexes. The 45.35 men is an **expected frequency** for the men-approve cell. The same process may be used to obtain the expected frequency of each cell in the table. The computation of expected frequencies is most often stated in the following terms:

> To compute the **expected frequency** for any cell, multiply the marginal total for the row that contains the cell by the marginal total for the column that contains the cell, and divide this product by the total number of cases in the table.

For the men-approve cell the expected frequency is given by

$$\frac{(93)(79)}{162} = 45.35$$

Since $\frac{93}{162}$ is the proportion of the total sample approving and 79 is the number of males in the sample, this method uses the same logic as the percentage method described above.

If the expected frequencies are computed in this manner for each cell, the observed frequencies (symbolized by O) and the expected frequencies (symbolized by E) for the sample data are as follows:

	Men	Women
Approve	$O = 58$ $E = 45.35$	$O = 35$ $E = 47.65$
Neutral	$O = 11$ $E = 17.56$	$O = 25$ $E = 18.44$
Disapprove	$O = 10$ $E = 16.09$	$O = 23$ $E = 16.91$

If the calculation of the expected frequencies is correct, the row and column sums of the expected frequencies will equal the marginal sums of the observed frequencies.

We now have a set of expected frequencies, given that the null hypothesis of the equivalence of population distributions within the categories for the two groups is true and given the observed marginal values. Now the task is to determine some index of the extent to which the observed frequencies are consonant with the null hypothesis. The following expression represents such an index, and its sampling distribution approximates a theoretical distribution called **chi square** (χ^2):

$$\chi_{obs}^{2} = \sum_{j=1}^{r} \sum_{k=1}^{c} \frac{(O_{jk} - E_{jk})^2}{E_{jk}}$$

in which

$\quad O_{jk}$ = the observed frequency in the cell corresponding to the intersection of the jth row and kth column
$\quad E_{jk}$ = the expected frequency in the cell corresponding to the intersection of the jth row and kth column
$\quad r$ = the number of rows
$\quad c$ = the number of columns

The formula directs one to take the difference between the observed and expected frequencies for each cell, square it, divide by the expected frequency, and sum these values over all cells. The resulting total is distributed as chi square with degrees of freedom

$$df = (r - 1)(c - 1)$$

where r is the number of rows and c the number of columns in the table of frequencies.

The percentiles of the chi-square distribution for each number of degrees of freedom are known, and Table F in Appendix II lists values of chi square for selected significance levels at several df. In the present case, the table of data has three rows and two columns; thus, there are $(r - 1)(c - 1)$ $= (3 - 1)(2 - 1) = 2$ degrees of freedom. The critical value for a nondirectional test at the .05 level of significance is 5.99. If the observed value of chi square, symbolized by χ_{obs}^{2}, exceeds the critical value, the null hypothesis, H_0, is rejected.

For the data presented, the calculation of χ proceeds as follows:

$$\chi_{obs}^{2} = \frac{(58 - 45.35)^2}{45.35} + \frac{(35 - 47.65)^2}{47.65} + \frac{(11 - 17.56)^2}{17.56}$$

$$+ \frac{(25 - 18.44)^2}{18.44} + \frac{(10 - 16.09)^2}{16.09} + \frac{(23 - 16.91)^2}{16.91}$$

$$\chi_{obs}^{2} = 16.17$$

Since the observed value 16.17 exceeds the critical value 5.99 and, in fact, is beyond the critical value for a test at the .001 level, the null hypothesis is rejected. It can be concluded from the data that the distribution of responses to the statement regarding woman's role in the family is different for the two sexes. Apparently, college men are more likely to endorse such a statement than are college women. (See Table 13–2 for a formal summary of this example.)

When the assumptions were listed for performing such a test, it was noted that a large sample is required. The approximation to the theoretical chi-square distribution is not very good for small samples and thus the probabilities are somewhat inaccurate. Unfortunately, the closeness of the approximation is a function of many factors, and a single rule of thumb concerning sample size is not totally adequate. However, a conservative guideline would be not to have any expected frequency less than 5. If the table is only 2×2 in size, then no expected frequency should be less than 10. Notice that the requirement is based upon the *expected* frequencies, not the observed frequencies.

13–2 Summary of $r \times c$ Chi-Square Test
(Sex $\times$ Approval Example)

Hypotheses

H_0: Given the observed marginals, the distributions of frequencies in the population are not different for the groups.

H_1: Given the observed marginals, these distributions are different for the groups (nondirectional).

Assumptions and Conditions

1. The subjects for each group are **randomly** and **independently** selected.
2. The groups are **independent**.
3. Each observation must qualify for **one and only one category**.
4. The sample size must be fairly large such that **no expected frequency is less than 5 for r or c greater than 2, or less than 10 if $r = c = 2$.**

Decision Rules (from Table F)

Given: .05, a nondirectional test, and $df = (r - 1)(c - 1) = 2$

If $\chi_{obs}^2 < 5.99$, do not reject H_0.
If $\chi_{obs}^2 \geq 5.99$, reject H_0.

Computation (see text for data and details)

$$\chi_{\text{obs}}^2 = \sum_{j=1}^{r} \sum_{k=1}^{c} \frac{(O_{jk} - E_{jk})^2}{E_{jk}}$$

in which

O_{jk} = the observed frequency in the cell corresponding to the intersection of the jth row and kth column

E_{jk} = the expected frequency in the cell corresponding to the intersection of the jth row and kth column, determined by multiplying the marginal totals of frequencies for the row and column which contain the cell and dividing by the number of frequencies in the table

r = the number of rows

c = the number of columns

$$\chi_{\text{obs}}^2 = 16.17$$

Decision

Reject H_0.

Mann-Whitney U test for the difference between two independent samples

If two samples are randomly and independently selected, if there is an underlying continuous distribution, and if there is at least an ordinal scale of measurement, then the **Mann-Whitney U test** may be used to evaluate the difference between the two population distributions. The U test is one of the most popular alternatives to the parametric t test.

Hypotheses   The null hypothesis is that the populations from which the two samples have been drawn are identical. The alternative hypothesis is that these two populations are not identical. Note that this is not equivalent to testing the difference between two population means, since two distributions could be quite different in form but have identical means. Therefore, it is theoretically possible to obtain a statistically significant result with the U test when in fact the means are identical. From a practical standpoint, the general shapes of the distributions of two groups within a single experiment are not often markedly different. When the forms of the distributions are similar, the U test does compare the central tendencies of the groups. Thus, if the forms of the sample distributions are similar, the results of the U test are often interpreted in terms of differences in central tendency; if the forms of the distributions are not similar, the results must be viewed in terms of the difference between the distributions in general.

Assumptions and conditions   The assumptions required for the U test are few. **Random** and **independent** sampling with **independent groups** is required. It is also assumed that the dependent variable is **continuous**. This implies that an infinite number of values theoretically exists between any two measured values (see page 27). Last, the measurement scale must be at least **ordinal** in character.

Rationale and computation   The rationale for the test is based upon the premise that if two distributions of equal size are identical, then it follows that listing the observations from both groups together in rank order (i.e., smallest first) should yield a sequence in which the scores from the two groups are well mixed. If they are well mixed, then the number of scores in group A that precede scores in group B should equal, within sampling error, the number of scores in B that precede scores in A. If the scores in group A tend to be smaller in value than those in B, then more of the A scores will precede B scores when all the scores are rank ordered.

The statistic based upon an extension of this type of rationale is the Mann-Whitney U. Suppose that the sets of scores for groups A and B are as follows:

$$A = \{5, 9, 17, 3\}$$
$$B = \{1, 8, 28, 20, 18\}$$

Arranging the scores in order beginning with the smallest score (retain the group identity of each observation) and assigning each score a rank beginning with rank 1 for the smallest score, the following table results:

Score	1	3	5	8	9	17	18	20	28
Group	B	A	A	B	A	A	B	B	B
Rank	1	2	3	4	5	6	7	8	9

Although the value of U can be obtained by counting the number of scores of one group that precede those of another, a more convenient method requires that the total (T_j) of the ranks for one group be obtained. For the example just presented,

$$T_A = 2 + 3 + 5 + 6 = 16$$

The statistic U is then given by

$$U_{obs} = n_A n_B + \frac{n_A(n_A + 1)}{2} - T_A$$

in which n_A and n_B are the numbers of cases in groups A and B, respectively. For the present example,

$$U_{obs} = 4(5) + \frac{4(4 + 1)}{2} - 16$$

$$U_{obs} = 14$$

Size of n   If n for each group is 20 or less, Table G in Appendix II gives the critical values for U. Notice that Table G presents several different tables for different significance levels. In each case, the rows and columns represent different numbers of cases for the two groups. (It makes no difference which group is used for the rows and which for the columns.) At the intersection of the appropriate row and column, two values of U are given. These two values are a type of critical value such that if the observed U falls *between* these two values, H_0 is not rejected. If the observed value is less than or equal to the lower value in the table or greater than or equal to the larger value in the table, H_0 is rejected.

If n for either group is greater than 20, then Table G cannot be used. It happens that with such a large sample, the observed value of U approaches a normal distribution with

$$\text{mean} = \frac{n_A n_B}{2}$$

and

$$\text{standard deviation} = \sqrt{\frac{(n_A)(n_B)(n_A + n_B + 1)}{12}}$$

Consequently, if the size of a group is greater than 20 and the difference in sample sizes of the two groups is not too great, the significance of U_{obs} may be determined by calculating

$$z_{obs} = \frac{U_{obs} - n_A n_B / 2}{\sqrt{\frac{(n_A)(n_B)(n_A + n_B + 1)}{12}}}$$

which approaches the standard normal distribution. In that event, critical values of z may be obtained by consulting Table A of Appendix II.

Ties   Occasionally, the values of scores are tied.[3] For example, consider the set of scores {13, 15, 15, 18}. In this case, the score of 13 receives rank 1. The two scores of 15 are given the average of the next two ranks. The ranks are 2 and 3, and their average is 2.5. Consequently, each of the two scores of 15 receives a rank of 2.5. The score of 18 is then assigned a rank of 4. It is not given a rank of 3 because this rank was used in the previous averaging. If three or more scores are tied, each score receives the average rank that these scores would have received if they were distinct. The next score(s) begin with the next unused rank. An example of ranking with several ties follows:

Score	13	13	16	19	19	22	22	22	28	28	30
Rank	1.5	1.5	3	4.5	4.5	7	7	7	9.5	9.5	11

Small-sample illustration   One premise of Freudian theory is that a person is born with certain innate drives and needs, such as an oral need. Another important concept is that the energy system of an organism is closed, so that if a need is blocked from satisfaction the need does not dissipate but will express itself in other ways. For example, during the early months of life if oral needs (e.g., the need for sucking) are not exercised, there should be more oral activity later in order to make up for this frustration. However, from the standpoint of learning theory, a person who is not allowed to suck as an infant may suck less on other objects later in life. Which is true, Freudian or learning theory? An experiment is run in which four infants are

---

[3]The method of handling tied observations suggested in this chapter is a common and convenient one, but it has certain technical liabilities. Most rank-order statistics assume an underlying continuous distribution. If this is true, ties in scores result from imprecision of measurement. If more precise methods were available, no ties would exist. Consequently, in this instance the occurrence of ties is not a true reflection of what really exists. The issue of just how to handle ties is not firmly resolved, and Bradley (*Statistical Tests*) presents a good though sophisticated discussion of the alternatives. One approach advocated by Siegel (*Nonparametric Statistics*) is to use the average-rank technique presented in the text and correct for ties with special formulas. However, Siegel points out, the corrections do not change the result very much even when a large proportion of the scores are tied. Bradley notes that under some conditions the average-rank approach actually biases the outcome in one direction or another rather than yielding a result that is itself a type of average or medium approximation. One alternative to this problem is to treat the tied scores as if they were not tied scores, selecting the ranks in the manner *least* favorable to rejecting the null hypothesis. Then, rerank the observations, this time treating the ties in a manner *most* favorable to rejecting the null hypothesis. As a result, one obtains two test statistics, one most and one least favorable toward rejecting the null hypothesis. If both statistics fall into the region of rejection, the null hypothesis may be unambiguously rejected. If both fall outside the rejection region, the null hypothesis may be not rejected without ambiguity. However, if one value does and the other does not fall within the rejection region, no clear decision can be made. In short, most of the methods discussed in this chapter technically assume that no ties exist. Since ties are common in social science, some procedures must be followed to handle them, and the choice of which method to follow rests on technical considerations largely beyond the scope of this text. However, the careful student will note that the method of handling ties is an issue and will be sensitive to the possible qualifications their presence may make upon the accuracy of conclusions.

fed from special cups as soon as possible after birth.[4] Another six infants are fed on bottles throughout the course of infancy, in the manner of many American infants. After eight months of these experiences, samples are made of the infants' behavior to determine how much of the time the infants suck their thumbs. These measures are expressed in terms of the percentage of time spent sucking during the observations.

The null hypothesis is that the population distribution for cup-fed babies is the same as for bottle-fed babies. The alternative hypothesis is that these distributions are different in some way.

The assumptions are that the subjects are **randomly** and **independently** sampled, the dependent variable is **continuous**, and the measurement scale is at least **ordinal**. The .05 level will be adopted.

In the following table, A indicates cup feeding and B bottle feeding. The scores are percentages.

Score	3	5	6	9	10	12	13	13	16	24
Group	A	A	A	B	B	A	B	B	B	B
Rank	1	2	3	4	5	6	7.5	7.5	9	10

Ranking scores is an important aspect of this and other nonparametric tests, so it is helpful to test the accuracy of the ranks assigned with two kinds of checks. First, the highest score should receive, or share with a tied score, a rank equal to that of the total number of scores (N). In the table above, the highest score, 24, has a rank of 10, which equals the total number of cases to be ranked. Second, the sum of N rankings should always be as follows:

$$\text{Sum of } N \text{ ranks} = \frac{N(N+1)}{2}$$

For $N = 10$ subjects, the sum of the ranks should equal

$$\frac{10(10+1)}{2} = \frac{10(11)}{2} = 55$$

The sum of the ranks in the bottom row of the above table is also 55. Notice that these checks determine whether all the correct rankings have been used, but they do not test whether the ranks have been correctly assigned to individuals or groups of subjects.

[4]Inspired by, but not identical to, R. R. Sears and G. W. Wise, "Relation of Cup Feeding in Infancy to Thumbsucking and the Oral Drive," *American Journal of Orthopsychiatry* 20 (1950): 123–38.

13–3 Data and Computation for Large-Sample Mann-Whitney U Test
(Cup vs. Bottle Example)

Cup-Fed		Bottle-Fed	
Score	Rank	Score	Rank
8	2	3	1
10	3	22	9
11	4.5	27	10
11	4.5	29	11.5
12	6	30	13
18	7	36	14
21	8	51	23
29	11.5	51	23
45	15	51	23
46	16	59	26
49	18	65	27
49	18	74	30
49	18	76	31.5
50	20.5	76	31.5
50	20.5	81	33.5
57	25	83	35
71	28	96	40.5
73	29	98	42
81	33.5	122	44
89	36	135	45
90	37	142	46
93	38	159	47
94	39	183	48
96	40.5	190	49
109	43		
$n_A = 25$	$T_A = 521.5$	$n_B = 24$	$T_B = 703.5$

Check: $T_A + T_B = 521.5 + 703.5 \stackrel{?}{=} \dfrac{N(N+1)}{2} = \dfrac{49(49+1)}{2} = 1225$

$$U_{obs} = n_A n_B + \frac{n_A(n_A+1)}{2} - T_A$$

$$= 25(24) + \frac{25(25+1)}{2} - 521.5$$

$$U_{obs} = 403.5$$

13–4 Summary of Large-Sample Mann-Whitney U Test (Cup vs. Bottle Example)

Hypotheses

H_0: The population distributions from which the samples are drawn are identical.

H_1: These populations are different in some way (nondirectional).

Assumptions and Conditions

1. The observations are **randomly** and **independently** sampled.
2. The groups are **independent**.
3. The dependent variable is **continuous** and the measurement scale is at least **ordinal**.

Decision Rules (from Table A)

Since the sample is large ($n > 20$), the standard normal approximation will be used. For a nondirectional test at .05, the decision rules are

If $-1.96 < z_{obs} < 1.96$, do not reject H_0.
If $z_{ob} \leq -1.96$ or $z_{obs} \geq 1.96$, reject H_0.

Computation

$$U_{obs} = n_A n_B + \frac{n_A(n_A + 1)}{2} - T_A$$

in which n_A and n_B = the number of subjects in samples A and B, respectively, and T_A = the sum of the ranks for sample A. U_{obs} is calculated to be 403.5 in Table 13–3 for the illustrative data. The approximation to the standard normal distribution is given by

$$z_{obs} = \frac{U_{obs} - n_A n_B / 2}{\sqrt{\dfrac{(n_A)(n_B)(n_A + n_B + 1)}{12}}}$$

$$= \frac{403.5 - 24(25)/2}{\sqrt{\dfrac{(24)(25)(24 + 25 + 1)}{12}}}$$

$$z_{obs} = 2.07$$

Decision

Reject H_0.

U_{obs} is computed as follows:

$$U_{\text{obs}} = n_A n_B + \frac{n_A(n_A + 1)}{2} - T_A$$

$$= 4(6) + \frac{4(5)}{2} - 12$$

$$U_{\text{obs}} = 22$$

Looking at Table G of Appendix II for a nondirectional test at the .05 level, one finds the critical values of U for $n = (4, 6)$ are 2 and 22. The observed value of U is 22 and since this equals the critical value for U, H_0 is rejected. The distributions are probably different, and it would appear that cup-fed infants suck their thumbs less, not more, than bottle-fed infants. According to the learning theory, they have not learned to need or to want to suck.

Large-sample illustration   Suppose a similar experiment is carried out with a much larger sample, 25 cup-fed and 24 bottle-fed infants. The assumptions and hypotheses are the same as before, but since the sample sizes are large (greater than 20), the standard normal approximation will be used. For a nondirectional alternative at the .05 level, the observed z must be less than or equal to -1.96 or greater than or equal to 1.96 in order to reject H_0.

The data and computation are presented in Table 13–3 and a formal summary of the procedure is presented in Table 13–4. The data are arranged in increasing order of score value within each group. Then, disregarding group affiliation, the scores are rank ordered, and the total of the ranks for the two groups is determined. The value of U_{obs} is computed in Table 13–3. The formula approximating the standard normal is given in Table 13–4 and the corresponding values entered. The resulting observed z is 2.07, which complies with the second decision rule, and the null hypothesis is rejected. The interpretation is that the groups probably do differ in the population, and an examination of the data indicate that learning theory is favored.

Kruskal-Wallis test for k independent samples

The Mann-Whitney U test for two samples may be generalized to several independent samples. The most common approach to this kind of analysis is known as the **Kruskal-Wallis test**. It is analogous to the simple parametric analysis of variance but does not make all of the same assumptions. In Chapter 12, a study was described in which children were shown a movie of an adult striking a large doll. The adult was either rewarded or punished for that behavior. Suppose there had been three rather than two different films shown to the children in the experiment.[5] All three movies showed an adult

[5]Similar to A. Bandura, D. Ross, and S. A. Ross, "Imitation of Film-Mediated Aggressive Models," *Journal of Abnormal and Social Psychology* 66 (1963): 3–11.

striking and otherwise assaulting a large doll, Bobo. Each film depicted one of three consequences to the aggressor: The person was rewarded with praise, punished with reprimands, or nothing at all happened. The question is whether vicarious experience with different consequences would determine the number of imitative aggressive behaviors in a ten-minute test situation in which each child is left in a room with Bobo and other toys.

Hypotheses   The null hypothesis is that the three samples have identical population distributions, while the alternative hypothesis is that their population distributions are different. If the distributions have the same form, then H_1 implies that the mean scores are higher or lower in some of the groups than in others.

Assumptions and conditions   The three **independent** groups of subjects are **randomly** selected. Assume also that the dependent variable is **continuous** and that it is measured with at least an **ordinal** scale. The conceptual dependent variable—which is continuous—is the extent to which the child imitates the adult's behavior, and it is measured by counting the number of imitative acts, which is at least an ordinal measurement. Finally, there should be **at least five observations per group** for an accurate estimation of the probability.[6]

Rationale and computation   The data and computation for this example are presented in Table 13–5. The method is similar to that for the Mann-Whitney test. The scores are arranged into groups in ascending order of score value. Then, without regard to group affiliation, all the observations are ranked, the smallest score value being assigned a rank of 1. If any scores are tied, each tied score receives the average of the ranks available for those scores. The ranks are totaled within each group, and this total is signified by T_j for the jth group. Once again, the accuracy of the rankings can be partly checked by making sure that the highest score receives (or shares) a rank equivalent to the total number of subjects in the analysis (N) and that the sum of the ranks in the entire analysis equals $N(N + 1)/2$. That is, $\sum_{j=1}^{k} T_j = N(N + 1)/2$.

The general rationale is that if the groups are distributed in the population in the same way and with the same central tendency, then the total of the ranks for the several samples should be approximately equal. To the extent that the scores in one group are higher than those in the others, the several totals will be unequal. Obviously, the totals are not likely to be precisely equal, but under the null hypothesis they will differ only because of sampling error. When the differences in average ranks become so great that it is

---

[6]For procedures to handle the case in which there are fewer than five observations in a group, see Siegel, *Nonparametric Statistics*, pp. 185–88.

13–5 Data and Computation for Kruskal-Wallis Test
(Imitation Example)

Punished		Ignored		Rewarded	
Imitations	Rank	Imitations	Rank	Imitations	Rank
0	1	2	3	8	13.5
1	2	3	5	12	16.5
3	5	4	7.5	13	18
3	5	6	10.5	16	20.5
4	7.5	7	12	19	22
5	9	10	15	21	23
6	10.5	12	16.5	22	24
8	13.5	14	19	23	25
		16	20.5		
$n_1 = 8$	$T_1 = 53.5$	$n_2 = 9$	$T_2 = 109$	$n_3 = 8$	$T_3 = 162.5$

Check: $\sum_{j=1}^{k} T_j = 53.5 + 109 + 162.5 \overset{?}{=} \dfrac{N(N+1)}{2} = \dfrac{25(25+1)}{2} = 325$

$$H_{obs} = \left[\frac{12}{N(N+1)}\right]\left[\sum_{j=1}^{k}\frac{T_j^2}{n_j}\right] - 3(N+1)$$

$$= \left[\frac{12}{25(25+1)}\right]\left[\frac{(53.5)^2}{8} + \frac{(109)^2}{9} + \frac{(162.5)^2}{8}\right] - 3(25+1)$$

$$H_{obs} = 13.91$$
$$df = k - 1 = 3 - 1 = 2$$

implausible to attribute them to sampling error, the tentative assumption of the null hypothesis is rejected and the existence of treatment effects in the population is suspected.

This logic is translated into a formula for the statistic H that reflects the extent to which the sum of the ranks for the several groups differ from one another. The formula for H_{obs} is

$$H_{obs} = \left[\frac{12}{N(N+1)}\right]\left[\sum_{j=1}^{k}\frac{T_j^2}{n_j}\right] - 3(N+1)$$

13–6 Summary of Kruskal-Wallis Test (Imitation Example)

Hypotheses

H_0: The population distributions from which the groups are sampled are identical.

H_1: These distributions are different in some way.

Assumptions and Conditions

1. Subjects are **randomly** and **independently** sampled and divided into k **independent** groups with all $n_j \geq 5$.
2. The dependent variable is **continuous** and the measurement scale is at least **ordinal**.

Decision Rules (from Table F)

Given: the .05 level of significance, a nondirectional test, and that H is distributed as chi square with $k - 1 = 3 - 1 = 2$ df

If $H_{obs} < 5.99$, do not reject H_0.
If $H_{obs} \geq 5.99$, reject H_0.

Computation (see Table 13–5)

$$H_{obs} = \left[\frac{12}{N(N + 1)} \right] \left[\sum_{j=1}^{k} \frac{T_j^2}{n_j} \right] - 3(N + 1)$$

in which

k = the number of groups in the analysis
n_j = the number of subjects in group j
N = the total number of subjects in the analysis
T_j^2 = the square of the total of the ranks for scores in group j

$H_{obs} = 13.91$

Decision

Reject H_0.

in which

k = the number of groups in the analysis
n_j = the number of subjects in group j
N = the total number of subjects in the analysis
T_j^2 = the square of the total of the ranks for scores in group j

The formula directs one to sum the ranks and square this total separately for each group. Divide each squared total by the number of observations in that particular group, and sum across all groups to obtain

$$\sum_{j=1}^{k} \frac{T_j^2}{n_j}$$

Enter this and N into the above expression for H_{obs}.

The sampling distribution of the statistic H has approximately the same form as chi square with $k - 1$ degrees of freedom (k = the number of groups). This approximation is close only when there are at least five observations per group, and the accuracy of the approximation improves as N increases.

In the present case there are three groups, so the degrees of freedom for H is $k - 1 = 3 - 1 = 2$. The critical value of chi square for a nondirectional test with $df = 2$ is 5.99, as obtained from Table F in Appendix II. Thus, if H_{obs} is greater than or equal to 5.99, the null hypothesis of equivalence of population distributions will be rejected. The computation in Table 13–5 reveals H_{obs} to be 13.91, which exceeds the critical value at .05 (and at .01 as well), so H_0 is rejected. The analysis suggests that the populations from which these three groups are drawn do differ in some way, and an examination of the data implies that a child is more likely to imitate if positive consequences are perceived to derive from the aggressor's behavior. A formal summary of this analysis is presented in Table 13–6.

TESTS ON CORRELATED SAMPLES

All tests described thus far in this chapter have been for independent samples —groups of observations on separate or unmatched subjects. Just as special procedures are required for the t test when the sets of observations are made on the same or matched subjects, so too a different nonparametric analysis must be made when the samples are not independent. One of the most common nonparametric tests for this situation is the **Wilcoxon test** for two correlated or matched samples.

Wilcoxon test for two correlated samples

If the same subjects are measured under two conditions (or if matched pairs of subjects provide the scores) and if the measurement is ordinal for both within-pair and between-pair differences, then the Wilcoxon test may be used to test the null hypothesis that the population distributions corresponding to the two sets of observations are identical.

In the sensorimotor development of organisms there seems to be a close relationship between visual experiences and the opportunity for the organism to interact physically with its environment. Some theorists have suggested that the development of normal visually guided behavior requires not only experience with visual stimuli but also feedback obtained through physical interaction with the stimulus environment. One application of this theory was made in the following experiment.[7] Eight pairs of kittens, each pair from the same litter, were reared in darkness until they were 10 weeks of age. They then received visual stimulation for only a short period of time each day under very special conditions. The two kittens of each pair were placed into a vertically oriented cylindrical apparatus with stripes painted on the side. One member of each pair was allowed to move about at will, except that it wore a harness attached by means of pulleys and gears to a gondola-like carriage that held the other kitten. A post in the center of the apparatus prevented one member of the pair from seeing the other, and the circumstances were arranged so that the visual experience of the two kittens was quite similar. If the theory is correct, the active kitten should have shown faster development on visual-motor tasks than the kitten carried passively through the environment in the gondola. A series of tasks that measure visual-motor development was used, and each subject received a score indicating proficiency on the test battery.

Hypotheses   The null hypothesis is that the population distributions under the two conditions are identical. The alternative hypothesis is that they are not identical. Usually, if the distributions are symmetrical, the alternative hypothesis is taken to mean that the central tendency of one distribution is higher than that of the other.

Assumptions and conditions   The assumptions required for the Wilcoxon test are less restrictive than those required for the analogous t test. First, the Wilcoxon test assumes **correlated measurements**. In the present example, pairs of measurements from related subjects—littermates—are the basic data. Littermates are assumed to be more similar to one another than

---

[7]Inspired by, but not identical to, R. Held and A. Hein, "Movement-Produced Stimulation in the Development of Visually-Guided Behavior," *Journal of Comparative and Physiological Psychology* 56 (1963): 872–76.

non-littermates, so measurements on littermates will be correlated. In other experiments, the same variable might be measured twice on the same subjects —before and after an experimental treatment, for example. These two sets of measures will also be correlated, because they are made on the same subjects. In addition to correlated groups, the Wilcoxon test assumes that the matched pairs or individual subjects who are measured twice are **randomly** and **independently** sampled. In the present example, each pair of kittens must be randomly and independently selected from its litter, and only one pair may come from a given litter. Moreover, both the measurement scale and the differences between related measurements must be at least **ordinal**, so that it will be possible to rank order the differences between related pairs of scores.

Rationale and computation   The procedure requires that one score value in each pair be subtracted from the other in order to obtain the difference between them, and that the algebraic sign of the difference be retained. Following this, the *absolute* values of the differences are assigned ranks starting with a rank of 1 for the smallest difference. Recall that the signs of the differences were retained but that the rankings were made without regard to sign. The ranks may now be attributed to positive and negative differences between the two groups. If there is no difference between the groups, then the sum of the ranks associated with positive differences between the groups should be about equal to the sum of the ranks associated with negative differences between the groups. Since under this circumstance the total of all ranks would be divided relatively evenly between positive and negative differences, the smaller of these two sums would take on its largest value under the null hypothesis of no difference. If there is a difference between groups, then either the positive or negative sum of ranks will be quite a bit smaller than the other. As the difference between groups increases, the smaller of the two sums gets smaller and smaller. The sampling distribution of this smaller sum is known, and when the observed total becomes sufficiently small as to be unlikely to have arisen solely because of sampling error, the null hypothesis of no difference is rejected.

The eight pairs of subjects and their scores are listed in the computation section of Table 13–7. The difference between these scores (d_i) is computed. If a difference is zero (see pair D), it is dropped from the analysis, and the total N (the number of *paired* observations) is reduced by the number of such zero differences. Then the absolute values of these differences ($|d_i|$) are taken and ranked, assigning the rank of 1 to the lowest absolute difference. If some of the absolute differences are of equal value (see pairs A, E, G), these ties are assigned the average of the ranks that they would have received if they had been distinct.

After the absolute differences have been ranked, the last column of the table merely repeats the rank that each matched pair has received but with the algebraic sign of the difference in score values attached to the rank. Now

13–7 Summary of Wilcoxon Test of Correlated Groups (Sensorimotor Experience Example)

Hypotheses

H_0: The population distributions for the two correlated groups of observations are identical.

H_1: These two distributions are different in some way (nondirectional).

Assumptions and Conditions

1. The pairs of observations are **randomly** and **independently** selected, but the two observations of a pair are made on the **same** or **matched** subjects.
2. At least **ordinal measurement** is available both between pairs and between members' scores within each pair (i.e., d_i).

Decision Rules (from Table H)

Given: the .05 level, a nondirectional alternative, and $N = 7$ (see below)

$$\text{If } W_{obs} > 2, \text{ do not reject } H_0.$$
$$\text{If } W_{obs} \leq 2, \text{ reject } H_0.$$

Computation

Pair	Active Kitten	Passive Kitten	d_i	$|d_i|$	Rank of $|d_i|$	Signed Rank of $|d_i|$
A	4	2	2	2	5	5
B	3	0	3	3	7	7
C	1	0	1	1	2	2
D	2	2	0	0	eliminated	
E	5	3	2	2	5	5
F	1	2	−1	1	2	−2
G	3	1	2	2	5	5
H	5	4	1	1	2	2

N = number of pairs minus the number of zero differences $\quad T_+ = 26$

$N = 8 - 1 = 7$ $\qquad\qquad\qquad\qquad\qquad\qquad\qquad\qquad T_- = 2$

$\qquad\qquad\qquad\qquad\qquad\qquad\qquad\qquad\qquad\qquad\qquad\qquad W_{obs} = 2$

$$\text{Check: } T_+ + T_- = 26 + 2 \overset{?}{=} \frac{N(N+1)}{2} = \frac{7(7+1)}{2} = 28$$

Decision

Reject H_0.

two quantities are computed. T_+ is the total of the ranks having positive signs associated with them in the last column of the table, and T_- is the sum of the ranks having negative signs. The rankings may be checked by noticing whether the largest d_i has been given (or shares) a rank equal to the total number of pairs having a nonzero difference (7 for this particular test); in addition, $T_+ + T_-$ should equal $N(N+1)/2$.

The statistic of interest is W_{obs}, which is simply the smaller of T_+ and T_-. If the two groups are similar, these two totals will be about equal. As the difference between groups increases so does the difference between T_+ and T_-, and W_{obs} (the smaller of the two) takes on a smaller and smaller value. When W_{obs} becomes sufficiently small relative to its sampling distribution, H_0 is rejected.

For N (the number of *paired* observations sampled less the number of pairs having zero differences in scores) between 5 and 50, the critical values for the sampling distribution of W as a function of N for several levels of significance and for directional and nondirectional tests are presented in Table H of Appendix II. If W_{obs} is *less than or equal to* the tabled critical value, H_0 is rejected and a difference in the population is presumed. In the present illustration $N = 7$, and if the test is taken to be nondirectional at the .05 level, the critical value is 2. Since W_{obs} equals this critical value, H_0 is rejected and the observed differences between the freely moving and gondola kittens are not likely to result from sampling error. A formal summary is presented in Table 13–7.

When N is greater than 50, an approximation to the standard normal may be used (actually the approximation is sufficiently good so that it can be employed with $N \geq 10$). The conversion to a standard normal deviate is given by the following:

$$z_{obs} = \frac{W_{obs} - N(N+1)/4}{\sqrt{\dfrac{N(N+1)(2N+1)}{24}}}$$

RANK–ORDER CORRELATION

Spearman rank–order correlation coefficient

The Pearson product-moment correlation coefficient, introduced in Chapter 6, may be applied to ordinal as well as interval or ratio data. Often, if the data are markedly skewed, measurements made with an interval or ratio scale are transformed to ranks before the correlation is computed. When the Pearson product-moment correlation is applied to rank orderings, it is called **Spearman's rank-order correlation**, symbolized by r_S.

Suppose that a group of 15 nursery-school children is observed by two judges, who rank the children on their social aggressiveness. Each judge rank orders the 15 children, assigning a rank of 1 to the child presumed to have the least aggressive behavior and a rank of 15 to the child presumed to have the most aggressive behavior. The question is, to what extent do the two judges agree in their rankings? That is, how reliable are the two judges? The requirements for computing such an index are that the subjects be **randomly** sampled and that the measurement be at least **ordinal**.

13–8 Rankings of Social Aggressiveness by Two Judges

Pupil	Judge I	Judge II	d_i	d_i^2
A	1	3	−2	4
B	4	4	0	0
C	5	8	−3	9
D	10	5	5	25
E	8	2	6	36
F	14	15	−1	1
G	7	9	−2	4
H	2	6	−4	16
I	12	14	−2	4
J	9	7	2	4
K	15	13	2	4
L	3	1	2	4
M	13	12	1	1
N	11	10	1	1
O	6	11	−5	25
	$N = 15$		$\Sigma d_i = 0$	$\Sigma d_i^2 = 138$

Suppose the data are those presented in Table 13–8. In the present case, the raw data are themselves rank orderings of the subjects by two different judges. If the data were numbers of aggressive acts, for example, these measurements would have to be rank ordered before proceeding. Notice that *observations are ranked only within, not across, a condition*. That is, scores for each variable (judge, in this case) are ranked separately. Ties may be handled as before, by assigning the average of the ranks that the tied observations would otherwise have received.

The method proceeds by computing the difference (d_i) between the ranks for each subject and then squaring each of these differences. The sum of the squared differences in ranks $\left(\sum\limits_{i=1}^{N} d_i^2 \right)$ and the number of **pairs of observations (N) including those with zero differences**, are entered into the following formula for r_S:

$$r_S = 1 - \left[\frac{6 \left(\sum\limits_{i=1}^{N} d_i^2 \right)}{N^3 - N} \right]$$

For the data presented in Table 13–8:

$$r_S = 1 - \left[\frac{6(138)}{15^3 - 15} \right]$$

$$r_S = .75$$

The formula given above for the Spearman rank-order correlation is a simplification of the formula for the Pearson product-moment correlation as applied to ranked data. It can be shown that when the Pearson formula is applied to data that have been ranked, the expression can be reduced to the formula given above for r_s. However, the Pearson formula applied to ranked data will not yield exactly the same value as that obtained when the formula is applied to the same data before ranking takes place.

Testing the significance of r_S

As was mentioned in the discussion of the Pearson r, it is desirable to be able to test the hypothesis that the observed value of r_S is computed on a sample drawn from a population in which the correlation is actually zero—in other words, that the observed value is merely a function of sampling error.

Using the data presented above for the similarity of the two judges' rankings, the null hypothesis would be that the observed Spearman correlation of .75 is based upon a sample from a population in which the correlation is actually zero (that is, H_0: $\rho_S = 0$). The alternative hypothesis is that a

13-9 Summary of Test of the Significance of the Spearman
Rank-Order Correlation Coefficient
(Reliability of Two Judges Example)

Hypotheses

H_0: $\rho_S \leq 0$
H_1: $\rho_S > 0$ (directional)

Assumptions and Conditions

1. The sample is **randomly** and **independently** selected.
2. The measurement scale is at least **ordinal**.

Decision Rules (from Table I)
Given: the .05 level, a directional alternative, and $N = 15$

If $r_S < .447$, do not reject H_0.
If $r_S \geq .447$, reject H_0.

Computation
Given: $N = 15$ and $\sum\limits_{i=1}^{N} d_i^2 = 138$ (from Table 13-8)

$$r_S = 1 - \left[\frac{6 \sum\limits_{i=1}^{N} d_i^2}{N^3 - N} \right]$$

$$= 1 - \left[\frac{6(138)}{15^3 - 15} \right]$$

$$r_S = .75$$

in which

N = the number of pairs of scores including zero differences
d_i^2 = the squared difference in ranks for the ith pair of
scores

Decision
Reject H_0.

nonzero correlation actually does exist in the population (H_1: $\rho_S \neq 0$). The assumptions are that the subjects were **randomly** and **independently** selected and that the measurement was at least **ordinal**.

Table I[8] in Appendix II lists critical values for the Spearman correlation for samples of size $N = 5$ to $N = 30$ for several levels of significance and for directional and nondirectional tests. If the observed correlation equals or exceeds the value in the table for the specified N, α, and type of test (directional or nondirectional), the null hypothesis is rejected. For the present case, $N = 15$, $\alpha = .05$, and the alternative is directional since one would predict that the judges would tend to agree. Thus, the critical value of r_S is .447. Since the observed correlation of .75 is greater than the critical value, H_0 is rejected and it is concluded that the two judges do show some degree of agreement in their evaluations of the children. Table 13–9 presents a formal summary of this example. For $N > 30$, the following expression translates r_S into an approximation to Student's t distribution with $N - 2$ degrees of freedom:

$$t = \frac{r_S \sqrt{N - 2}}{\sqrt{1 - r_S^2}}$$

Because the approximations to theoretical sampling distributions for samples of intermediate size are somewhat less accurate than might be desired, some statisticians and researchers prefer another method of assessing rank correlation. This statistic is called Kendall's tau, and the procedures for testing its significance are more precise than those for the Spearman coefficient. The Kendall procedure reflects a slightly different rationale, and thus the two coefficients are not precisely equivalent. Procedures for the use of tau are outlined in Siegel (*Nonparametric Statistics*) and Bradley (*Statistical Tests*).

---

[8]The values in this table are exact for $N \leq 10$ but approximate for other values of N. Note also that the entries in the table may be regarded as + (positive) or − (negative).

FORMULAS

1. Chi Square for $r \times c$ tables

$$\chi_{\text{obs}}^2 = \sum_{j=1}^{r} \sum_{k=1}^{c} \frac{(O_{jk} - E_{jk})^2}{E_{jk}}$$

in which

O_{jk} = the observed frequency of the jkth cell
E_{jk} = the expected frequency of the jkth cell
r = the number of rows
c = the number of columns

Refer to the chi-square distribution in Table F of Appendix II, with

$$df = (r - 1)(c - 1)$$

2. Mann-Whitney U test for two independent samples
a. For $n \leq 20$

$$U_{obs} = n_A n_B + \frac{n_A(n_A + 1)}{2} - T_A$$

in which

n_A = the number of observations in group A
n_B = the number of observations in group B
T_A = the total of the ranks for group A

Refer to the U distribution in Table G of Appendix II.
b. For $n > 20$

$$z_{obs} = \frac{U_{obs} - n_A n_B / 2}{\sqrt{\dfrac{(n_A)(n_B)(n_A + n_B + 1)}{12}}}$$

in which U_{obs}, n_A, and n_B are as defined above.
Refer to the standard normal distribution in Table A of Appendix II.
3. Kruskal-Wallis test for k independent samples

$$H_{obs} = \left[\frac{12}{N(N + 1)} \right] \left[\sum_{j=1}^{k} \frac{T_j^2}{n_j} \right] - 3(N + 1)$$

in which

k = the number of groups in the analysis
n_j = the number of subjects in group j
N = the total number of subjects in the analysis
T_j^2 = the square of the total of the ranks for scores in group j

Refer H_{obs} to the chi-square distribution in Table F of Appendix II, with

$$df = k - 1$$

4. Wilcoxon test for two correlated samples
a. For $N \leq 50$

W_{obs} = the smaller of the sum of the ranks associated with positive differences in pairs of scores (T_+) and the sum of the ranks associated with negative differences (T_-)
N = the number of pairs of observations having nonzero differences
Refer to the W distribution in Table H of Appendix II.

b. For $N > 50$

$$z_{obs} = \frac{W_{obs} - N(N+1)/4}{\sqrt{\dfrac{N(N+1)(2N+1)}{24}}}$$

in which W_{obs} is defined as above

Refer to the standard normal distribution in Table A of Appendix II.

5. Spearman rank-order correlation coefficient

$$r_S = 1 - \left[\frac{6\left(\sum_{i=1}^{N} d_i^2\right)}{N^3 - N} \right]$$

in which

N = the number of pairs of observations including zero differences
d_i = the difference in ranks for the ith pair of scores

a. For $N \leq 30$
Refer to critical values for r_S in Table I of Appendix II.

b. For $N > 30$

$$t = \frac{r_S\sqrt{N-2}}{\sqrt{1 - r_S^2}}$$

Refer to Student's t distribution in Table B of Appendix II, with

$$df = N - 2$$

EXERCISES

1. Under what conditions might a researcher use a nonparametric rather than a parametric statistical technique?
2. To what does power-efficiency refer? For a given N, which are more powerful: parametric tests or analogous nonparametric tests?
3. A sample of children was observed from 3 to 12 years of age and their IQ was tested periodically. It was found that 56 children showed increases in the general trend of their IQs over this age period, while 55 displayed essentially declining trends. The mothers of these children were seen in their homes during this period and the home visitor rated each mother on the extent to which she expected and encouraged intellectual achievement and success. The data follow at the top of the next page.

Test the hypothesis that there is no difference in the distribution of maternal encouragement for intellectual success for

		Amount of Maternal Encouragement		
		Low	Medium	High
IQ Trend over Age	Increased	12	15	29
	Decreased	29	16	10

the children who evidenced IQ increases and for those showing declines.[9]

4. Consider the following practical situation. Suppose that a father comes home from work after a hard day and sits down with his wife and five-year-old child for dinner. The wife asks the husband, "How was your day, dear?" The retort is "It was . . . ," in which " . . . " is a set of expletives not in the best social and child-rearing traditions. The mother may censor her husband in front of the child by saying, "Jim, you really shouldn't swear like that in front of Junior," or she could continue the conversation without commenting on the profanity. The question is, under which of these maternal response conditions is the child more likely to remember the profanity? Suppose it is possible to perform such a study in which some fathers are verbally censored while others are not.[10] Later, the child is asked what his father has said in response to the question "How was your day?" The responses are scored as either not remembering or remembering the father's profanity. The numbers of children responding at each level are presented in the table below. Test the hypothesis that the mother's response to the father's profanity does not influence the ability of the child to remember the profanity.

5. One hypothesis about thumbsucking in infants and children is that the behavior is a learned habit. An interesting question is how children come to learn it. It happens that all normal infants have a rooting reflex: Stroking the side of the infant's mouth with a finger (for example) elicits a widening of the mouth, a turning of the head toward the stimulus, and a propensity to suck the finger. This reflex is useful in helping the infant find the mother's breast. However, sleeping infants are usually laid on their stomachs so that their arms are on either side of the head. The thumb may easily touch the side of the mouth, and infants root and suck their thumb. Perhaps infants learn the habit in this context. If

		Child's Memory of Profanity	
		No	Yes
Mother's Response to Profanity	Censors Father	16	28
	No Response	26	17

[9]Inspired by, but not identical to, R. B. McCall, M. I. Appelbaum, and P. S. Hogarty, *Developmental Changes in Mental Performance*, Monographs of the Society for Research in Child Development, no. 150 (Chicago: University of Chicago Press, 1973).

[10]Inspired by, but not identical to, R. M. Liebert and L. E. Fernandez, "Effects of Vicarious Consequences on Imitative Performance," *Child Development* 41 (1970): 847–52.

infants could be prevented from rooting and sucking their thumbs when they go to sleep and wake up, presumably they would not develop as strong a habit for this behavior.[11] Suppose nylon mittens are put on

[11]Inspired by, but not identical to, L. S. Benjamin, "The Beginning of Thumbsucking," *Child Development* 38 (1967): 1079–88.

the hands of nine infants just before they go to sleep and are removed when they wake up, thus preventing thumbsucking. Ten other infants are reared without the gloves. In an observation period of several hours when the infants are nine months of age, the amount of time each infant spends thumbsucking is noted. The data follow. With the appropriate nonparametric technique, test the hypothesis that the wearing of mittens does not alter the amount of thumbsucking.

Mittens	Control
0	6
5	12
0	6
1	5
7	10
4	11
2	9
3	4
8	8
	10

6. Given the following data, calculate the means and medians for the two groups and use the parametric t test for independent groups to compare the two groups. Then compute a Mann-Whitney U test on the same data. Compare the results of the two tests. Explain any differences and attempt to draw some conclusions about when nonparametric tests might be more appropriate than parametric tests. In what way is the difference between the mean and the median similar to the difference between a parametric and a nonparametric test in this example?

A	B
10	16
11	17
12	18
13	19
14	20
15	21
62	24

7. The concept of reminiscence in learning refers to the improvement in performance of a learned task after a period of rest following an initial practice period. The effect often appears in tasks requiring motor learning. For example, a pursuit rotor is a machine on which a spot of light moves in a circular pattern, and the subject's task is to keep the point of a pencil-like stylus on the moving dot. The apparatus is such that it can monitor the length of time the point of the stylus is actually on the spot of light. Each of several randomly selected subjects is allowed 25 seconds of practice on the pursuit rotor and then given a rest of either .5, 1, or 3 minutes. This is followed by a test period on the rotor. The number of seconds in contact with the spot during the test is given below for the three groups.[12]

Using the appropriate nonparametric technique, test the hypothesis that the amount of rest does not alter postrest performance.

Rest in Minutes		
.5	1.0	3.0
19	27	39
25	32	45
34	61	61
10	29	65
45	41	52
47	37	52
21	49	51
28		57
		49

8. Twelve culturally deprived children are given an intensive preschool enrichment program in an attempt to raise their IQs. They are tested first after one month of training and then at the end of the first year. The IQs of the children for these two

[12]Inspired by, but not identical to, A. L. Irion, "Reminiscence in Pursuit-Rotor Learning as a Function of Length of Rest and Amount of Pre-Test Practice," *Journal of Experimental Psychology* 39 (1949): 492–99.

testing follow. By using the appropriate nonparametric technique, test the hypothesis that the program has had no effect on the IQs of the youngsters. (Would such a program likely retard IQ?)

Child	One Month	One Year
A	92	101
B	91	95
C	86	92
D	103	98
E	87	92
F	92	90
G	99	96
H	93	105
I	83	89
J	97	103
K	93	101
L	87	87

9. For the data in exercise 8, using r_S, determine the degree of relationship between the first and the second testings for the 12 children. Test the hypothesis that in the population the relationship is zero. On psychological and educational grounds, how do you interpret the result observed in exercise 8 in view of this additional information? (Would such correlations likely be negative?)

10. People who try to stop smoking must overcome a need for nicotine. Psychologist Stanley Schachter once suggested that smokers whose physiological systems have high acid levels excrete more unmetabolized nicotine and therefore have a greater need for nicotine; those who have low acid levels retain the nicotine from cigarettes and have less need. Presumably, smokers given a substance to make their systems more basic, such as bicarbonate, should be able to give up smoking more easily than those given a substance that would make

their systems more acidic, such as vitamin C. A group of researchers[13] decided to test this notion. They randomly selected smokers and arbitrarily assigned them to three groups. The three groups were treated alike for the first three weeks of the stop-smoking program. Then one group was given vitamin C supplements, another group was furnished with bicarbonate, and a third group was given no pills. The two supplement groups did not know which substance they were receiving. The number of cigarettes smoked per day during the third and the fifth weeks of the experiment are given below. Using nonparametric tests, assess the following questions.

a. Did the vitamin C group improve between the third and fifth week?

b. What is the correlation between the third and fifth week for the bicarbonate group? Is it significant?

c. Were the three treatment groups different in the number of cigarettes smoked at three weeks, that is, before the supplements were given? Afterward, at five weeks?

Vitamin C (Acid)		Nothing		Bicarbonate (Base)	
Third	Fifth	Third	Fifth	Third	Fifth
42	10	39	9	23	3
25	6	21	3	38	1
15	19	17	10	41	2
28	9	30	7	16	7
33	12	35	8	23	3
40	12	41	13	15	5
26	5	25	2	39	3
29	6	31	9	40	0

[13]Based on, but not identical to, research by A. James Fix, Irving Kass, Joseph Shipp, and Jack Smith reported in *Science News*, 14 April 1979, p. 244.

review of basic mathematics and algebra

A CLASS in elementary statistics almost always includes students whose mathematical backgrounds range from a knowledge of high school algebra to facility with differential equations. It is difficult to begin a course in statistics unless some common level of mathematical experience can be assumed. This appendix reviews basic mathematical concepts and operations, including symbols, fractions, factorials, exponents, and factoring. Although some students will not need to spend much time on this section, others would profit from studying it fairly seriously.[1] Students should check themselves by working the exercises at the end of the section.

Most students have hand calculators that will add, subtract, multiply, divide, and take square roots. Nonetheless, a complete understanding of the material in this text requires that students be able to follow simple algebraic operations. Hand calculators provide numerically accurate answers quickly, but they do not help the student understand how a formula is derived, what it means, or how to interpret the numerical result. Therefore, students are strongly encouraged to become thoroughly familiar with the material in this appendix, and *only then* to use their calculators.

Symbols

Some students feel that their basic problem with mathematical material is using the symbols. Usually, this problem occurs in part because the students do not take time to learn what the symbols mean. It will be helpful if readers do not proceed until they can easily read and interpret all the symbols presented to that point.

Most symbols will be introduced as they arise in the text. However, students are assumed to be familiar with the signs for equality ($=$) and inequality ($\neq$); and with the signs for addition ($+$), subtraction ($-$), multiplication (2×3, $2 \cdot 3$, $(2)(3)$, or ab), and division ($5 \div 2$ or $5/2$). The signs for "greater than" and "less than" are reviewed below.

The symbol $>$ means "is greater than." The expression $5 > 3$ is read "5 is greater than 3" and $a > b$ is read "a is greater than b." Conversely, the symbol $<$ means "is less than." The expression $3 < 5$ is read "3 is less than 5" and $b < a$ is read "b is less than a." Some students remember the difference between these two symbols by observing that the open, or larger, end of the symbol is always next to the larger quantity.

Sometimes an expression of the following type will be encountered:

$$-1.96 < t < 1.96$$

This is a short way of writing two inequalities, specifically $-1.96 < t$ and $t < 1.96$. Therefore, the statement means that the value of t is greater than -1.96 but less than 1.96. More simply, t lies between -1.96 and 1.96.

Occasionally, it is desirable to write that some quantity is "greater than or equal to" some other quantity. This fact can be stated with the symbol $\geq$, which is simply a combination of the signs $>$ and $=$. To show that z assumes values of 2.56 or greater,

[1]Students who need a more thorough review of the concepts presented in this section or additional material relevant to this text are referred to H. M. Walker, *Mathematic Essentials in Elementary Statistics* (New York: Holt, Rinehart & Winston, 1951); C. L. Johnston and A. T. Willis, *Essential Algebra*, 2nd ed. (Belmont, Calif.: Wadsworth, 1978); and T. Carnevale and R. Schloming, *Encounters with Algebra* (New York: Harcourt Brace Jovanovich, 1980).

write

$$z \geq 2.56$$

Similarly, the symbol $\leq$ means "is less than or equal to." If z is less than or equal to zero, one can write

$$z \leq 0$$

In a few places, we will use the expression $|X|$, which is read "the absolute value of X." This means that regardless of whether X is a positive or negative number, $|X|$ equals its positive, or absolute, value. Thus,

$$|-5| = 5 \quad \text{and} \quad |5| = 5$$

and if $W = 4$ and $Y = 7$,

$$|W - Y| = |4 - 7| = |-3| = 3$$

Signed numbers

Perhaps the first real problem students encounter in algebra is negative numbers and how to perform simple mathematical operations (e.g., addition, subtraction, multiplication, division) with them. It helps to think of the number scale as running from large negative numbers at the left, through zero, to large positive numbers at the right.

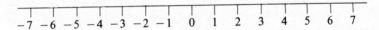

We use such a scale every day to measure temperature. Notice that any given number (except zero) has both quantity and sign. If you must locate 4 on the scale, you need to know if it is a $+4$ or a -4. The value 4 is its quantity, or distance from zero, while the $+$ or $-$ is its sign, or direction from zero. When a number is given no sign it is assumed to be positive. Thus, 4 is understood to be $+4$.

Addition and subtraction   Adding and subtracting are *operations*, and mathematical operations, like other kinds of operations, require a sort of movement. Adding or subtracting can be seen as moving up or down the scale a certain number of steps. Since positive numbers are to the right of zero while negative numbers are to the left of zero, adding 2 and 4 means starting at $+2$ and taking 4 steps to the right, thus ending at $+6$. Similarly, if you start at -3 and add 2—that is, move two steps to the right—you wind up at -1. If you start at -3 and add 5, you move five steps to the right, passing through zero to $+2$.

Finally, consider $(-4) + (7)$. The parentheses are included to help you distinguish between signs that describe the location of a number and signs that instruct you to perform a given operation. This expression says to start at -4 on the scale and move 7 steps to the right. $(-4) + (7) = +3$.

To subtract, you move to the left. Therefore, $6 - 4$ directs one to start at $+6$ and move four steps to the left, stopping at $+2$. The problem $5 - 8$ is solved by starting at $+5$ and moving eight steps to the left, passing through zero and stopping at -3.

Signs indicating location and signs indicating operations have essentially the same meaning—plus means to the right while minus means to the left. Because they are similar, they oppose each other when they occur in the same expression. As we have seen, adding a positive number is "ordinary" addition:

$$4 + (+2) = 6$$
$$3 + (+4) = 7$$
$$-5 + (+3) = -2$$

But adding a negative number is the same as subtracting that quantity:

$$4 + (-2) = 2$$
$$3 + (-4) = -1$$
$$-5 - (+3) = -8$$

But—and here is the special new case—subtracting a negative number is the same as adding that quantity:

$$4 - (-2) = 6$$
$$3 - (-4) = 7$$
$$-5 - (-3) = -2$$

Notice that the answers in the first and last sets are identical—adding a positive number is the same as subtracting a negative number. "Minus a minus is a plus."

Multiplication   When a string of numbers that have different signs is multiplied, the product is positive if there is an even number of negative terms to be multiplied:

$$(-5)(-2) = 10$$
$$(-4)(1)(2)(-3) = 24$$
$$(-a)(-b)(-c)(-d) = abcd$$

The product is negative if there is an odd number of negative terms:

$$(-4)(3) = -12$$
$$(-3)(2)(5) = -30$$
$$(-a)(b)(-c)(-d) = -abcd$$

Division   The same rule applies to division. If there is an even number of negative terms, the result is positive:

$$\frac{-4}{-3} = 1.33$$

$$\frac{(-a)(b)}{-c} \quad \text{or} \quad \frac{(-a)(-b)}{c} = \frac{ab}{c}$$

If there is an odd number of negative terms in the division, the result is negative:

$$\frac{6}{-3} = -2$$

$$\frac{c}{(-b)(d)} \quad \text{or} \quad \frac{-c}{(b)(d)} = -\frac{c}{bd}$$

Students very often wonder if the statistical formulas presented in the text will work if some or all of the data (original numbers) are negative. Yes, they all work, even with negative numbers. However, the negative signs must be handled properly, as described above.

Fractions

Multiplication   The product of two or more fractions equals the product of the numerators divided by the product of the denominators:

$$\frac{1}{2} \cdot \frac{3}{5} = \frac{(1)(3)}{(2)(5)} = \frac{3}{10}$$

In general, multiply $\frac{a}{b}$ times $\frac{c}{d}$ as follows:

$$\frac{a}{b} \cdot \frac{c}{d} = \frac{ac}{bd}$$

Division   To divide one fraction by another, invert the divisor (the fraction you want to divide by) and multiply:

$$\frac{1}{2} \div \frac{1}{3} = \frac{1}{2} \cdot \frac{3}{1} = \frac{3}{2} = 1.5$$

In general,

$$\frac{a}{b} \div \frac{c}{d} = \frac{a}{b} \cdot \frac{d}{c} = \frac{ad}{bc}$$

Reducing fractions   If large numbers are involved in fractions or if one must multiply several fractions together, one can frequently simplify the computations by reducing the fractions. To do this, divide both numerator and denominator by the same number, selecting one that goes evenly into each.

Dividing by 2,

$$\frac{2}{4} \text{ becomes } \frac{2^1}{4_2} = \frac{1}{2}$$

Dividing by 3,

$$\frac{6}{15} \text{ becomes } \frac{6^2}{15_5} = \frac{2}{5}$$

Sometimes you need to divide several times before obtaining a numerator and denominator that cannot be divided further by the same number.

Dividing by 2,

$$\frac{12}{20} \text{ becomes } \frac{12^6}{20_{15}} = \frac{6}{15}$$

which, dividing by 3, becomes

$$\frac{6^2}{15_5} = \frac{2}{5}$$

Of course, if you see that 12 and 30 can both be divided by 6, the answer will be obtained sooner:

Dividing by 6,

$$\frac{12}{30} \text{ becomes } \frac{\cancel{12}^2}{\cancel{30}_5} = \frac{2}{5}$$

Essentially the same process is involved when there is a string of fractions to be multiplied:

$$\left(\frac{\cancel{2}^1}{\cancel{3}}\right)\left(\frac{\cancel{3}^1}{\cancel{4}}\right) = \frac{1}{2}$$

$$\left(\frac{\cancel{3}^1}{\cancel{5}}\right)\left(\frac{2}{\cancel{3}}\right)\left(\frac{\cancel{5}^2}{\cancel{6}}\right) = \frac{1}{3}$$

Occasionally, the reduction can become rather complicated:

$$\left(\frac{1}{\cancel{3}}\right)\left(\frac{\cancel{3}}{\cancel{8}}\right)\left(\frac{\cancel{12}}{\cancel{35}}\right)\left(\frac{\cancel{7}}{\cancel{3}}\right)\left(\frac{\cancel{5}}{\cancel{8}}\right)\left(\frac{2}{\cancel{3}}\right) = \frac{1}{4}$$

Addition and subtraction   Before fractional quantities can be added, the denominators of the two fractions must be made equal. Thus, to add $\frac{1}{2}$ and $\frac{1}{3}$ it is necessary to change both fractions to sixths. This is accomplished by multiplying the first fraction by $\frac{3}{3}$ and the second by $\frac{2}{2}$:

$$\frac{1}{2} + \frac{1}{3} = ?$$

$$\frac{1}{2}\left(\frac{3}{3}\right) + \frac{1}{3}\left(\frac{2}{2}\right) = \frac{3}{6} + \frac{2}{6} = \frac{5}{6}$$

More generally, to add $\frac{a}{b} + \frac{c}{d}$,

$$\frac{a}{b} + \frac{c}{d} = \frac{a}{b}\left(\frac{d}{d}\right) + \left(\frac{b}{b}\right)\frac{c}{d} = \frac{ad}{bd} + \frac{bc}{bd} = \frac{ad + bc}{bd}$$

Working with zero   In some formulas requiring multiplication or division, one or more terms might be zero. It is helpful to remember that

$$\frac{0}{n} = 0$$

that

$$0 \cdot n = 0$$

Thus,

$$\frac{(0)(4)(7)(2)}{(6)(14)} = 0$$

In other words, any number multiplied by 0 equals 0, and 0 divided by any nonzero number equals 0.

Factorials

In some probability problems it is necessary to multiply a positive integer (a whole number) by each of the integers having a value less than that integer, ending with 1. This string of multiplications of integers is called a *factorial* and the sign "!" following the largest integer in the string is used to indicate this operation. For example, 3!, read "three factorial," would equal

$$3! = 3 \cdot 2 \cdot 1 = 6$$

In general, $n!$ means

$$n! = n(n - 1)(n - 2)(n - 3) \cdots (1)$$

Note also that, by custom,

$$0! = 1$$

Exponents

An exponent is a number written as a superscript to a base number and signifies that the base number should be multiplied by itself as many times as the exponent states. Therefore,

$$2^3 = 2 \cdot 2 \cdot 2 = 8$$

and more generally

$$n^r = \underbrace{n \cdot n \cdot n \cdots n}_{r \text{ times}}$$

Note also that

$$n^1 = n$$

and

$$n^0 = 1$$

Addition and subtraction   Generally, numbers with exponents cannot be added or subtracted without first carrying out the exponentiation of each quantity. This is true even if the numbers have the same base. For example, $2^2 + 2^3$ is handled by performing the indicated exponentiation and *then* adding:

$$2^2 + 2^3 = 2 \cdot 2 + 2 \cdot 2 \cdot 2 = 4 + 8 = 12$$

Multiplication   The product of two exponential quantities with the *same base number* equals that base number raised to the sum of the two exponents. For example,

$$(2^2)(2^3) = 2^{2+3} = 2^5$$

because

$$(2^2)(2^3) = (2 \cdot 2)(2 \cdot 2 \cdot 2) = 2^5$$

More generally,

$$(n^r)(n^s) = n^{r+s}$$

If the exponents *do not have the same base number* (e.g., $2^4 \times 3^4$), this procedure does not apply and the exponentiation should be carried out first.

Division   The quotient of two exponential quantities with the *same base number* is that base raised to the difference between the exponent of the numerator and that of the denominator. For example,

$$\frac{2^3}{2^2} = 2^{3-2} = 2^1 = 2$$

More generally,

$$\frac{n^r}{n^s} = n^{r-s}$$

It is helpful to remember that if s is larger than r in the expression n^r/n^s, then the result has a negative exponent.

Whenever a number has a negative exponent, a reciprocal $\left(\dfrac{1}{\text{number}}\right)$ is taken before the exponentiation is carried out. For example,

$$2^{-3} = \frac{1}{2^3} = \frac{1}{2 \cdot 2 \cdot 2} = \frac{1}{8}$$

$$\frac{3^2}{3^5} = 3^{2-5} = \frac{1}{3^3} = \frac{1}{27}$$

Fractions   A fraction raised to a power equals the numerator raised to that power divided by the denominator raised to that power. For example,

$$\left(\frac{3}{5}\right)^2 = \frac{3^2}{5^2} = \frac{9}{25}$$

More generally,

$$\left(\frac{r}{s}\right)^n = \frac{r^n}{s^n}$$

Binomial expansion   A *binomial* is an expression involving the addition or subtraction of two quantities. Many of the algebraic manipulations presented in this text require the student to understand the expansion of a binomial, such as $(a + b)^2$:

$$(a + b)^2 = a^2 + 2ab + b^2$$

This result is obtained by taking the square of the first term in the binomial (here, a^2), plus 2 times the product of the two terms in the binomial ($2ab$), plus the square of the second term (b^2). A more common problem is to expand $(a - b)^2$. This is accomplished in the same way but with attention to the algebraic signs:

$$(a - b)^2 = a^2 + 2a(-b) + (-b)^2 = a^2 - 2ab + b^2$$

Again, the result equals the square of the first term (a^2), plus 2 times the product of the two terms $[2(a)(-b) = -2ab]$, plus the square of the last term $[(-b)^2 = b^2]$. The student must remember that these procedures apply to any binomial regardless of the specific terms.

For example, quite frequently it will be necessary to expand $(X_i - \bar{X})^2$. In this expression, X_i stands for any particular score in a group of scores; $\bar{X}$ refers to the average of all the scores in the group. The result is generated just as in the examples above:

$$\left(X_i - \bar{X}\right)^2 = X_i^2 - 2X_i\bar{X} + \bar{X}^2$$

Square roots

Many statistical formulas require taking a square root, the reverse operation to squaring. Since it is assumed that students have calculators for this purpose, the way to extract a square root by hand will not be reviewed here. However, notice that in addition to symbolizing a mathematical operation, the square root sign (i.e., the radical) is treated as if it were parentheses enclosing a quantity. For example,

$$\sqrt{a^2} = a \text{ (technically, either } + a \text{ or } - a)$$

$$3\sqrt{4} = 3(\sqrt{4}) = 3(2) = 6$$

$$2\sqrt{4 + 21} = 2(\sqrt{25}) = 2(5) = 10$$

The radical can be applied to each number in a complex term that calls for multiplication or division but not to the numbers in a term that calls for addition or subtraction. For example:

$$\sqrt{4a} = \sqrt{4}\,\sqrt{a} = 2\sqrt{a}$$

$$\sqrt{\frac{a + b}{c}} = \frac{\sqrt{a + b}}{\sqrt{c}}$$

But

$$\sqrt{a + b} \text{ does not equal } \sqrt{a} + \sqrt{b}$$

and

$$\sqrt{a - b} \text{ does not equal } \sqrt{a} - \sqrt{b}$$

Factoring and simplification

Factoring is the process of breaking down a number into parts which, when multiplied together, equal the number. *Simplification* is the process of reducing the number of quantities in an expression. Sometimes an expression can be simplified by factoring it first.

Removing parentheses   Occasionally, algebraic expressions may be simplified by removing parentheses. When no multiplications or divisions are involved, one simply removes the parentheses and performs any required addition or subtraction:

$$4 + (2) = 4 + 2 = 6$$
$$6 + (-5) = 6 - 5 = 1$$
$$3 - (-2) + 7 = 3 + 2 + 7 = 12$$
$$a + (-b) - (-c) = a - b + c$$

Sometimes squaring or taking a square root is combined with multiplying or dividing by another quantity. In such cases, one must be very careful which operations are performed on which numbers. For example,

$$2s^2 = 2ss$$
$$(2s)^2 = (2s)(2s) = 4s^2$$
$$4(2x)^2 = 4(2x)(2x) = 16x^2$$

Similarly,

$$\sqrt{4s^2} = 2s$$
$$\sqrt{2p^2} = p\sqrt{2}$$
$$\tfrac{1}{2}\sqrt{4x} = \left(\tfrac{1}{2}\right)(2)\sqrt{x} = \sqrt{x}$$

The sequence in which quantities are added or subtracted does not matter. For example,

$$4 + 3 = 3 + 4 = 7$$
$$4 - 3 = -3 + 4 = 1$$
$$3 + (a - 2) = (a - 2) + 3 = a + 1$$

Similarly, the sequence in which quantities are multiplied or divided does not matter:

$$4(3) = 3(4) = 12$$
$$2b = b2$$
$$2b3a = 6ab$$
$$3(4 - 5) = (-5 + 4)3 = -3$$

But when expressions become more complicated, the sequence of operations does matter. Sometimes an expression to be simplified is quite complex, involving several different terms. A term is a quantity to be added to or subtracted from another quantity. For example, each of the following expressions contains two terms:

$$4 + 3$$
$$2x + y$$
$$6\sqrt{5y} \; - \; \frac{4(x + y)^2}{3}$$

Notice that multiplication and division as well as exponentiation and taking roots can be required within a term. In a complex expression, it helps to perform the mathematical operations in a given sequence, specifically:

1. Perform any addition or subtraction *within* terms, parentheses, or radicals.
2. Calculate any exponents or roots.
3. Compute any multiplications or divisions.
4. Perform any addition or subtraction *between* terms.

For example, suppose you must simplify the expression

$$ab + 3[a - b(4 - 2)]^2$$

First, perform any addition or subtraction within terms, parentheses, or radicals:

$$ab + 3[a - b(2)]^2$$
$$ab + 3[a - 2b]^2$$

Second, calculate any exponents or roots:

$$ab + 3\left[a^2 - 2(a2b) + (2b)^2 \right]$$
$$ab + 3[a^2 - 4ab + 4b^2]$$

Third, compute any multiplications or divisions:

$$ab + 3a^2 - 3(4ab) + 3(4b^2)$$
$$ab + 3a^2 - 12ab + 12b^2$$

Fourth, perform any addition or subtraction between terms:

$$3a^2 - 11ab + 12b^2$$

Transposing   Sometimes it is convenient to transpose a term from one side of an equation to the other. When a term is to be transposed, it is added or subtracted on both sides of the equation:

$$
\begin{array}{r}
a - b = c \\
\text{plus:} \quad -a = -a \\
\hline
a - b - a = c - a \\
-b = c - a
\end{array}
$$

The operation amounts to placing the term (here, a) on the other side of the equation with a change of sign.

If the term to be transposed is involved in multiplication or division, the opposite operation (i.e., division or multiplication) must be applied to both sides of the equation:

$$\frac{a}{b} = c$$

$$\left(\frac{a}{b}\right)\left(\frac{b}{1}\right) = c\left(\frac{b}{1}\right)$$

$$a = cb$$

Complex factoring   Factoring an algebraic expression involves determining the simplest set of quantities which, when multiplied together, will yield the original quantities:

$$ab + ac = a(b + c)$$

$$- ab - ac = -a(b + c)$$

Factoring is often used to simplify algebraic expressions. Consider the following example:

$$\frac{-a(b - c) - (c - ab)}{a - 1}$$

Carrying out the appropriate multiplications in the numerator, one obtains

$$\frac{- ab - a(-c) - (c - ab)}{a - 1}$$

and simplifying the signs of certain expressions, one obtains

$$\frac{- ab + ac - c + ab}{a - 1}$$

By subtracting,

$$\frac{ac - c}{a - 1}$$

factoring,

$$\frac{c(a - 1)}{a - 1}$$

and dividing, one finds the expression reduces to

$$c$$

Sometimes in the course of simplifying an expression it helps to divide each term in the numerator by the denominator:

$$\frac{a - b}{c} = \frac{a}{c} - \frac{b}{c}$$

However, the student must be cautious. The above manipulation is correct but the following manipulations are not correct.

$$\frac{c}{a - b} \quad \textit{does not equal} \quad \frac{c}{a} - \frac{c}{b}$$

$$\frac{a - b}{c - d} \quad \textit{does not equal} \quad \frac{a}{c} - \frac{b}{d}$$

EXERCISES

1. Perform the indicated operations.

   a. $\dfrac{2}{5} + \dfrac{1}{5}$

   b. $\dfrac{4}{6} + \dfrac{3}{6}$

   c. $\dfrac{3}{8} + \dfrac{5}{16}$

   d. $\dfrac{3}{5} + \dfrac{4}{7}$

   e. $\dfrac{5}{7} - \dfrac{2}{7}$

   f. $\dfrac{3}{4} - \dfrac{1}{8}$

   g. $\dfrac{2}{5} - \dfrac{1}{9}$

   h. $\dfrac{2}{7} - \dfrac{4}{11}$

2. Perform the indicated operations.

   a. $\dfrac{1}{4} \cdot \dfrac{2}{4}$

   b. $\dfrac{4}{5} \cdot \dfrac{1}{3}$

   c. $\dfrac{3}{5} \cdot \dfrac{10}{21} \cdot \dfrac{7}{12} \cdot \dfrac{6}{7}$

   d. $\dfrac{1}{2} \div \dfrac{2}{3}$

   e. $\dfrac{3}{4} \div \dfrac{7}{8}$

   f. $\dfrac{9}{11} \div \dfrac{3}{22}$

3. Simplify the following.

   a. $5!$ b. $\dfrac{4!}{2!}$ c. $\dfrac{3!5!}{4!}$

4. Simplify the following.

   a. 2^3

   b. $3^2 + 3^3$

   c. $2^2 + 3^2$

   d. $4^3 - 2^3$

   e. $2^2 \cdot 2^3$

   f. $3^2 \cdot 2^3$

   g. $2^3 \div 2^2$

   h. $3^3 \div 3^4$

   i. $4^3 \div 2^3$

   j. $\left(\dfrac{3}{5}\right)^2$

   k. $\left(\dfrac{4}{5}\right)^3 \div \dfrac{2}{25^2}$

   l. $\left(\dfrac{2}{6}\right)^2 \cdot \left(\dfrac{1}{4}\right)^3$

5. Expand and simplify the following expressions.

a. $(x + y)^2$

f. $\dfrac{ab - c}{b}$

b. $(p - q)^2$

g. $\dfrac{abc + bdc}{cb(a - b)}$

c. $(X - \bar{X})^2$

h. $\sqrt{4X^2}$

d. $(ab - bc)^2$

i. $\sqrt{3Y^2}$

e. $\dfrac{a(b - c) + (ac - ab)}{a - 1}$

j. $\sqrt{X(3X^2)}$

6. For each of the following sets, determine the sum, the square of the sum, and the sum of the squared numbers (called sum of squares).

a.	**b.**	**c.**
2	7	3
5	4	−4
1	0	8
	3	−1
		0
		5

tables

appendix

Table A. Proportions of Area under the Standard Normal Curve

z			z			z		
0.00	.0000	.5000	0.55	.2088	.2912	1.10	.3643	.1357
0.01	.0040	.4960	0.56	.2123	.2877	1.11	.3665	.1335
0.02	.0080	.4920	0.57	.2157	.2843	1.12	.3686	.1314
0.03	.0120	.4880	0.58	.2190	.2810	1.13	.3708	.1292
0.04	.0160	.4840	0.59	.2224	.2776	1.14	.3729	.1271
0.05	.0199	.4801	0.60	.2257	.2743	1.15	.3749	.1251
0.06	.0239	.4761	0.61	.2291	.2709	1.16	.3770	.1230
0.07	.0279	.4721	0.62	.2324	.2676	1.17	.3790	.1210
0.08	.0319	.4681	0.63	.2357	.2643	1.18	.3810	.1190
0.09	.0359	.4641	0.64	.2389	.2611	1.19	.3830	.1170
0.10	.0398	.4602	0.65	.2422	.2578	1.20	.3849	.1151
0.11	.0438	.4562	0.66	.2454	.2546	1.21	.3869	.1131
0.12	.0478	.4522	0.67	.2486	.2514	1.22	.3888	.1112
0.13	.0517	.4483	0.68	.2517	.2483	1.23	.3907	.1093
0.14	.0557	.4443	0.69	.2549	.2451	1.24	.3925	.1075
0.15	.0596	.4404	0.70	.2580	.2420	1.25	.3944	.1056
0.16	.0636	.4364	0.71	.2611	.2389	1.26	.3962	.1038
0.17	.0675	.4325	0.72	.2642	.2358	1.27	.3980	.1020
0.18	.0714	.4286	0.73	.2673	.2327	1.28	.3997	.1003
0.19	.0753	.4247	0.74	.2704	.2296	1.29	.4015	.0985
0.20	.0793	.4207	0.75	.2734	.2266	1.30	.4032	.0968
0.21	.0832	.4168	0.76	.2764	.2236	1.31	.4049	.0951
0.22	.0871	.4129	0.77	.2794	.2206	1.32	.4066	.0934
0.23	.0910	.4090	0.78	.2823	.2177	1.33	.4082	.0918
0.24	.0948	.4052	0.79	.2852	.2148	1.34	.4099	.0901
0.25	.0987	.4013	0.80	.2881	.2119	1.35	.4115	.0885
0.26	.1026	.3974	0.81	.2910	.2090	1.36	.4131	.0869
0.27	.1064	.3936	0.82	.2939	.2061	1.37	.4147	.0853
0.28	.1103	.3897	0.83	.2967	.2033	1.38	.4162	.0838
0.29	.1141	.3859	0.84	.2995	.2005	1.39	.4177	.0823
0.30	.1179	.3821	0.85	.3023	.1977	1.40	.4192	.0808
0.31	.1217	.3783	0.86	.3051	.1949	1.41	.4207	.0793
0.32	.1255	.3745	0.87	.3078	.1922	1.42	.4222	.0778
0.33	.1293	.3707	0.88	.3106	.1894	1.43	.4236	.0764
0.34	.1331	.3669	0.89	.3133	.1867	1.44	.4251	.0749
0.35	.1368	.3632	0.90	.3159	.1841	1.45	.4265	.0735
0.36	.1406	.3594	0.91	.3186	.1814	1.46	.4279	.0721
0.37	.1443	.3557	0.92	.3212	.1788	1.47	.4292	.0708
0.38	.1480	.3520	0.93	.3238	.1762	1.48	.4306	.0694
0.39	.1517	.3483	0.94	.3264	.1736	1.49	.4319	.0681
0.40	.1554	.3446	0.95	.3289	.1711	1.50	.4332	.0668
0.41	.1591	.3409	0.96	.3315	.1685	1.51	.4345	.0655
0.42	.1628	.3372	0.97	.3340	.1660	1.52	.4357	.0643
0.43	.1664	.3336	0.98	.3365	.1635	1.53	.4370	.0630
0.44	.1700	.3300	0.99	.3389	.1611	1.54	.4382	.0618
0.45	.1736	.3264	1.00	.3413	.1587	1.55	.4394	.0606
0.46	.1772	.3228	1.01	.3438	.1562	1.56	.4406	.0594
0.47	.1808	.3192	1.02	.3461	.1539	1.57	.4418	.0582
0.48	.1844	.3156	1.03	.3485	.1515	1.58	.4429	.0571
0.49	.1879	.3121	1.04	.3508	.1492	1.59	.4441	.0559
0.50	.1915	.3085	1.05	.3531	.1469	1.60	.4452	.0548
0.51	.1950	.3050	1.06	.3554	.1446	1.61	.4463	.0537
0.52	.1985	.3015	1.07	.3577	.1423	1.62	.4474	.0526
0.53	.2019	.2981	1.08	.3599	.1401	1.63	.4484	.0516
0.54	.2054	.2946	1.09	.3621	.1379	1.64	.4495	.0505

Source: P. Runyon and Audrey Haber, *Fundamentals of Behavioral Statistics*, 3rd ed., © 1976. Addison-Wesley, Reading, Massachusetts. Table A. Reprinted with permission.

Table A (continued)

z	0 z	0 z	z	0 z	0 z	z	0 z	0 z
1.65	.4505	.0495	2.22	.4868	.0132	2.79	.4974	.0026
1.66	.4515	.0485	2.23	.4871	.0129	2.80	.4974	.0026
1.67	.4525	.0475	2.24	.4875	.0125	2.81	.4975	.0025
1.68	.4535	.0465	2.25	.4878	.0122	2.82	.4976	.0024
1.69	.4545	.0455	2.26	.4881	.0119	2.83	.4977	.0023
1.70	.4554	.0446	2.27	.4884	.0116	2.84	.4977	.0023
1.71	.4564	.0436	2.28	.4887	.0113	2.85	.4978	.0022
1.72	.4573	.0427	2.29	.4890	.0110	2.86	.4979	.0021
1.73	.4582	.0418	2.30	.4893	.0107	2.87	.4979	.0021
1.74	.4591	.0409	2.31	.4896	.0104	2.88	.4980	.0020
1.75	.4599	.0401	2.32	.4898	.0102	2.89	.4981	.0019
1.76	.4608	.0392	2.33	.4901	.0099	2.90	.4981	.0019
1.77	.4616	.0384	2.34	.4904	.0096	2.91	.4982	.0018
1.78	.4625	.0375	2.35	.4906	.0094	2.92	.4982	.0018
1.79	.4633	.0367	2.36	.4909	.0091	2.93	.4983	.0017
1.80	.4641	.0359	2.37	.4911	.0089	2.94	.4984	.0016
1.81	.4649	.0351	2.38	.4913	.0087	2.95	.4984	.0016
1.82	.4656	.0344	2.39	.4916	.0084	2.96	.4985	.0015
1.83	.4664	.0336	2.40	.4918	.0082	2.97	.4985	.0015
1.84	.4671	.0329	2.41	.4920	.0080	2.98	.4986	.0014
1.85	.4678	.0322	2.42	.4922	.0078	2.99	.4986	.0014
1.86	.4686	.0314	2.43	.4925	.0075	3.00	.4987	.0013
1.87	.4693	.0307	2.44	.4927	.0073	3.01	.4987	.0013
1.88	.4699	.0301	2.45	.4929	.0071	3.02	.4987	.0013
1.89	.4706	.0294	2.46	.4931	.0069	3.03	.4988	.0012
1.90	.4713	.0287	2.47	.4932	.0068	3.04	.4988	.0012
1.91	.4719	.0281	2.48	.4934	.0066	3.05	.4989	.0011
1.92	.4726	.0274	2.49	.4936	.0064	3.06	.4989	.0011
1.93	.4732	.0268	2.50	.4938	.0062	3.07	.4989	.0011
1.94	.4738	.0262	2.51	.4940	.0060	3.08	.4990	.0010
1.95	.4744	.0256	2.52	.4941	.0059	3.09	.4990	.0010
1.96	.4750	.0250	2.53	.4943	.0057	3.10	.4990	.0010
1.97	.4756	.0244	2.54	.4945	.0055	3.11	.4991	.0009
1.98	.4761	.0239	2.55	.4946	.0054	3.12	.4991	.0009
1.99	.4767	.0233	2.56	.4948	.0052	3.13	.4991	.0009
2.00	.4772	.0228	2.57	.4949	.0051	3.14	.4992	.0008
2.01	.4778	.0222	2.58	.4951	.0049	3.15	.4992	.0008
2.02	.4783	.0217	2.59	.4952	.0048	3.16	.4992	.0008
2.03	.4788	.0212	2.60	.4953	.0047	3.17	.4992	.0008
2.04	.4793	.0207	2.61	.4955	.0045	3.18	.4993	.0007
2.05	.4798	.0202	2.62	.4956	.0044	3.19	.4993	.0007
2.06	.4803	.0197	2.63	.4957	.0043	3.20	.4993	.0007
2.07	.4808	.0192	2.64	.4959	.0041	3.21	.4993	.0007
2.08	.4812	.0188	2.65	.4960	.0040	3.22	.4994	.0006
2.09	.4817	.0183	2.66	.4961	.0039	3.23	.4994	.0006
2.10	.4821	.0179	2.67	.4962	.0038	3.24	.4994	.0006
2.11	.4826	.0174	2.68	.4963	.0037	3.25	.4994	.0006
2.12	.4830	.0170	2.69	.4964	.0036	3.30	.4995	.0005
2.13	.4834	.0166	2.70	.4965	.0035	3.35	.4996	.0004
2.14	.4838	.0162	2.71	.4966	.0034	3.40	.4997	.0003
2.15	.4842	.0158	2.72	.4967	.0033	3.45	.4997	.0003
2.16	.4846	.0154	2.73	.4968	.0032	3.50	.4998	.0002
2.17	.4850	.0150	2.74	.4969	.0031	3.60	.4998	.0002
2.18	.4854	.0146	2.75	.4970	.0030	3.70	.4999	.0001
2.19	.4857	.0143	2.76	.4971	.0029	3.80	.4999	.0001
2.20	.4861	.0139	2.77	.4972	.0028	3.90	.49995	.00005
2.21	.4864	.0136	2.78	.4973	.0027	4.00	.49997	.00003

Table **B**. Critical Values of *t*

	Level of significance for a directional (one-tailed) test					
	.10	.05	.025	.01	.005	.0005
	Level of significance for a non-directional (two-tailed) test					
df	.20	.10	.05	.02	.01	.001
1	3.078	6.314	12.706	31.821	63.657	636.619
2	1.886	2.920	4.303	6.965	9.925	31.598
3	1.638	2.353	3.182	4.541	5.841	12.941
4	1.533	2.132	2.776	3.747	4.604	8.610
5	1.476	2.015	2.571	3.365	4.032	6.859
6	1.440	1.943	2.447	3.143	3.707	5.959
7	1.415	1.895	2.365	2.998	3.499	5.405
8	1.397	1.860	2.306	2.896	3.355	5.041
9	1.383	1.833	2.262	2.821	3.250	4.781
10	1.372	1.812	2.228	2.764	3.169	4.587
11	1.363	1.796	2.201	2.718	3.106	4.437
12	1.356	1.782	2.179	2.681	3.055	4.318
13	1.350	1.771	2.160	2.650	3.012	4.221
14	1.345	1.761	2.145	2.624	2.977	4.140
15	1.341	1.753	2.131	2.602	2.947	4.073
16	1.337	1.746	2.120	2.583	2.921	4.015
17	1.333	1.740	2.110	2.567	2.898	3.965
18	1.330	1.734	2.101	2.552	2.878	3.922
19	1.328	1.729	2.093	2.539	2.861	3.883
20	1.325	1.725	2.086	2.528	2.845	3.850
21	1.323	1.721	2.080	2.518	2.831	3.819
22	1.321	1.717	2.074	2.508	2.819	3.792
23	1.319	1.714	2.069	2.500	2.807	3.767
24	1.318	1.711	2.064	2.492	2.797	3.745
25	1.316	1.708	2.060	2.485	2.787	3.725
26	1.315	1.706	2.056	2.479	2.779	3.707
27	1.314	1.703	2.052	2.473	2.771	3.690
28	1.313	1.701	2.048	2.467	2.763	3.674
29	1.311	1.699	2.045	2.462	2.756	3.659
30	1.310	1.697	2.042	2.457	2.750	3.646
40	1.303	1.684	2.021	2.423	2.704	3.551
60	1.296	1.671	2.000	2.390	2.660	3.460
120	1.289	1.658	1.980	2.358	2.617	3.373
∞	1.282	1.645	1.960	2.326	2.576	3.291

The value listed in the table is the critical value of *t* for the number of degrees of freedom listed in the left column for a directional (one-tailed) or nondirectional (two-tailed) test at the significance level indicated at the top of each column. If the observed *t* is *greater than or equal to* the tabled value, reject H_0. Since the *t* distribution is symmetrical about $t = 0$, these critical values represent both + and − values for nondirectional tests.

Source: Table B is taken from Table III of Fisher and Yates, *Statistical Tables for Biological, Agricultural and Medical Research*, published by Longman Group Ltd., London (previously published by Oliver and Boyd, Ltd., Edinburgh), and by permission of the authors and publishers.

Table **C.**   Critical Values of the Pearson Product–Moment Correlation Coefficient

	Level of significance for a directional (one-tailed) test				
	.05	.025	.01	.005	.0005
	Level of significance for a non-directional (two-tailed) test†				
$df = N-2$	.10	.05	.02	.01	.001
1	.9877	.9969	.9995	.9999	1.0000
2	.9000	.9500	.9800	.9900	.9990
3	.8054	.8783	.9343	.9587	.9912
4	.7293	.8114	.8822	.9172	.9741
5	.6694	.7545	.8329	.8745	.9507
6	.6215	.7067	.7887	.8343	.9249
7	.5822	.6664	.7498	.7977	.8982
8	.5494	.6319	.7155	.7646	.8721
9	.5214	.6021	.6851	.7348	.8471
10	.4973	.5760	.6581	.7079	.8233
11	.4762	.5529	.6339	.6835	.8010
12	.4575	.5324	.6120	.6614	.7800
13	.4409	.5139	.5923	.6411	.7603
14	.4259	.4973	.5742	.6226	.7420
15	.4124	.4821	.5577	.6055	.7246
16	.4000	.4683	.5425	.5897	.7084
17	.3887	.4555	.5285	.5751	.6932
18	.3783	.4438	.5155	.5614	.6787
19	.3687	.4329	.5034	.5487	.6652
20	.3598	.4227	.4921	.5368	.6524
25	.3233	.3809	.4451	.4869	.5974
30	.2960	.3494	.4093	.4487	.5541
35	.2746	.3246	.3810	.4182	.5189
40	.2573	.3044	.3578	.3932	.4896
45	.2428	.2875	.3384	.3721	.4648
50	.2306	.2732	.3218	.3541	.4433
60	.2108	.2500	.2948	.3248	.4078
70	.1954	.2319	.2737	.3017	.3799
80	.1829	.2172	.2565	.2830	.3568
90	.1726	.2050	.2422	.2673	.3375
100	.1638	.1946	.2301	.2540	.3211

If the observed value of r is *greater than or equal to* the tabled value for the appropriate level of significance (columns) and degrees of freedom (rows), reject H_0. The degrees of freedom are the number of pairs of scores minus two, or $N - 2$. The critical values in the table are both $+$ and $-$ for nondirectional (two-tailed) tests.

Source: Table C is taken from Table VII of Fisher and Yates, *Statistical Tables for Biological, Agricultural, and Medical Research*, published by Longman Group Ltd., London (previously published by Oliver and Boyd, Ltd., Edinburgh), and by permission of the authors and publishers.

Table D.   Transformation of r to z_r

r	z_r	r	z_r	r	z_r	r	z_r	r	z_r
.000	.000	.200	.203	.400	.424	.600	.693	.800	1.099
.005	.005	.205	.208	.405	.430	.605	.701	.805	1.113
.010	.010	.210	.213	.410	.436	.610	.709	.810	1.127
.015	.015	.215	.218	.415	.442	.615	.717	.815	1.142
.020	.020	.220	.224	.420	.448	.620	.725	.820	1.157
.025	.025	.225	.229	.425	.454	.625	.733	.825	1.172
.030	.030	.230	.234	.430	.460	.630	.741	.830	1.188
.035	.035	.235	.239	.435	.466	.635	.750	.835	1.204
.040	.040	.240	.245	.440	.472	.640	.758	.840	1.221
.045	.045	.245	.250	.445	.478	.645	.767	.845	1.238
.050	.050	.250	.255	.450	.485	.650	.775	.850	1.256
.055	.055	.255	.261	.455	.491	.655	.784	.855	1.274
.060	.060	.260	.266	.460	.497	.660	.793	.860	1.293
.065	.065	.265	.271	.465	.504	.665	.802	.865	1.313
.070	.070	.270	.277	.470	.510	.670	.811	.870	1.333
.075	.075	.275	.282	.475	.517	.675	.820	.875	1.354
.080	.080	.280	.288	.480	.523	.680	.829	.880	1.376
.085	.085	.285	.293	.485	.530	.685	.838	.885	1.398
.090	.090	.290	.299	.490	.536	.690	.848	.890	1.422
.095	.095	.295	.304	.495	.543	.695	.858	.895	1.447
.100	.100	.300	.310	.500	.549	.700	.867	.900	1.472
.105	.105	.305	.315	.505	.556	.705	.877	.905	1.499
.110	.110	.310	.321	.510	.563	.710	.887	.910	1.528
.115	.116	.315	.326	.515	.570	.715	.897	.915	1.557
.120	.121	.320	.332	.520	.576	.720	.908	.920	1.589
.125	.126	.325	.337	.525	.583	.725	.918	.925	1.623
.130	.131	.330	.343	.530	.590	.730	.929	.930	1.658
.135	.136	.335	.348	.535	.597	.735	.940	.935	1.697
.140	.141	.340	.354	.540	.604	.740	.950	.940	1.738
.145	.146	.345	.360	.545	.611	.745	.962	.945	1.783
.150	.151	.350	.365	.550	.618	.750	.973	.950	1.832
.155	.156	.355	.371	.555	.626	.755	.984	.955	1.886
.160	.161	.360	.377	.560	.633	.760	.996	.960	1.946
.165	.167	.365	.383	.565	.640	.765	1.008	.965	2.014
.170	.172	.370	.388	.570	.648	.770	1.020	.970	2.092
.175	.177	.375	.394	.575	.655	.775	1.033	.975	2.185
.180	.182	.380	.400	.580	.662	.780	1.045	.980	2.298
.185	.187	.385	.406	.585	.670	.785	1.058	.985	2.443
.190	.192	.390	.412	.590	.678	.790	1.071	.990	2.647
.195	.198	.395	.418	.595	.685	.795	1.085	.995	2.994

Source: From *Statistical Methods*, Second Edition, by Allen L. Edwards. Copyright © 1967 by Allen L. Edwards. First edition copyright 1954 by Allen L. Edwards under the title *Statistical Methods for the Behavioral Sciences*. Reprinted by permission of Holt, Rinehart and Winston.

Table E. Critical Values of F (.05 level in roman type, .01 level in boldface)

Degrees of freedom for the numerator (columns) / Degrees of freedom for the denominator (rows)

den \ num	1	2	3	4	5	6	7	8	9	10	11	12	14	16	20	24	30	40	50	75	100	200	500	∞
1	161 / 4,052	200 / 4,999	216 / 5,403	225 / 5,625	230 / 5,764	234 / 5,859	237 / 5,928	239 / 5,981	241 / 6,022	242 / 6,056	243 / 6,082	244 / 6,106	245 / 6,142	246 / 6,169	248 / 6,208	249 / 6,234	250 / 6,261	251 / 6,286	252 / 6,302	253 / 6,323	253 / 6,334	254 / 6,352	254 / 6,361	254 / 6,366
2	18.51 / 98.49	19.00 / 99.00	19.16 / 99.17	19.25 / 99.25	19.30 / 99.30	19.33 / 99.33	19.36 / 99.36	19.37 / 99.37	19.38 / 99.39	19.39 / 99.40	19.40 / 99.41	19.41 / 99.42	19.42 / 99.43	19.43 / 99.44	19.44 / 99.45	19.45 / 99.46	19.46 / 99.47	19.47 / 99.48	19.47 / 99.48	19.48 / 99.49	19.49 / 99.49	19.49 / 99.49	19.50 / 99.50	19.50 / 99.50
3	10.13 / 34.12	9.55 / 30.82	9.28 / 29.46	9.12 / 28.71	9.01 / 28.24	8.94 / 27.91	8.88 / 27.67	8.84 / 27.49	8.81 / 27.34	8.78 / 27.23	8.76 / 27.13	8.74 / 27.05	8.71 / 26.92	8.69 / 26.83	8.66 / 26.69	8.64 / 26.60	8.62 / 26.50	8.60 / 26.41	8.58 / 26.35	8.57 / 26.27	8.56 / 26.23	8.54 / 26.18	8.54 / 26.14	8.53 / 26.12
4	7.71 / 21.20	6.94 / 18.00	6.59 / 16.69	6.39 / 15.98	6.26 / 15.52	6.16 / 15.21	6.09 / 14.98	6.04 / 14.80	6.00 / 14.66	5.96 / 14.54	5.93 / 14.45	5.91 / 14.37	5.87 / 14.24	5.84 / 14.15	5.80 / 14.02	5.77 / 13.93	5.74 / 13.83	5.71 / 13.74	5.70 / 13.69	5.68 / 13.61	5.66 / 13.57	5.65 / 13.52	5.64 / 13.48	5.63 / 13.46
5	6.61 / 16.26	5.79 / 13.27	5.41 / 12.06	5.19 / 11.39	5.05 / 10.97	4.95 / 10.67	4.88 / 10.45	4.82 / 10.29	4.78 / 10.15	4.74 / 10.05	4.70 / 9.96	4.68 / 9.89	4.64 / 9.77	4.60 / 9.68	4.56 / 9.55	4.53 / 9.47	4.50 / 9.38	4.46 / 9.29	4.44 / 9.24	4.42 / 9.17	4.40 / 9.13	4.38 / 9.07	4.37 / 9.04	4.36 / 9.02
6	5.99 / 13.74	5.14 / 10.92	4.76 / 9.78	4.53 / 9.15	4.39 / 8.75	4.28 / 8.47	4.21 / 8.26	4.15 / 8.10	4.10 / 7.98	4.06 / 7.87	4.03 / 7.79	4.00 / 7.72	3.96 / 7.60	3.92 / 7.52	3.87 / 7.39	3.84 / 7.31	3.81 / 7.23	3.77 / 7.14	3.75 / 7.09	3.72 / 7.02	3.71 / 6.99	3.69 / 6.94	3.68 / 6.90	3.67 / 6.88
7	5.59 / 12.25	4.74 / 9.55	4.35 / 8.45	4.12 / 7.85	3.97 / 7.46	3.87 / 7.19	3.79 / 7.00	3.73 / 6.84	3.68 / 6.71	3.63 / 6.62	3.60 / 6.54	3.57 / 6.47	3.52 / 6.35	3.49 / 6.27	3.44 / 6.15	3.41 / 6.07	3.38 / 5.98	3.34 / 5.90	3.32 / 5.85	3.29 / 5.78	3.28 / 5.75	3.25 / 5.70	3.24 / 5.67	3.23 / 5.65
8	5.32 / 11.26	4.46 / 8.65	4.07 / 7.59	3.84 / 7.01	3.69 / 6.63	3.58 / 6.37	3.50 / 6.19	3.44 / 6.03	3.39 / 5.91	3.34 / 5.82	3.31 / 5.74	3.28 / 5.67	3.23 / 5.56	3.20 / 5.48	3.15 / 5.36	3.12 / 5.28	3.08 / 5.20	3.05 / 5.11	3.03 / 5.06	3.00 / 5.00	2.98 / 4.96	2.96 / 4.91	2.94 / 4.88	2.93 / 4.86
9	5.12 / 10.56	4.26 / 8.02	3.86 / 6.99	3.63 / 6.42	3.48 / 6.06	3.37 / 5.80	3.29 / 5.62	3.23 / 5.47	3.18 / 5.35	3.13 / 5.26	3.10 / 5.18	3.07 / 5.11	3.02 / 5.00	2.98 / 4.92	2.93 / 4.80	2.90 / 4.73	2.86 / 4.64	2.82 / 4.56	2.80 / 4.51	2.77 / 4.45	2.76 / 4.41	2.73 / 4.36	2.72 / 4.33	2.71 / 4.31
10	4.96 / 10.04	4.10 / 7.56	3.71 / 6.55	3.48 / 5.99	3.33 / 5.64	3.22 / 5.39	3.14 / 5.21	3.07 / 5.06	3.02 / 4.95	2.97 / 4.85	2.94 / 4.78	2.91 / 4.71	2.86 / 4.60	2.82 / 4.52	2.77 / 4.41	2.74 / 4.33	2.70 / 4.25	2.67 / 4.17	2.64 / 4.12	2.61 / 4.05	2.59 / 4.01	2.56 / 3.96	2.55 / 3.93	2.54 / 3.91
11	4.84 / 9.65	3.98 / 7.20	3.59 / 6.22	3.36 / 5.67	3.20 / 5.32	3.09 / 5.07	3.01 / 4.88	2.95 / 4.74	2.90 / 4.63	2.86 / 4.54	2.82 / 4.46	2.79 / 4.40	2.74 / 4.29	2.70 / 4.21	2.65 / 4.10	2.61 / 4.02	2.57 / 3.94	2.53 / 3.86	2.50 / 3.80	2.47 / 3.74	2.45 / 3.70	2.42 / 3.66	2.41 / 3.62	2.40 / 3.60
12	4.75 / 9.33	3.88 / 6.93	3.49 / 5.95	3.26 / 5.41	3.11 / 5.06	3.00 / 4.82	2.92 / 4.65	2.85 / 4.50	2.80 / 4.39	2.76 / 4.30	2.72 / 4.22	2.69 / 4.16	2.64 / 4.05	2.60 / 3.98	2.54 / 3.86	2.50 / 3.78	2.46 / 3.70	2.42 / 3.61	2.40 / 3.56	2.36 / 3.49	2.35 / 3.46	2.32 / 3.41	2.31 / 3.38	2.30 / 3.36
13	4.67 / 9.07	3.80 / 6.70	3.41 / 5.74	3.18 / 5.20	3.02 / 4.86	2.92 / 4.62	2.84 / 4.44	2.77 / 4.30	2.72 / 4.19	2.67 / 4.10	2.63 / 4.02	2.60 / 3.96	2.55 / 3.85	2.51 / 3.78	2.46 / 3.67	2.42 / 3.59	2.38 / 3.51	2.34 / 3.42	2.32 / 3.37	2.28 / 3.30	2.26 / 3.27	2.24 / 3.21	2.22 / 3.18	2.21 / 3.16

The values in the table are the critical values of F for the degrees of freedom listed over the columns (the degrees of freedom for the numerator of the F ratio) and the degrees of freedom listed for the rows (the degrees of freedom for the denominator of the F ratio). The critical value for the .05 level of significance is presented first (roman type) followed by the critical value at the .01 level (boldface). If the observed value is *greater than or equal to* the tabled value, reject H_0. F values are always positive.

Source: Reprinted by permission from *Statistical Methods* by George B. Snedecor and William G. Cochran, Sixth Edition © 1967 by The Iowa State University Press, Ames, Iowa 50010.

Table E (continued)

Degrees of freedom for the denominator	\1	2	3	4	5	6	7	8	9	10	11	12	14	16	20	24	30	40	50	75	100	200	500	∞
14	4.60 / 8.86	3.74 / 6.51	3.34 / 5.56	3.11 / 5.03	2.96 / 4.69	2.85 / 4.46	2.77 / 4.28	2.70 / 4.14	2.65 / 4.03	2.60 / 3.94	2.56 / 3.86	2.53 / 3.80	2.48 / 3.70	2.44 / 3.62	2.39 / 3.51	2.35 / 3.43	2.31 / 3.34	2.27 / 3.26	2.24 / 3.21	2.21 / 3.14	2.19 / 3.11	2.16 / 3.06	2.14 / 3.02	2.13 / 3.00
15	4.54 / 8.68	3.68 / 6.36	3.29 / 5.42	3.06 / 4.89	2.90 / 4.56	2.79 / 4.32	2.70 / 4.14	2.64 / 4.00	2.59 / 3.89	2.55 / 3.80	2.51 / 3.73	2.48 / 3.67	2.43 / 3.56	2.39 / 3.48	2.33 / 3.36	2.29 / 3.29	2.25 / 3.20	2.21 / 3.12	2.18 / 3.07	2.15 / 3.00	2.12 / 2.97	2.10 / 2.92	2.08 / 2.89	2.07 / 2.87
16	4.49 / 8.53	3.63 / 6.23	3.24 / 5.29	3.01 / 4.77	2.85 / 4.44	2.74 / 4.20	2.66 / 4.03	2.59 / 3.89	2.54 / 3.78	2.49 / 3.69	2.45 / 3.61	2.42 / 3.55	2.37 / 3.45	2.33 / 3.37	2.28 / 3.25	2.24 / 3.18	2.20 / 3.10	2.16 / 3.01	2.13 / 2.96	2.09 / 2.98	2.07 / 2.86	2.04 / 2.80	2.02 / 2.77	2.01 / 2.75
17	4.45 / 8.40	3.59 / 6.11	3.20 / 5.18	2.96 / 4.67	2.81 / 4.34	2.70 / 4.10	2.62 / 3.93	2.55 / 3.79	2.50 / 3.68	2.45 / 3.59	2.41 / 3.52	2.38 / 3.45	2.33 / 3.35	2.29 / 3.27	2.23 / 3.16	2.19 / 3.08	2.15 / 3.00	2.11 / 2.92	2.08 / 2.86	2.04 / 2.79	2.02 / 2.76	1.99 / 2.70	1.97 / 2.67	1.96 / 2.65
18	4.41 / 8.28	3.55 / 6.01	3.16 / 5.09	2.93 / 4.58	2.77 / 4.25	2.66 / 4.01	2.58 / 3.85	2.51 / 3.71	2.46 / 3.60	2.41 / 3.51	2.37 / 3.44	2.34 / 3.37	2.29 / 3.27	2.25 / 3.19	2.19 / 3.07	2.15 / 3.00	2.11 / 2.91	2.07 / 2.83	2.04 / 2.78	2.00 / 2.71	1.98 / 2.68	1.95 / 2.62	1.93 / 2.59	1.92 / 2.57
19	4.38 / 8.18	3.52 / 5.93	3.13 / 5.01	2.90 / 4.50	2.74 / 4.17	2.63 / 3.94	2.55 / 3.77	2.48 / 3.63	2.43 / 3.52	2.38 / 3.43	2.34 / 3.36	2.31 / 3.30	2.26 / 3.19	2.21 / 3.12	2.15 / 3.00	2.11 / 2.92	2.07 / 2.84	2.02 / 2.76	2.00 / 2.70	1.96 / 2.63	1.94 / 2.60	1.91 / 2.54	1.90 / 2.51	1.88 / 2.49
20	4.35 / 8.10	3.49 / 5.85	3.10 / 4.94	2.87 / 4.43	2.71 / 4.10	2.60 / 3.87	2.52 / 3.71	2.45 / 3.56	2.40 / 3.45	2.35 / 3.37	2.31 / 3.30	2.28 / 3.23	2.23 / 3.13	2.18 / 3.05	2.12 / 2.94	2.08 / 2.86	2.04 / 2.77	1.99 / 2.69	1.96 / 2.63	1.92 / 2.56	1.90 / 2.53	1.87 / 2.47	1.85 / 2.44	1.84 / 2.42
21	4.32 / 8.02	3.47 / 5.78	3.07 / 4.87	2.84 / 4.37	2.68 / 4.04	2.57 / 3.81	2.49 / 3.65	2.42 / 3.51	2.37 / 3.40	2.32 / 3.31	2.28 / 3.24	2.25 / 3.17	2.20 / 3.07	2.15 / 2.99	2.09 / 2.88	2.05 / 2.80	2.00 / 2.72	1.96 / 2.63	1.93 / 2.58	1.89 / 2.51	1.87 / 2.47	1.84 / 2.42	1.82 / 2.38	1.81 / 2.36
22	4.30 / 7.94	3.44 / 5.72	3.05 / 4.82	2.82 / 4.31	2.66 / 3.99	2.55 / 3.76	2.47 / 3.59	2.40 / 3.45	2.35 / 3.35	2.30 / 3.26	2.26 / 3.18	2.23 / 3.12	2.18 / 3.02	2.13 / 2.94	2.07 / 2.83	2.03 / 2.75	1.98 / 2.67	1.93 / 2.58	1.91 / 2.53	1.87 / 2.46	1.84 / 2.42	1.81 / 2.37	1.80 / 2.33	1.78 / 2.31
23	4.28 / 7.88	3.42 / 5.66	3.03 / 4.76	2.80 / 4.26	2.64 / 3.94	2.53 / 3.71	2.45 / 3.54	2.38 / 3.41	2.32 / 3.30	2.28 / 3.21	2.24 / 3.14	2.20 / 3.07	2.14 / 2.97	2.10 / 2.89	2.04 / 2.78	2.00 / 2.70	1.96 / 2.62	1.91 / 2.53	1.88 / 2.48	1.84 / 2.41	1.82 / 2.37	1.79 / 2.32	1.77 / 2.28	1.76 / 2.26
24	4.26 / 7.82	3.40 / 5.61	3.01 / 4.72	2.78 / 4.22	2.62 / 3.90	2.51 / 3.67	2.43 / 3.50	2.36 / 3.36	2.30 / 3.25	2.26 / 3.17	2.22 / 3.09	2.18 / 3.03	2.13 / 2.93	2.09 / 2.85	2.02 / 2.74	1.98 / 2.66	1.94 / 2.58	1.89 / 2.49	1.86 / 2.44	1.82 / 2.36	1.80 / 2.33	1.76 / 2.27	1.74 / 2.23	1.73 / 2.21
25	4.24 / 7.77	3.38 / 5.57	2.99 / 4.68	2.76 / 4.18	2.60 / 3.86	2.49 / 3.63	2.41 / 3.46	2.34 / 3.32	2.28 / 3.21	2.24 / 3.13	2.20 / 3.05	2.16 / 2.99	2.11 / 2.89	2.06 / 2.81	2.00 / 2.70	1.96 / 2.62	1.92 / 2.54	1.87 / 2.45	1.84 / 2.40	1.80 / 2.32	1.77 / 2.29	1.74 / 2.23	1.72 / 2.19	1.71 / 2.17
26	4.22 / 7.72	3.37 / 5.53	2.98 / 4.64	2.74 / 4.14	2.59 / 3.82	2.47 / 3.59	2.39 / 3.42	2.32 / 3.29	2.27 / 3.17	2.22 / 3.09	2.18 / 3.02	2.15 / 2.96	2.10 / 2.86	2.05 / 2.77	1.99 / 2.66	1.95 / 2.58	1.90 / 2.50	1.85 / 2.41	1.82 / 2.36	1.78 / 2.28	1.76 / 2.25	1.72 / 2.19	1.70 / 2.15	1.69 / 2.13

Degrees of freedom for the numerator

The function, $F = e$ with exponent $2z$, is computed in part from Fisher's table VI. Additional entries are by interpolation, mostly graphical.

Table E (continued)

Degrees of freedom for the numerator

df (denom.)	1	2	3	4	5	6	7	8	9	10	11	12	14	16	20	24	30	40	50	75	100	200	500	∞
27	4.21 / 7.68	3.35 / 5.49	2.96 / 4.60	2.73 / 4.11	2.57 / 3.79	2.46 / 3.56	2.37 / 3.39	2.30 / 3.26	2.25 / 3.14	2.20 / 3.06	2.16 / 2.98	2.13 / 2.93	2.08 / 2.83	2.03 / 2.74	1.97 / 2.63	1.93 / 2.55	1.88 / 2.47	1.84 / 2.38	1.80 / 2.33	1.76 / 2.25	1.74 / 2.21	1.71 / 2.16	1.68 / 2.12	1.67 / 2.10
28	4.20 / 7.64	3.34 / 5.45	2.95 / 4.57	2.71 / 4.07	2.56 / 3.76	2.44 / 3.53	2.36 / 3.36	2.29 / 3.23	2.24 / 3.11	2.19 / 3.03	2.15 / 2.95	2.12 / 2.90	2.06 / 2.80	2.02 / 2.71	1.96 / 2.60	1.91 / 2.52	1.87 / 2.44	1.81 / 2.35	1.78 / 2.30	1.75 / 2.22	1.72 / 2.18	1.69 / 2.13	1.67 / 2.09	1.65 / 2.06
29	4.18 / 7.60	3.33 / 5.42	2.93 / 4.54	2.70 / 4.04	2.54 / 3.73	2.43 / 3.50	2.35 / 3.33	2.28 / 3.20	2.22 / 3.08	2.18 / 3.00	2.14 / 2.92	2.10 / 2.87	2.05 / 2.77	2.00 / 2.68	1.94 / 2.57	1.90 / 2.49	1.85 / 2.41	1.80 / 2.32	1.77 / 2.27	1.73 / 2.19	1.71 / 2.15	1.68 / 2.10	1.65 / 2.06	1.64 / 2.03
30	4.17 / 7.56	3.32 / 5.39	2.92 / 4.51	2.69 / 4.02	2.53 / 3.70	2.42 / 3.47	2.34 / 3.30	2.27 / 3.17	2.21 / 3.06	2.16 / 2.98	2.12 / 2.90	2.09 / 2.84	2.04 / 2.74	1.99 / 2.66	1.93 / 2.55	1.89 / 2.47	1.84 / 2.38	1.79 / 2.29	1.76 / 2.24	1.72 / 2.16	1.69 / 2.13	1.66 / 2.07	1.64 / 2.03	1.62 / 2.01
32	4.15 / 7.50	3.30 / 5.34	2.90 / 4.46	2.67 / 3.97	2.51 / 3.66	2.40 / 3.42	2.32 / 3.25	2.25 / 3.12	2.19 / 3.01	2.14 / 2.94	2.10 / 2.86	2.07 / 2.80	2.02 / 2.70	1.97 / 2.62	1.91 / 2.51	1.86 / 2.42	1.82 / 2.34	1.76 / 2.25	1.74 / 2.20	1.69 / 2.12	1.67 / 2.08	1.64 / 2.02	1.61 / 1.98	1.59 / 1.96
34	4.13 / 7.44	3.28 / 5.29	2.88 / 4.42	2.65 / 3.93	2.49 / 3.61	2.38 / 3.38	2.30 / 3.21	2.23 / 3.08	2.17 / 2.97	2.12 / 2.89	2.08 / 2.82	2.05 / 2.76	2.00 / 2.66	1.95 / 2.58	1.89 / 2.47	1.84 / 2.38	1.80 / 2.30	1.74 / 2.21	1.71 / 2.15	1.67 / 2.08	1.64 / 2.04	1.61 / 1.98	1.59 / 1.94	1.57 / 1.91
36	4.11 / 7.39	3.26 / 5.25	2.86 / 4.38	2.63 / 3.89	2.48 / 3.58	2.36 / 3.35	2.28 / 3.18	2.21 / 3.04	2.15 / 2.94	2.10 / 2.86	2.06 / 2.78	2.03 / 2.72	1.98 / 2.62	1.93 / 2.54	1.87 / 2.43	1.82 / 2.35	1.78 / 2.26	1.72 / 2.17	1.69 / 2.12	1.65 / 2.04	1.62 / 2.00	1.59 / 1.94	1.56 / 1.90	1.55 / 1.87
38	4.10 / 7.35	3.25 / 5.21	2.85 / 4.34	2.62 / 3.86	2.46 / 3.54	2.35 / 3.32	2.26 / 3.15	2.19 / 3.02	2.14 / 2.91	2.09 / 2.82	2.05 / 2.75	2.02 / 2.69	1.96 / 2.59	1.92 / 2.51	1.85 / 2.40	1.80 / 2.32	1.76 / 2.22	1.71 / 2.14	1.67 / 2.08	1.63 / 2.00	1.60 / 1.97	1.57 / 1.90	1.54 / 1.86	1.53 / 1.84
40	4.08 / 7.31	3.23 / 5.18	2.84 / 4.31	2.61 / 3.83	2.45 / 3.51	2.34 / 3.29	2.25 / 3.12	2.18 / 2.99	2.12 / 2.88	2.07 / 2.80	2.04 / 2.73	2.00 / 2.66	1.95 / 2.56	1.90 / 2.49	1.84 / 2.37	1.79 / 2.29	1.74 / 2.20	1.69 / 2.11	1.66 / 2.05	1.61 / 1.97	1.59 / 1.94	1.55 / 1.88	1.53 / 1.84	1.51 / 1.81
42	4.07 / 7.27	3.22 / 5.15	2.83 / 4.29	2.59 / 3.80	2.44 / 3.49	2.32 / 3.26	2.24 / 3.10	2.17 / 2.96	2.11 / 2.86	2.06 / 2.77	2.02 / 2.70	1.99 / 2.64	1.94 / 2.54	1.89 / 2.46	1.82 / 2.35	1.78 / 2.26	1.73 / 2.17	1.68 / 2.08	1.64 / 2.02	1.60 / 1.94	1.57 / 1.91	1.54 / 1.85	1.51 / 1.80	1.49 / 1.78
44	4.06 / 7.24	3.21 / 5.12	2.82 / 4.26	2.58 / 3.78	2.43 / 3.46	2.31 / 3.24	2.23 / 3.07	2.16 / 2.94	2.10 / 2.84	2.05 / 2.75	2.01 / 2.68	1.98 / 2.62	1.92 / 2.52	1.88 / 2.44	1.81 / 2.32	1.76 / 2.24	1.72 / 2.15	1.66 / 2.06	1.63 / 2.00	1.58 / 1.92	1.56 / 1.88	1.52 / 1.82	1.50 / 1.78	1.48 / 1.75
46	4.05 / 7.21	3.20 / 5.10	2.81 / 4.24	2.57 / 3.76	2.42 / 3.44	2.30 / 3.22	2.22 / 3.05	2.14 / 2.92	2.09 / 2.82	2.04 / 2.73	2.00 / 2.66	1.97 / 2.60	1.91 / 2.50	1.87 / 2.42	1.80 / 2.30	1.75 / 2.22	1.71 / 2.13	1.65 / 2.04	1.62 / 1.98	1.57 / 1.90	1.54 / 1.86	1.51 / 1.80	1.48 / 1.76	1.46 / 1.72
48	4.04 / 7.19	3.19 / 5.08	2.80 / 4.22	2.56 / 3.74	2.41 / 3.42	2.30 / 3.20	2.21 / 3.04	2.14 / 2.90	2.08 / 2.80	2.03 / 2.71	1.99 / 2.64	1.96 / 2.58	1.90 / 2.48	1.86 / 2.40	1.79 / 2.28	1.74 / 2.20	1.70 / 2.11	1.64 / 2.02	1.61 / 1.96	1.56 / 1.88	1.53 / 1.84	1.50 / 1.78	1.47 / 1.73	1.45 / 1.70

Degrees of freedom for the denominator

Table E (continued)

Degrees of freedom for the numerator

df (denom)	1	2	3	4	5	6	7	8	9	10	11	12	14	16	20	24	30	40	50	75	100	200	500	∞
50	4.03 / 7.17	3.18 / 5.06	2.79 / 4.20	2.56 / 3.72	2.40 / 3.41	2.29 / 3.18	2.20 / 3.02	2.13 / 2.88	2.07 / 2.78	2.02 / 2.70	1.98 / 2.62	1.95 / 2.56	1.90 / 2.46	1.85 / 2.39	1.78 / 2.26	1.74 / 2.18	1.69 / 2.10	1.63 / 2.00	1.60 / 1.94	1.55 / 1.86	1.52 / 1.82	1.48 / 1.76	1.46 / 1.71	1.44 / 1.68
55	4.02 / 7.12	3.17 / 5.01	2.78 / 4.16	2.54 / 3.68	2.38 / 3.37	2.27 / 3.15	2.18 / 2.98	2.11 / 2.85	2.05 / 2.75	2.00 / 2.66	1.97 / 2.59	1.93 / 2.53	1.88 / 2.43	1.83 / 2.35	1.76 / 2.23	1.72 / 2.15	1.67 / 2.06	1.61 / 1.96	1.58 / 1.90	1.52 / 1.82	1.50 / 1.78	1.46 / 1.71	1.43 / 1.66	1.41 / 1.64
60	4.00 / 7.08	3.15 / 4.98	2.76 / 4.13	2.52 / 3.65	2.37 / 3.34	2.25 / 3.12	2.17 / 2.95	2.10 / 2.82	2.04 / 2.72	1.99 / 2.63	1.95 / 2.56	1.92 / 2.50	1.86 / 2.40	1.81 / 2.32	1.75 / 2.20	1.70 / 2.12	1.65 / 2.03	1.59 / 1.93	1.56 / 1.87	1.50 / 1.79	1.48 / 1.74	1.44 / 1.68	1.41 / 1.63	1.39 / 1.60
65	3.99 / 7.04	3.14 / 4.95	2.75 / 4.10	2.51 / 3.62	2.36 / 3.31	2.24 / 3.09	2.15 / 2.93	2.08 / 2.79	2.02 / 2.70	1.98 / 2.61	1.94 / 2.54	1.90 / 2.47	1.85 / 2.37	1.80 / 2.30	1.73 / 2.18	1.68 / 2.09	1.63 / 2.00	1.57 / 1.90	1.54 / 1.84	1.49 / 1.76	1.46 / 1.71	1.42 / 1.64	1.39 / 1.60	1.37 / 1.56
70	3.98 / 7.01	3.13 / 4.92	2.74 / 4.08	2.50 / 3.60	2.35 / 3.29	2.23 / 3.07	2.14 / 2.91	2.07 / 2.77	2.01 / 2.67	1.97 / 2.59	1.93 / 2.51	1.89 / 2.45	1.84 / 2.35	1.79 / 2.28	1.72 / 2.15	1.67 / 2.07	1.62 / 1.98	1.56 / 1.88	1.53 / 1.82	1.47 / 1.74	1.45 / 1.69	1.40 / 1.62	1.37 / 1.56	1.35 / 1.53
80	3.96 / 6.96	3.11 / 4.88	2.72 / 4.04	2.48 / 3.56	2.33 / 3.25	2.21 / 3.04	2.12 / 2.87	2.05 / 2.74	1.99 / 2.64	1.95 / 2.55	1.91 / 2.48	1.88 / 2.41	1.82 / 2.32	1.77 / 2.24	1.70 / 2.11	1.65 / 2.03	1.60 / 1.94	1.54 / 1.84	1.51 / 1.78	1.45 / 1.70	1.42 / 1.65	1.38 / 1.57	1.35 / 1.52	1.32 / 1.49
100	3.94 / 6.90	3.09 / 4.82	2.70 / 3.98	2.46 / 3.51	2.30 / 3.20	2.19 / 2.99	2.10 / 2.82	2.03 / 2.69	1.97 / 2.59	1.92 / 2.51	1.88 / 2.43	1.85 / 2.36	1.79 / 2.26	1.75 / 2.19	1.68 / 2.06	1.63 / 1.98	1.57 / 1.89	1.51 / 1.79	1.48 / 1.73	1.42 / 1.64	1.39 / 1.59	1.34 / 1.51	1.30 / 1.46	1.28 / 1.43
125	3.92 / 6.84	3.07 / 4.78	2.68 / 3.94	2.44 / 3.47	2.29 / 3.17	2.17 / 2.95	2.08 / 2.79	2.01 / 2.65	1.95 / 2.56	1.90 / 2.47	1.86 / 2.40	1.83 / 2.33	1.77 / 2.23	1.72 / 2.15	1.65 / 2.03	1.60 / 1.94	1.55 / 1.85	1.49 / 1.75	1.45 / 1.68	1.39 / 1.59	1.36 / 1.54	1.31 / 1.46	1.27 / 1.40	1.25 / 1.37
150	3.91 / 6.81	3.06 / 4.75	2.67 / 3.91	2.43 / 3.44	2.27 / 3.14	2.16 / 2.92	2.07 / 2.76	2.00 / 2.62	1.94 / 2.53	1.89 / 2.44	1.85 / 2.37	1.82 / 2.30	1.76 / 2.20	1.71 / 2.12	1.64 / 2.00	1.59 / 1.91	1.54 / 1.83	1.47 / 1.72	1.44 / 1.66	1.37 / 1.56	1.34 / 1.51	1.29 / 1.43	1.25 / 1.37	1.22 / 1.33
200	3.89 / 6.76	3.04 / 4.71	2.65 / 3.88	2.41 / 3.41	2.26 / 3.11	2.14 / 2.90	2.05 / 2.73	1.98 / 2.60	1.92 / 2.50	1.87 / 2.41	1.83 / 2.34	1.80 / 2.28	1.74 / 2.17	1.69 / 2.09	1.62 / 1.97	1.57 / 1.88	1.52 / 1.79	1.45 / 1.69	1.42 / 1.62	1.35 / 1.53	1.32 / 1.48	1.26 / 1.39	1.22 / 1.33	1.19 / 1.28
400	3.86 / 6.70	3.02 / 4.66	2.62 / 3.83	2.39 / 3.36	2.23 / 3.06	2.12 / 2.85	2.03 / 2.69	1.96 / 2.55	1.90 / 2.46	1.85 / 2.37	1.81 / 2.29	1.78 / 2.23	1.72 / 2.12	1.67 / 2.04	1.60 / 1.92	1.54 / 1.84	1.49 / 1.74	1.42 / 1.64	1.38 / 1.57	1.32 / 1.47	1.28 / 1.42	1.22 / 1.32	1.16 / 1.24	1.13 / 1.19
1000	3.85 / 6.66	3.00 / 4.62	2.61 / 3.80	2.38 / 3.34	2.22 / 3.04	2.10 / 2.82	2.02 / 2.66	1.95 / 2.53	1.89 / 2.43	1.84 / 2.34	1.80 / 2.26	1.76 / 2.20	1.70 / 2.09	1.65 / 2.01	1.58 / 1.89	1.53 / 1.81	1.47 / 1.71	1.41 / 1.51	1.36 / 1.54	1.30 / 1.44	1.26 / 1.38	1.19 / 1.28	1.13 / 1.19	1.08 / 1.11
∞	3.84 / 6.64	2.99 / 4.60	2.60 / 3.78	2.37 / 3.32	2.21 / 3.02	2.09 / 2.80	2.01 / 2.64	1.94 / 2.51	1.88 / 2.41	1.83 / 2.32	1.79 / 2.24	1.75 / 2.18	1.69 / 2.07	1.64 / 1.99	1.57 / 1.87	1.52 / 1.79	1.46 / 1.69	1.40 / 1.59	1.35 / 1.52	1.28 / 1.41	1.24 / 1.36	1.17 / 1.25	1.11 / 1.15	1.00 / 1.00

Degrees of freedom for the denominator

Table **F.**   Critical Values of Chi Square

	Level of significance for a directional test					
	.10	.05	.025	.01	.005	.0005
	Level of significance for a non-directional test					
df	.20	.10	.05	.02	.01	.001
1	1.64	2.71	3.84	5.41	6.64	10.83
2	3.22	4.60	5.99	7.82	9.21	13.82
3	4.64	6.25	7.82	9.84	11.34	16.27
4	5.99	7.78	9.49	11.67	13.28	18.46
5	7.29	9.24	11.07	13.39	15.09	20.52
6	8.56	10.64	12.59	15.03	16.81	22.46
7	9.80	12.02	14.07	16.62	18.48	24.32
8	11.03	13.36	15.51	18.17	20.09	26.12
9	12.24	14.68	16.92	19.68	21.67	27.88
10	13.44	15.99	18.31	21.16	23.21	29.59
11	14.63	17.28	19.68	22.62	24.72	31.26
12	15.81	18.55	21.03	24.05	26.22	32.91
13	16.98	19.81	22.36	25.47	27.69	34.53
14	18.15	21.06	23.68	26.87	29.14	36.12
15	19.31	22.31	25.00	28.26	30.58	37.70
16	20.46	23.54	26.30	29.63	32.00	39.29
17	21.62	24.77	27.59	31.00	33.41	40.75
18	22.76	25.99	28.87	32.35	34.80	42.31
19	23.90	27.20	30.14	33.69	36.19	43.82
20	25.04	28.41	31.41	35.02	37.57	45.32
21	26.17	29.62	32.67	36.34	38.93	46.80
22	27.30	30.81	33.92	37.66	40.29	48.27
23	28.43	32.01	35.17	38.97	41.64	49.73
24	29.55	33.20	36.42	40.27	42.98	51.18
25	30.68	34.38	37.65	41.57	44.31	52.62
26	31.80	35.56	38.88	42.86	45.64	54.05
27	32.91	36.74	40.11	44.14	46.96	55.48
28	34.03	37.92	41.34	45.42	48.28	56.89
29	35.14	39.09	42.69	46.69	49.59	58.30
30	36.25	40.26	43.77	47.96	50.89	59.70
32	38.47	42.59	46.19	50.49	53.49	62.49
34	40.68	44.90	48.60	53.00	56.06	65.25
36	42.88	47.21	51.00	55.49	58.62	67.99
38	45.08	49.51	53.38	57.97	61.16	70.70
40	47.27	51.81	55.76	60.44	63.69	73.40
44	51.64	56.37	60.48	65.34	68.71	78.75
48	55.99	60.91	65.17	70.20	73.68	84.04
52	60.33	65.42	69.83	75.02	78.62	89.27
56	64.66	69.92	74.47	79.82	83.51	94.46
60	68.97	74.40	79.08	84.58	88.38	99.61

The table lists the critical values of chi square for the degrees of freedom shown at the left for tests corresponding to those significance levels which head each column. If the observed value of χ_{obs}^2 is *greater than or equal to* the tabled value, reject H_0. All chi squares are positive.

Source: Table F is taken from Table IV of Fisher and Yates, *Statistical Tables for Biological, Agricultural and Medical Research*, published by Longman Group Ltd., London (previously published by Oliver and Boyd, Ltd., Edinburgh), and by permission of the authors and publishers.

Table **G.** Critical Values of the Mann-Whitney U for a Directional Test at .005 or a Nondirectional Test at .01

n_B \ n_A	1	2	3	4	5	6	7	8	9	10	11	12	13	14	15	16	17	18	19	20
1	--	--	--	--	--	--	--	--	--	--	--	--	--	--	--	--	--	--	0/19	0/20
2	--	--	--	--	0/10	0/12	0/14	1/15	1/17	1/19	1/21	2/22	2/24	2/26	3/27	3/29	3/31	4/32	4/34	4/36
3	--	--	0/9	0/12	1/14	2/16	2/19	3/21	3/24	4/26	5/28	5/31	6/33	7/35	7/38	8/40	9/42	9/45	10/47	11/49
4	--	--	0/12	1/15	2/18	3/21	4/24	5/27	6/30	7/33	8/36	9/39	10/42	11/45	12/48	14/50	15/53	16/56	17/59	18/62
5	--	0/10	1/14	2/18	4/21	5/25	6/29	8/32	9/36	11/39	12/43	13/47	15/50	16/54	18/57	19/61	20/65	22/68	23/72	25/75
6	--	0/12	2/16	3/21	5/25	7/29	8/34	10/38	12/42	14/46	16/50	17/55	19/59	21/63	23/67	25/71	26/76	28/80	30/84	32/88
7	--	0/14	2/19	4/24	6/29	8/34	11/38	13/43	15/48	17/53	19/58	21/63	24/67	26/72	28/77	30/82	33/86	35/91	37/96	39/101
8	--	1/15	3/21	5/27	8/32	10/38	13/43	15/49	18/54	20/60	23/65	26/70	28/76	31/81	33/87	36/92	39/97	41/103	44/108	47/113
9	--	1/17	3/24	6/30	9/36	12/42	15/48	18/54	21/60	24/66	27/72	30/78	33/84	36/90	39/96	42/102	45/108	48/114	51/120	54/126
10	--	1/19	4/26	7/33	11/39	14/46	17/53	20/60	24/66	27/73	31/79	34/86	37/93	41/99	44/106	48/112	51/119	55/125	58/132	62/138
11	--	1/21	5/28	8/36	12/43	16/50	19/58	23/65	27/72	31/79	34/87	38/94	42/101	46/108	50/115	54/122	57/130	61/137	65/144	69/151
12	--	2/22	5/31	9/39	13/47	17/55	21/63	26/70	30/78	34/86	38/94	42/102	47/109	51/117	55/125	60/132	64/140	68/148	72/156	77/163
13	--	2/24	6/33	10/42	15/50	19/59	24/67	28/76	33/84	37/93	42/101	47/109	51/118	56/126	61/134	65/143	70/151	75/159	80/167	84/176
14	--	2/26	7/35	11/45	16/54	21/63	26/72	31/81	36/90	41/99	46/108	51/117	56/126	61/135	66/144	71/153	77/161	82/170	87/179	92/188
15	--	3/27	7/38	12/48	18/57	23/67	28/77	33/87	39/96	44/106	50/115	55/125	61/134	66/144	72/153	77/163	83/172	88/182	94/191	100/200
16	--	3/29	8/40	14/50	19/61	25/71	30/82	36/92	42/102	48/112	54/122	60/132	65/143	71/153	77/163	83/173	89/183	95/193	101/203	107/213
17	--	3/31	9/42	15/53	20/65	26/76	33/86	39/97	45/108	51/119	57/130	64/140	70/151	77/161	83/172	89/183	96/193	102/204	109/214	115/225
18	--	4/32	9/45	16/56	22/68	28/80	35/91	41/103	48/114	55/123	61/137	68/148	75/159	82/170	88/182	95/193	102/204	109/215	116/226	123/237
19	0/19	4/34	10/47	17/59	23/72	30/84	37/96	44/108	51/120	58/132	65/144	72/156	80/167	87/179	94/191	101/203	109/214	116/226	123/238	130/250
20	0/20	4/36	11/49	18/62	25/75	32/88	39/101	47/113	54/126	62/138	69/151	77/163	84/176	92/188	100/200	107/213	115/225	123/237	130/250	138/262

(Dashes in the body of the table indicate that no decision is possible at the stated level of significance.)

If the observed value of U falls between the two values presented in the table for n_A and n_B, do not reject H_0. Otherwise, reject H_0.

Table G. Critical Values of the Mann-Whitney U for a Directional Test at .01 or a Nondirectional Test at .02

n_B \ n_A	1	2	3	4	5	6	7	8	9	10	11	12	13	14	15	16	17	18	19	20
1	--	--	--	--	--	--	--	--	--	--	--	--	--	--	--	--	--	--	--	--
2	--	--	--	--	--	--	--	0/16	0/18	0/20	0/22	1/23	1/25	1/27	1/29	1/31	2/32	2/34	2/36	2/38
3	--	--	--	--	0/15	1/17	1/20	2/22	2/25	3/27	3/30	4/32	4/35	5/37	5/40	6/42	6/45	7/47	7/50	8/52
4	--	--	--	0/16	1/19	2/22	3/25	4/28	4/32	5/35	6/38	7/41	8/44	9/47	10/50	11/53	11/57	12/60	13/63	13/67
5	--	--	0/15	1/19	2/23	3/27	5/30	6/34	7/38	8/42	9/46	11/49	12/53	13/57	14/61	15/65	17/68	18/72	19/76	20/80
6	--	--	1/17	2/22	3/27	5/31	6/36	8/40	10/44	11/49	13/53	14/58	16/62	17/67	19/71	21/75	22/80	24/84	25/89	27/93
7	--	--	1/20	3/25	5/30	6/36	8/41	10/46	12/51	14/56	16/61	18/66	20/71	22/76	24/81	26/86	28/91	30/96	32/101	34/106
8	--	0/16	2/22	4/28	6/34	8/40	10/46	13/51	15/57	17/63	19/69	22/74	24/80	26/86	29/91	31/97	34/102	36/108	38/111	41/119
9	--	0/18	2/25	4/32	7/38	10/44	12/51	15/57	17/64	20/70	23/76	26/82	28/89	31/95	34/101	37/107	39/114	42/120	45/126	48/132
10	--	0/20	3/27	5/35	8/42	11/49	14/56	17/63	20/70	23/77	26/84	29/91	33/97	36/104	39/111	42/118	45/125	48/132	52/138	55/145
11	--	0/22	3/30	6/38	9/46	13/53	16/61	19/69	23/76	26/84	30/91	33/99	37/106	40/114	44/121	47/129	51/136	55/143	58/151	62/158
12	--	1/23	4/32	7/41	11/49	14/58	18/66	22/74	26/82	29/91	33/99	37/107	41/115	45/123	49/131	53/139	57/147	61/155	65/163	69/171
13	--	1/25	4/35	8/44	12/53	16/62	20/71	24/80	28/89	33/97	37/106	41/115	45/124	50/132	54/141	59/149	63/158	67/167	72/175	76/184
14	--	1/27	5/37	9/47	13/51	17/67	22/76	26/86	31/95	36/104	40/114	45/123	50/132	55/141	59/151	64/160	67/171	74/178	78/188	83/197
15	--	1/29	5/40	10/50	14/61	19/71	24/81	29/91	34/101	39/111	44/121	49/131	54/141	59/151	64/161	70/170	75/180	80/190	85/200	90/210
16	--	1/31	6/42	11/53	15/65	21/75	26/86	31/97	37/107	42/118	47/129	53/139	59/149	64/160	70/170	75/181	81/191	86/202	92/212	98/222
17	--	2/32	6/45	11/57	17/68	22/80	28/91	34/102	39/114	45/125	51/136	57/147	63/158	67/171	75/180	81/191	87/202	93/213	99/224	105/235
18	--	2/34	7/47	12/60	18/72	24/84	30/96	36/108	42/120	48/132	55/143	61/155	67/167	74/178	80/190	86/202	93/213	99/225	106/236	112/248
19	--	2/36	7/50	13/63	19/76	25/89	32/101	38/114	45/126	52/138	58/151	65/163	72/175	78/188	85/200	92/212	99/224	106/236	113/248	119/261
20	--	2/38	8/52	13/67	20/80	27/93	34/106	41/119	48/132	55/145	62/158	69/171	76/184	83/197	90/210	98/222	105/235	112/248	119/261	127/273

(Dashes in the body of the table indicate that no decision is possible at the stated level of significance.)

Source: From Mann, H. B., and Whitney, D. R., "On a Test of Whether One of Two Random Variables Is Stochastically Larger Than the Other," *Annals of Mathematical Statistics* 18 (1947): 50–60, and Auble, D., "Extended Tables for the Mann-Whitney Statistic," *Bulletin of the Institute of Educational Research at Indiana University*, vol. 1, no. 2 (1953), as used in Runyon and Haber, *Fundamentals of Behavioral Statistics*, 3rd ed., Addison-Wesley, Reading, Mass., 1976. Reprinted by permission.

Table **G.** Critical Values of the Mann-Whitney U for a Directional Test at .025 or a Nondirectional Test at .05

n_B \ n_A	1	2	3	4	5	6	7	8	9	10	11	12	13	14	15	16	17	18	19	20
1	--	--	--	--	--	--	--	--	--	--	--	--	--	--	--	--	--	--	--	--
2	--	--	--	--	--	--	--	--	--	--	--	--	0/26	0/28	0/30	0/32	0/34	0/36	1/37	1/39
3	--	--	--	--	--	--	0/21	0/24	1/26	1/29	1/32	2/34	2/37	2/40	3/42	3/45	4/47	4/50	4/52	5/55
4	--	--	--	--	0/20	1/23	1/27	2/30	3/33	3/37	4/40	5/43	5/47	6/50	7/53	7/57	8/60	9/63	9/67	10/70
5	--	--	--	0/20	1/24	2/28	3/32	4/36	5/40	6/44	7/48	8/52	9/56	10/60	11/64	12/68	13/72	14/76	15/80	16/84
6	--	--	--	1/23	2/28	3/33	4/38	6/42	7/47	8/52	9/57	11/61	12/66	13/71	15/75	16/80	18/84	19/89	20/94	22/93
7	--	--	0/21	1/27	3/32	4/38	6/43	7/49	9/54	11/59	12/65	14/70	16/75	17/81	19/86	21/91	23/96	24/102	26/107	28/112
8	--	--	0/24	2/30	4/36	6/42	7/49	9/55	11/61	13/67	15/73	17/79	20/84	22/90	24/96	26/102	28/108	30/114	32/120	34/126
9	--	--	1/26	3/33	5/40	7/47	9/54	11/61	14/67	16/74	18/81	21/87	23/94	26/100	28/107	31/113	33/120	36/126	38/133	40/140
10	--	--	1/29	3/37	6/44	8/52	11/59	13/67	16/74	19/81	22/88	24/96	27/103	30/110	33/117	36/124	38/132	41/139	44/146	47/153
11	--	--	1/32	4/40	7/48	9/57	12/65	15/73	18/81	22/88	25/96	28/104	31/112	34/120	37/128	41/135	44/143	47/151	50/159	53/167
12	--	--	2/34	5/43	8/52	11/61	14/70	17/79	21/87	24/96	28/104	31/113	35/121	38/130	42/138	46/146	49/155	53/163	56/172	60/180
13	--	0/26	2/37	5/47	9/56	12/66	16/75	20/84	23/94	27/103	31/112	35/121	39/130	43/139	47/148	51/157	55/166	59/175	63/184	67/193
14	--	0/28	2/40	6/50	10/60	13/71	17/81	22/90	26/100	30/110	34/120	38/130	43/139	47/149	51/159	56/168	60/178	65/187	69/197	73/207
15	--	0/30	3/42	7/53	11/64	15/75	19/86	24/96	28/107	33/117	37/128	42/138	47/148	51/159	56/169	61/179	66/189	70/200	75/210	80/220
16	--	0/32	3/45	7/57	12/68	16/80	21/91	26/102	31/113	36/124	41/135	46/146	51/157	56/168	61/179	66/190	71/201	76/212	82/222	87/233
17	--	0/34	4/47	8/60	13/72	18/84	23/96	28/108	33/120	38/132	44/143	49/155	55/166	60/178	66/189	71/201	77/212	82/224	88/234	93/247
18	--	0/36	4/50	9/63	14/76	19/89	24/102	30/114	36/126	41/139	47/151	53/163	59/175	65/187	70/200	76/212	82/224	88/236	94/248	100/260
19	--	1/37	4/53	9/67	15/80	20/94	26/107	32/120	38/133	44/146	50/159	56/172	63/184	69/197	75/210	82/222	88/235	94/248	101/260	107/273
20	--	1/39	5/55	10/70	16/84	22/98	28/112	34/126	40/140	47/153	53/167	60/180	67/193	73/207	80/220	87/233	93/247	100/260	107/273	114/286

(Dashes in the body of the table indicate that no decision is possible at the stated level of significance.)

Table **G.** Critical Values of the Mann-Whitney U for a Directional Test at .05 or a Nondirectional Test at .10

n_B \ n_A	1	2	3	4	5	6	7	8	9	10	11	12	13	14	15	16	17	18	19	20
1	--	--	--	--	--	--	--	--	--	--	--	--	--	--	--	--	--	--	--	--
2	--	--	--	--	--	--	--	--	--	--	--	--	--	--	--	--	--	--	0/38	0/40
3	--	--	--	--	--	--	--	--	0/27	0/30	0/33	1/35	1/38	1/41	2/43	2/46	2/49	2/52	3/54	3/57
4	--	--	--	--	--	0/24	0/28	1/31	1/35	2/38	2/42	3/45	3/49	4/52	5/55	5/59	6/62	6/66	7/69	8/72
5	--	--	--	--	0/25	1/29	1/34	2/38	3/42	4/46	5/50	6/54	7/58	7/63	8/67	9/71	10/75	11/79	12/83	13/87
6	--	--	--	0/24	1/29	2/34	3/39	4/44	5/49	6/54	7/59	9/63	10/68	11/73	12/78	13/83	15/87	16/92	17/97	18/102
7	--	--	--	0/28	1/34	3/39	4/45	6/50	7/56	9/61	10/67	12/72	13/78	15/83	16/89	18/94	19/100	21/105	22/111	24/116
8	--	--	--	1/31	2/38	4/44	6/50	7/57	9/63	11/69	13/75	15/81	17/87	18/94	20/100	22/106	24/112	26/118	28/124	30/130
9	--	--	0/27	1/35	3/42	5/49	7/56	9/63	11/70	13/77	16/83	18/90	20/97	22/104	24/111	27/117	29/124	31/131	33/138	36/144
10	--	--	0/30	2/38	4/46	6/54	9/61	11/69	13/77	16/84	18/92	21/99	24/106	26/114	29/121	31/129	34/136	37/143	39/151	42/158
11	--	--	0/33	2/42	5/50	7/59	10/67	13/75	16/83	18/92	21/100	24/108	27/116	30/124	33/132	36/140	39/148	42/156	45/164	48/172
12	--	--	1/35	3/45	6/54	9/63	12/72	15/81	18/90	21/99	24/108	27/117	31/125	34/134	37/143	41/151	44/160	47/169	51/177	54/186
13	--	--	1/38	3/49	7/58	10/68	13/78	17/87	20/97	24/106	27/116	31/125	34/125	38/144	42/153	45/163	49/172	53/181	56/191	60/200
14	--	--	1/41	4/52	7/63	11/73	15/83	18/94	22/104	26/114	30/124	34/134	38/144	42/154	46/164	50/174	54/184	58/194	63/203	67/213
15	--	--	2/43	5/55	8/67	12/78	16/89	20/100	24/111	29/121	33/132	37/143	42/153	46/164	51/174	55/185	60/195	64/206	69/216	73/227
16	--	--	2/46	5/59	9/71	13/83	18/94	22/106	27/117	31/129	36/140	41/151	45/163	50/174	55/185	60/196	65/207	70/218	74/230	79/241
17	--	--	2/49	6/62	10/75	15/87	19/100	24/112	29/124	34/148	39/148	44/160	49/172	54/184	60/195	65/207	70/219	75/231	81/242	86/254
18	--	--	2/52	6/66	11/79	16/92	21/105	26/118	31/131	37/143	42/156	47/169	53/181	58/194	64/206	70/218	75/231	81/243	87/255	92/268
19	--	0/38	3/54	7/69	12/83	17/97	22/111	28/124	33/138	39/151	45/164	51/177	56/191	63/203	69/216	74/230	81/242	87/255	93/268	99/281
20	--	0/40	3/57	8/72	13/87	18/102	24/116	30/130	36/144	42/158	48/172	54/186	60/200	67/213	73/227	79/241	86/254	92/268	99/281	105/295

(Dashes in the body of the table indicate that no decision is possible at the stated level of significance.)

Table **H.**   Critical Values of *W* for the Wilcoxon Test

	Level of significance for a directional test					Level of significance for a directional test			
	.05	.025	.01	.005		.05	.025	.01	.005
	Level of significance for a non-directional test					Level of significance for a non-directional test			
N	.10	.05	.02	.01	*N*	.10	.05	.02	.01
5	0	--	--	--	28	130	116	101	91
6	2	0	--	--	29	140	126	110	100
7	3	2	0	--	30	151	137	120	109
8	5	3	1	0	31	163	147	130	118
9	8	5	3	1	32	175	159	140	128
10	10	8	5	3	33	187	170	151	138
11	13	10	7	5	34	200	182	162	148
12	17	13	9	7	35	213	195	173	159
13	21	17	12	9	36	227	208	185	171
14	25	21	15	12	37	241	221	198	182
15	30	25	19	15	38	256	235	211	194
16	35	29	23	19	39	271	249	224	207
17	41	34	27	23	40	286	264	238	220
18	47	40	32	27	41	302	279	252	233
19	53	46	37	32	42	319	294	266	247
20	60	52	43	37	43	336	310	281	261
21	67	58	49	42	44	353	327	296	276
22	75	65	55	48	45	371	343	312	291
23	83	73	62	54	46	389	361	328	307
24	91	81	69	61	47	407	378	345	322
25	100	89	76	68	48	426	396	362	339
26	110	98	84	75	49	446	415	379	355
27	119	107	92	83	50	466	434	397	373

For a given *N* (the number of pairs of scores minus the pairs having zero differences), if the observed value is *less than or equal to* the value in the table for the appropriate level of significance, reject H_0.

Source: From F. Wilcoxon, S. Katte, and R. A. Wilcox, *Critical Values and Probability Levels for the Wilcoxon Rank Sum Test and the Wilcoxon Signed Rank Test*, New York, American Cyanamid Co., 1963, and F. Wilcoxon and R. A. Wilcox, *Some Rapid Approximate Statistical Procedures*, New York, Lederle Laboratories, 1964, as used in Runyon and Haber, *Fundamentals of Behavioral Statistics*, 3rd ed., Addison-Wesley, Reading, Mass., 1976. Reprinted by permission of the American Cyanamid Company.

Table I. Critical Values for the Spearman Rank-Order Correlation Coefficient

	Significance level for a directional test at			
	.05	.025	.005	.001
	Significance level for a non-directional test at			
N	.10	.05	.01	.002
5	.900	1.000		
6	.829	.886	1.000	
7	.715	.786	.929	1.000
8	.620	.715	.881	.953
9	.600	.700	.834	.917
10	.564	.649	.794	.879
11	.537	.619	.764	.855
12	.504	.588	.735	.826
13	.484	.561	.704	.797
14	.464	.539	.680	.772
15	.447	.522	.658	.750
16	.430	.503	.636	.730
17	.415	.488	.618	.711
18	.402	.474	.600	.693
19	.392	.460	.585	.676
20	.381	.447	.570	.661
21	.371	.437	.556	.647
22	.361	.426	.544	.633
23	.353	.417	.532	.620
24	.345	.407	.521	.608
25	.337	.399	.511	.597
26	.331	.391	.501	.587
27	.325	.383	.493	.577
28	.319	.376	.484	.567
29	.312	.369	.475	.558
30	.307	.363	.467	.549

If the observed value of r_S is *greater than or equal to* the tabled value for the appropriate level of significance, reject H_0. Note that the left-hand column is the number of pairs of scores, not the number of degrees of freedom. The critical values listed are both + and − for nondirectional tests.

Source: Glasser, G. J., and Winter, R. F., "Critical Values of the Coefficient of Rank Correlation for Testing the Hypothesis of Independence," *Biometrika* 48 (1961):444.

Table J.   Random Numbers

22 17 68 65 84	68 95 23 92 35	87 02 22 57 51	61 09 43 95 06	58 24 82 03 47
19 36 27 59 46	13 79 93 37 55	39 77 32 77 09	85 52 05 30 62	47 83 51 62 74
16 77 23 02 77	09 61 87 25 21	28 06 24 25 93	16 71 13 59 78	23 05 47 47 25
78 43 76 71 61	20 44 90 32 64	97 67 63 99 61	46 38 03 93 22	69 81 21 99 21
03 28 28 26 08	73 37 32 04 05	69 30 16 09 05	88 69 58 28 99	35 07 44 75 47
93 22 53 64 39	07 10 63 76 35	87 03 04 79 88	08 13 13 85 51	55 34 57 72 69
78 76 58 54 74	92 38 70 96 92	52 06 79 79 45	82 63 18 27 44	69 66 92 19 09
23 68 35 26 00	99 53 93 61 28	52 70 05 48 34	56 65 05 61 86	90 92 10 70 80
15 39 25 70 99	93 86 52 77 65	15 33 59 05 28	22 87 26 07 47	86 96 98 29 06
58 71 96 30 24	18 46 23 34 27	85 13 99 24 44	49 18 09 79 49	74 16 32 23 02
57 35 27 33 72	24 53 63 94 09	41 10 76 47 91	44 04 95 49 66	39 60 04 59 81
48 50 86 54 48	22 06 34 72 52	82 21 15 65 20	33 29 94 71 11	15 91 29 12 03
61 96 48 95 03	07 16 39 33 66	98 56 10 56 79	77 21 30 27 12	90 49 22 23 62
36 93 89 41 26	29 70 83 63 51	99 74 20 52 36	87 09 41 15 09	98 60 16 03 03
18 87 00 42 31	57 90 12 02 07	23 47 37 17 31	54 08 01 88 63	39 41 88 92 10
88 56 53 27 59	33 35 72 67 47	77 34 55 45 70	08 18 27 38 90	16 95 86 70 75
09 72 95 84 29	49 41 31 06 70	42 38 06 45 18	64 84 73 31 65	52 53 37 97 15
12 96 88 17 31	65 19 69 02 83	60 75 86 90 68	24 64 19 35 51	56 61 87 39 12
85 94 57 24 16	92 09 84 38 76	22 00 27 69 85	29 81 94 78 70	21 94 47 90 12
38 64 43 59 98	98 77 87 68 07	91 51 67 62 44	40 98 05 93 78	23 32 65 41 18
53 44 09 42 72	00 41 86 79 79	68 47 22 00 20	35 55 31 51 51	00 83 63 22 55
40 76 66 26 84	57 99 99 90 37	36 63 32 08 58	37 40 13 68 97	87 64 81 07 83
02 17 79 18 05	12 59 52 57 02	22 07 90 47 03	28 14 11 30 79	20 69 22 40 98
95 17 82 06 53	31 51 10 96 46	92 06 88 07 77	56 11 50 81 69	40 23 72 51 39
35 76 22 42 92	96 11 83 44 80	34 68 35 48 77	33 42 40 90 60	73 96 53 97 86
26 29 13 56 41	85 47 04 66 08	34 72 57 59 13	82 43 80 46 15	38 26 61 70 04
77 80 20 75 82	72 82 32 99 90	63 95 73 76 63	89 73 44 99 05	48 67 26 43 18
46 40 66 44 52	91 36 74 43 53	30 82 13 54 00	78 45 63 98 35	55 03 36 67 68
37 56 08 18 09	77 53 84 46 47	31 91 18 95 58	24 16 74 11 53	44 10 13 85 57
61 65 61 68 66	37 27 47 39 19	84 83 70 07 48	53 21 40 06 71	95 06 79 88 54
93 43 69 64 07	34 18 04 52 35	56 27 09 24 86	61 85 53 83 45	19 90 70 99 00
21 96 60 12 99	11 20 99 45 18	48 13 93 55 34	18 37 79 49 90	65 97 38 20 46
95 20 47 97 97	27 37 83 28 71	00 06 41 41 74	45 89 09 39 84	51 67 11 52 49
97 86 21 78 73	10 65 81 92 59	58 76 17 14 97	04 76 62 16 17	17 95 70 45 80
69 92 06 34 13	59 71 74 17 32	27 55 10 24 19	23 71 82 13 74	63 52 52 01 41
04 31 17 21 56	33 73 99 19 87	26 72 39 27 67	53 77 57 68 93	60 61 97 22 61
61 06 98 03 91	87 14 77 43 96	43 00 65 98 50	45 60 33 01 07	98 99 46 50 47
85 93 85 86 88	72 87 08 62 40	16 06 10 89 20	23 21 34 74 97	76 38 03 29 63
21 74 32 47 45	73 96 07 94 52	09 65 90 77 47	25 76 16 19 33	53 05 70 53 30
15 69 53 82 80	79 96 23 53 10	65 39 07 16 29	45 33 02 43 70	02 87 40 41 45
02 89 08 04 49	20 21 14 68 86	87 63 93 95 17	11 29 01 95 80	35 14 97 35 33
87 18 15 89 79	85 43 01 72 73	08 61 74 51 69	89 74 39 82 15	94 51 33 41 67
98 83 71 94 22	59 97 50 99 52	08 52 85 08 40	87 80 61 65 31	91 51 80 32 44
10 08 58 21 66	72 68 49 29 31	89 85 84 46 06	59 73 19 85 23	65 09 29 75 63
47 90 56 10 08	88 02 84 27 83	42 29 72 23 19	66 56 45 65 79	20 71 53 20 25
22 85 61 68 90	49 64 92 85 44	16 40 12 89 88	50 14 49 81 06	01 82 77 45 12
67 80 43 79 33	12 83 11 41 16	25 58 19 68 70	77 02 54 00 52	53 43 37 15 26
27 62 50 96 72	79 44 61 40 15	14 53 40 65 39	27 31 58 50 28	11 39 03 34 25
33 78 80 87 15	38 30 06 38 21	14 47 47 07 26	54 96 87 53 32	40 36 40 96 76
13 13 92 66 99	47 24 49 57 74	32 25 43 62 17	10 97 11 69 84	99 63 22 32 98

Source: Table J is taken from Table XXXIII of Fisher and Yates,
Statistical Tables for Biological, Agricultural and Medical Research, published by
Longman Group Ltd., London (previously published by Oliver and Boyd,
Ltd., Edinburgh), and by permission of the authors and publishers.

Table J (continued)

```
10 27 53 96 23    71 50 54 36 23    54 31 04 82 98    04 14 12 15 09    26 78 25 47 47
28 41 50 61 88    64 85 27 20 18    83 36 36 05 56    39 71 65 09 62    94 76 62 11 89
34 21 42 57 02    59 19 18 97 48    80 30 03 30 98    05 24 67 70 07    84 97 50 87 46
61 81 77 23 23    82 82 11 54 08    53 28 70 58 96    44 07 39 55 43    42 34 43 39 28
61 15 18 13 54    16 86 20 26 88    90 74 80 55 09    14 53 90 51 17    52 01 63 01 59

91 76 21 64 64    44 91 13 32 97    75 31 62 66 54    84 80 32 75 77    56 08 25 70 29
00 97 79 08 06    37 30 28 59 85    53 56 68 53 40    01 74 39 59 73    30 19 99 85 48
36 46 18 34 94    75 20 80 27 77    78 91 69 16 00    08 43 18 73 68    67 69 61 34 25
88 98 99 60 50    65 95 79 42 94    93 62 40 89 96    43 56 47 71 66    46 76 29 67 02
04 37 59 87 21    05 02 03 24 17    47 97 81 56 51    92 34 86 01 82    55 51 33 12 91

63 62 06 34 41    94 21 78 55 09    72 76 45 16 94    29 95 81 83 83    79 88 01 97 30
78 47 23 53 90    34 41 92 45 71    09 23 70 70 07    12 38 92 79 43    14 85 11 47 23
87 68 62 15 43    53 14 36 59 25    54 47 33 70 15    59 24 48 40 35    50 03 42 99 36
47 60 92 10 77    88 59 53 11 52    66 25 69 07 04    48 68 64 71 06    61 65 70 22 12
56 88 87 59 41    65 28 04 67 53    95 79 88 37 31    50 41 06 94 76    81 83 17 16 33

02 57 45 86 67    73 43 07 34 48    44 26 87 93 29    77 09 61 67 84    06 69 44 77 75
31 54 14 13 17    48 62 11 90 60    68 12 93 64 28    46 24 79 16 76    14 60 25 51 01
28 50 16 43 36    28 97 85 58 99    67 22 52 76 23    24 70 36 54 54    59 28 61 71 96
63 29 62 66 50    02 63 45 52 38    67 63 47 54 75    83 24 78 43 20    92 63 13 47 48
45 65 58 26 51    76 96 59 38 72    86 57 45 71 46    44 67 76 14 55    44 88 01 62 12

39 65 36 63 70    77 45 85 50 51    74 13 39 35 22    30 53 36 02 95    49 34 88 73 61
73 71 98 16 04    29 18 94 51 23    76 51 94 84 86    79 93 96 38 63    08 58 25 58 94
72 20 56 20 11    72 65 71 08 86    79 57 95 13 91    97 48 72 66 48    09 71 17 24 89
75 17 26 99 76    89 37 20 70 01    77 31 61 95 46    26 97 05 73 51    53 33 18 72 87
37 48 60 82 29    81 30 15 39 14    48 38 75 93 29    06 87 37 78 48    45 56 00 84 47

68 08 02 80 72    83 71 46 30 49    89 17 95 88 29    02 39 56 03 46    97 74 06 56 17
14 23 98 61 67    70 52 85 01 50    01 84 02 78 43    10 62 98 19 41    18 83 99 47 99
49 08 96 21 44    25 27 99 41 28    07 41 08 34 66    19 42 74 39 91    41 96 53 78 72
78 37 06 08 43    63 61 62 42 29    39 68 95 10 96    09 24 23 00 62    56 12 80 73 16
37 21 34 17 68    68 96 83 23 56    32 84 60 15 31    44 73 67 34 77    91 15 79 74 58

14 29 09 34 04    87 83 07 55 07    76 58 30 83 64    87 29 25 58 84    86 50 60 00 25
58 43 28 06 36    49 52 83 51 14    47 56 91 29 34    05 87 31 06 95    12 45 57 09 09
10 43 67 29 70    80 62 80 03 42    10 80 21 38 84    90 56 35 03 09    43 12 74 49 14
44 38 88 39 54    86 97 37 44 22    00 95 01 31 76    17 16 29 56 63    38 78 94 49 81
90 69 59 19 51    85 39 52 85 13    07 28 37 07 61    11 16 36 27 03    78 86 72 04 95

41 47 10 25 62    97 05 31 03 61    20 26 36 31 62    68 69 86 95 44    84 95 48 46 45
91 94 14 63 19    75 89 11 47 11    31 56 34 19 09    79 57 92 36 59    14 93 87 81 40
80 06 54 18 66    09 18 94 06 19    98 40 07 17 81    22 45 44 84 11    24 62 20 42 31
67 72 77 63 48    84 08 31 55 58    24 33 45 77 58    80 45 67 93 82    75 70 16 08 24
59 40 24 13 27    79 26 88 86 30    01 31 60 10 39    53 58 47 70 93    85 81 56 39 38

05 90 35 89 95    01 61 16 96 94    50 78 13 69 36    37 68 53 37 31    71 26 35 03 71
44 43 80 69 98    46 68 05 14 82    90 78 50 05 62    77 79 13 57 44    59 60 10 39 66
61 81 31 96 82    00 57 25 60 59    46 72 60 18 77    55 66 12 62 11    08 99 55 64 57
42 88 07 10 05    24 98 65 63 21    47 21 61 88 32    27 80 30 21 60    10 92 35 36 12
77 94 30 05 39    28 10 99 00 27    12 73 73 99 12    49 99 57 94 82    96 88 57 17 91

78 83 19 76 16    94 11 68 84 26    23 54 20 86 85    23 86 66 99 07    36 37 34 92 09
87 76 59 61 81    43 63 64 61 61    65 76 36 95 90    18 48 27 45 68    27 23 65 30 72
91 43 05 96 47    55 78 99 95 24    37 55 85 78 78    01 48 41 19 10    35 19 54 07 73
84 97 77 72 73    09 62 06 65 72    87 12 49 03 60    41 15 20 76 27    50 47 02 29 16
87 41 60 76 83    44 88 96 07 80    83 05 83 38 96    73 70 66 81 90    30 56 10 48 59
```

Table **K**. Table of Squares and Square Roots

N	N^2	$\sqrt{N}$	N	N^2	$\sqrt{N}$
1	1	1.0000	41	1681	6.4031
2	4	1.4142	42	1764	6.4807
3	9	1.7321	43	1849	6.5574
4	16	2.0000	44	1936	6.6332
5	25	2.2361	45	2025	6.7082
6	36	2.4495	46	2116	6.7823
7	49	2.6458	47	2209	6.9557
8	64	2.8284	48	2304	6.8282
9	81	3.0000	49	2401	7.0000
10	100	3.1623	50	2500	7.0711
11	121	3.3166	51	2601	7.1414
12	144	3.4641	52	2704	7.2111
13	169	3.6056	53	2809	7.2801
14	196	3.7417	54	2916	7.3485
15	225	3.8730	55	3025	7.4162
16	256	4.0000	56	3136	7.4833
17	289	4.1231	57	3249	7.5498
18	324	4.2426	58	3364	7.6158
19	361	4.3589	59	3481	7.6811
20	400	4.4721	60	3600	7.7460
21	441	4.5826	61	3721	7.8102
22	484	4.6904	62	3844	7.8740
23	529	4.7958	63	3969	7.9373
24	576	4.8990	64	4096	8.0000
25	625	5.0000	65	4225	8.0623
26	676	5.0990	66	4356	8.1240
27	729	5.1962	67	4489	8.1854
28	784	5.2915	68	4624	8.2462
29	841	5.3852	69	4761	8.3066
30	900	5.4772	70	4900	8.3666
31	961	5.5678	71	5041	8.4261
32	1024	5.6569	72	5184	8.4853
33	1089	5.7446	73	5329	8.5440
34	1156	5.8310	74	5476	8.6023
35	1225	5.9161	75	5625	8.6603
36	1296	6.0000	76	5776	8.7178
37	1369	6.0828	77	5929	8.7750
38	1444	6.1644	78	6084	8.8318
39	1521	6.2450	79	6241	8.8882
40	1600	6.3246	80	6400	8.9443

Source: J. W. Dunlap and A. K. Kurtz, *Handbook of Statistical Monographs, Tables and Formulas*. New York: World Book Company, 1932, as used in A. L. Edwards, *Statistical Methods*, 2nd ed., Holt, Rinehart and Winston, New York, 1967.

Table **K** (continued)

N	N^2	$\sqrt{N}$	N	N^2	$\sqrt{N}$
81	6561	9.0000	121	14641	11.0000
82	6724	9.0554	122	14884	11.0454
83	6889	9.1104	123	15129	11.0905
84	7056	9.1652	124	15376	11.1355
85	7225	9.2195	125	15625	11.1803
86	7396	9.2736	126	15876	11.2250
87	7569	9.3274	127	16129	11.2694
88	7744	9.3808	128	16384	11.3137
89	7921	9.4340	129	16641	11.3578
90	8100	9.4868	130	16900	11.4018
91	8281	9.5394	131	17161	11.4455
92	8464	9.5917	132	17424	11.4891
93	8649	9.6437	133	17689	11.5326
94	8836	9.6954	134	17956	11.5758
95	9025	9.7468	135	18225	11.6190
96	9216	9.7980	136	18496	11.6619
97	9409	9.8489	137	18769	11.7047
98	9604	9.8995	138	19044	11.7473
99	9801	9.9499	139	19321	11.7898
100	10000	10.0000	140	19600	11.8322
101	10201	10.0499	141	19881	11.8743
102	10404	10.0995	142	20164	11.9164
103	10609	10.1489	143	20449	11.9583
104	10816	10.1980	144	20736	12.0000
105	11025	10.2470	145	21025	12.0416
106	11236	10.2956	146	21316	12.0830
107	11449	10.3441	147	21609	12.1244
108	11664	10.3923	148	21904	12.1655
109	11881	10.4403	149	22201	12.2066
110	12100	10.4881	150	22500	12.2474
111	12321	10.5357	151	22801	12.2882
112	12544	10.5830	152	23104	12.3288
113	12769	10.6301	153	23409	12.3693
114	12996	10.6771	154	23716	12.4097
115	13225	10.7238	155	24025	12.4499
116	13456	10.7703	156	24336	12.4900
117	13689	10.8167	157	24649	12.5300
118	13924	10.8628	158	24964	12.5698
119	14161	10.9087	159	25281	12.6095
120	14400	10.9545	160	25600	12.6491

Table **K** (continued)

N	N²	√N	N	N²	√N
161	25921	12.6886	201	40401	14.1774
162	26244	12.7279	202	40804	14.2127
163	26569	12.7671	203	41209	14.2478
164	26896	12.8062	204	41616	14.2829
165	27225	12.8452	205	42025	14.3178
166	27556	12.8841	206	42436	14.3527
167	27889	12.9228	207	42849	14.3875
168	28224	12.9615	208	43264	14.4222
169	28561	13.0000	209	43681	14.4568
170	28900	13.0384	210	44100	14.4914
171	29241	13.0767	211	44521	14.5258
172	29584	13.1149	212	44944	14.5602
173	29929	13.1529	213	45369	14.5945
174	30276	13.1909	214	45796	14.6287
175	30625	13.2288	215	46225	14.6629
176	30976	13.2665	216	46656	14.6969
177	31329	13.3041	217	47089	14.7309
178	31684	13.3417	218	47524	14.7648
179	32041	13.3791	219	47961	14.7986
180	32400	13.4164	220	48400	14.8324
181	32761	13.4536	221	48841	14.8661
182	33124	13.4907	222	49284	14.8997
183	33489	13.5277	223	49729	14.9332
184	33856	13.5647	224	50176	14.9666
185	34225	13.6015	225	50625	15.0000
186	34596	13.6382	226	51076	15.0333
187	34969	13.6748	227	51529	15.0665
188	35344	13.7113	228	51984	15.0997
189	35721	13.7477	229	52441	15.1327
190	36100	13.7840	230	52900	15.1658
191	36481	13.8203	231	53361	15.1987
192	36864	13.8564	232	53824	15.2315
193	37249	13.8924	233	54289	15.2643
194	37636	13.9284	234	54756	15.2971
195	38025	13.9642	235	55225	15.3297
196	38416	14.0000	236	55696	15.3623
197	38809	14.0357	237	56169	15.3948
198	39204	14.0712	238	56644	15.4272
199	39601	14.1067	239	57121	15.4596
200	40000	14.1421	240	57600	15.4919

Table **K** (continued)

N	N^2	$\sqrt{N}$	N	N^2	$\sqrt{N}$
241	58081	15.5242	281	78961	16.7631
242	58564	15.5563	282	79524	16.7929
243	59049	15.5885	283	80089	16.8226
244	59536	15.6205	284	80656	16.8523
245	60025	15.6525	285	81225	16.8819
246	60516	15.6844	286	81796	16.9115
247	61009	15.7162	287	82369	16.9411
248	61504	15.7480	288	82944	16.9706
249	62001	15.7797	289	83521	17.0000
250	62500	15.8114	290	84100	17 0294
251	63001	15.8430	291	84681	17.0587
252	63504	15.8745	292	85264	17.0880
253	64009	15.9060	293	85849	17.1172
254	64516	15.9374	294	86436	17.1464
255	65025	15.9687	295	87025	17.1756
256	65536	16.0000	296	87616	17.2047
257	66049	16.0312	297	88209	17.2337
258	66564	16.0624	298	88804	17.2627
259	67081	16.0935	299	89401	17.2916
260	67600	16.1245	300	90000	17.3205
261	68121	16.1555	301	90601	17.3494
262	68644	16.1864	302	91204	17.3781
263	69169	16.2173	303	91809	17.4069
264	69696	16.2481	304	92416	17.4356
265	70225	16.2788	305	93025	17.4642
266	70756	16.3095	306	93636	17.4929
267	71289	16.3401	307	94249	17.5214
268	71824	16.3707	308	94864	17.5499
269	72361	16.4012	309	95481	17.5784
270	72900	16.4317	310	96100	17.6068
271	73441	16.4621	311	96721	17.6352
272	73984	16.4924	312	97344	17.6635
273	74529	16.5227	313	97969	17.6918
274	75076	16.5529	314	98596	17.7200
275	75625	16.5831	315	99225	17.7482
276	76176	16.6132	316	99856	17.7764
277	76729	16.6433	317	100489	17.8045
278	77284	16.6733	318	101124	17.8326
279	77841	16.7033	319	101761	17.8606
280	78400	16.7332	320	102400	17.8885

Table **K** (continued)

N	N^2	$\sqrt{N}$	N	N^2	$\sqrt{N}$
321	103041	17.9165	361	130321	19.0000
322	103684	17.9444	362	131044	19.0263
323	104329	17.9722	363	131769	19.0526
324	104976	18.0000	364	132496	19.0788
325	105625	18.0278	365	133225	19.1050
326	106276	18.0555	366	133956	19.1311
327	106929	18.0831	367	134689	19.1572
328	107584	18.1108	368	135424	19.1833
329	108241	18.1384	369	136161	19.2094
330	108900	18.1659	370	136900	19.2354
331	109561	18.1934	371	137641	19.2614
332	110224	18.2209	372	138384	19.2873
333	110889	18.2483	373	139129	19.3132
334	111556	18.2757	374	139876	19.3391
335	112225	18.3030	375	140625	19.3649
336	112896	18.3303	376	141376	19.3907
337	113569	18.3576	377	142129	19.4165
338	114244	18.3848	378	142884	19.4422
339	114921	18.4120	379	143641	19.4679
340	115600	18.4391	380	144400	19.4936
341	116281	18.4662	381	145161	19.5192
342	116964	18.4932	382	145924	19.5448
343	117649	18.5203	383	146689	19.5704
344	118336	18.5472	384	147456	19.5959
345	119025	18.5742	385	148225	19.6214
346	119716	18.6011	386	148996	19.6469
347	120409	18.6279	387	149769	19.6723
348	121104	18.6548	388	150544	19.6977
349	121801	18.6815	389	151321	19.7231
350	122500	18.7083	390	152100	19.7484
351	123201	18.7350	391	152881	19.7737
352	123904	18.7617	392	153664	19.7990
353	124609	18.7883	393	154449	19.8242
354	125316	18.8149	394	155236	19.8494
355	126025	18.8414	395	156025	19.8746
356	126736	18.8680	396	156816	19.8997
357	127449	18.8944	397	157609	19.9249
358	128164	18.9209	398	158404	19.9499
359	128881	18.9473	399	159201	19.9750
360	129600	18.9737	400	160000	20.0000

Table **K** (continued)

N	N^2	$\sqrt{N}$	N	N^2	$\sqrt{N}$
401	160801	20.0250	441	194481	21.0000
402	161604	20.0499	442	195364	21.0238
403	162409	20.0749	443	196249	21.0476
404	163216	20.0998	444	197136	21.0713
405	164025	20.1246	445	198025	21.0950
406	164836	20.1494	446	198916	21.1187
407	165649	20.1742	447	199809	21.1424
408	166464	20.1990	448	200704	21.1660
409	167281	20.2237	449	201601	21.1896
410	168100	20.2485	450	202500	21.2132
411	168921	20.2731	451	203401	21.2368
412	169744	20.2978	452	204304	21.2603
413	170569	20.3224	453	205209	21.2838
414	171396	20.3470	454	206116	21.3073
415	172225	20.3715	455	207025	21.3307
416	173056	20.3961	456	207936	21.3542
417	173889	20.4206	457	208849	21.3776
418	174724	20.4450	458	209764	21.4009
419	175561	20.4695	459	210681	21.4243
420	176400	20.4939	460	211600	21.4476
421	177241	20.5183	461	212521	21.4709
422	178084	20.5426	462	213444	21.4942
423	178929	20.5670	463	214369	21.5174
424	179776	20.5913	464	215296	21.5407
425	180625	20.6155	465	216225	21.5639
426	181476	20.6398	466	217156	21.5870
427	182329	20.6640	467	218089	21.6102
428	183184	20.6882	468	219024	21.6333
429	184041	20.7123	469	219961	21.6564
430	184900	20.7364	470	220900	21.6795
431	185761	20.7605	471	221841	21.7025
432	186624	20.7846	472	222784	21.7256
433	187489	20.8087	473	223729	21.7486
434	188356	20.8327	474	224676	21.7715
435	189225	20.8567	475	225625	21.7945
436	190096	20.8806	476	226576	21.8174
437	190969	20.9045	477	227529	21.8403
438	191844	20.9284	478	228484	21.8632
439	192721	20.9523	479	229441	21.8861
440	193600	20.9762	480	230400	21.9089

Table K (continued)

N	N²	√N	N	N²	√N
481	231361	21.9317	521	271441	22.8254
482	232324	21.9545	522	272484	22.8473
483	233289	21.9773	523	273529	22.8692
484	234256	22.0000	524	274576	22.8910
485	235225	22.0227	525	275625	22.9129
486	236196	22.0454	526	276676	22.9347
487	237169	22.0681	527	277729	22.9565
488	238144	22.0907	528	278784	22.9783
489	239121	22.1133	529	279841	23.0000
490	240100	22.1359	530	280900	23.0217
491	241081	22.1585	531	281961	23.0434
492	242064	22.1811	532	283024	23.0651
493	243049	22.2036	533	284089	23.0868
494	244036	22.2261	534	285156	23.1084
495	245025	22.2486	535	286225	23.1301
496	246016	22.2711	536	287296	23.1517
497	247009	22.2935	537	288369	23.1733
498	248004	22.3159	538	289444	23.1948
499	249001	22.3383	539	290521	23.2164
500	250000	22.3607	540	291600	23.2379
501	251001	22.3830	541	292681	23.2594
502	252004	22.4054	542	293764	23.2809
503	253009	22.4277	543	294849	23.3024
504	254016	22.4499	544	295936	23.3238
505	255025	22.4722	545	297025	23.3452
506	256036	22.4944	546	298116	23.3666
507	257049	22.5167	547	299209	23.3880
508	258064	22.5389	548	300304	23.4094
509	259081	22.5610	549	301401	23.4307
510	260100	22.5832	550	302500	23.4521
511	261121	22.6053	551	303601	23.4734
512	262144	22.6274	552	304704	23.4947
513	263169	22.6495	553	305809	23.5160
514	264196	22.6716	554	306916	23.5372
515	265225	22.6936	555	308025	23.5584
516	266256	22.7156	556	309136	23.5797
517	267289	22.7376	557	310249	23.6008
518	268324	22.7596	558	311364	23.6220
519	269361	22.7816	559	312481	23.6432
520	270400	22.8035	560	313600	23.6643

Table K (continued)

N	N²	√N	N	N²	√N
561	314721	23.6854	601	361201	24.5153
562	315844	23.7065	602	362404	24.5357
563	316969	23.7276	603	363609	24.5561
564	318096	23.7487	604	364816	24.5764
565	319225	23.7697	605	366025	24.5967
566	320356	23.7908	606	367236	24.6171
567	321489	23.8118	607	368449	24.6374
568	322624	23.8328	608	369664	24.6577
569	323761	23.8537	609	370881	24.6779
570	324900	23.8747	610	372100	24.6982
571	326041	23.8956	611	373321	24.7184
572	327184	23.9165	612	374544	24.7386
573	328329	23.9374	613	375769	24.7588
574	329476	23.9583	614	376996	24.7790
575	330625	23.9792	615	378225	24.7992
576	331776	24.0000	616	379456	24.8193
577	332929	24.0208	617	380689	24.8395
578	334084	24.0416	618	381924	24.8596
579	335241	24.0624	619	383161	24.8797
580	336400	24.0832	620	384400	24.8998
581	337561	24.1039	621	385641	24.9199
582	338724	24.1247	622	386884	24.9399
583	339889	24.1454	623	388129	24.9600
584	341056	24.1661	624	389376	24.9800
585	342225	24.1868	625	390625	25.0000
586	343396	24.2074	626	391876	25.0200
587	344569	24.2281	627	393129	25.0400
588	345744	24.2487	628	394384	25.0599
589	346921	24.2693	629	395641	25.0799
590	348100	24.2899	630	396900	25.0998
591	349281	24.3105	631	398161	25.1197
592	350464	24.3311	632	399424	25.1396
593	351649	24.3516	633	400689	25.1595
594	352836	24.3721	634	401956	25.1794
595	354025	24.3926	635	403225	25.1992
596	355216	24.4131	636	404496	25.2190
597	356409	24.4336	637	405769	25.2389
598	357604	24.4540	638	407044	25.2587
599	358801	24.4745	639	408321	25.2784
600	360000	24.4949	640	409600	25.2982

Table **K** (continued)

N	N²	√N	N	N²	√N
641	410881	25.3180	681	463761	26.0960
642	412164	25.3377	682	465124	26.1151
643	413449	25.3574	683	466489	26.1343
644	414736	25.3772	684	467856	26.1534
645	416025	25.3969	685	469225	26.1725
646	417316	25.4165	686	470596	26.1916
647	418609	25.4362	687	471969	26.2107
648	419904	25.4558	688	473344	26.2298
649	421201	25.4755	689	474721	26.2488
650	422500	25.4951	690	476100	26.2679
651	423801	25.5147	691	477481	26.2869
652	425104	25.5343	692	478864	26.3059
653	426409	25.5539	693	480249	26.3249
654	427716	25.5734	694	481636	26.3439
655	429025	25.5930	695	483025	26.3629
656	430336	25.6125	696	484416	26.3818
657	431649	25.6320	697	485809	26.4008
658	432964	25.6515	698	487204	26.4197
659	434281	25.6710	699	488601	26.4386
660	435600	25.6905	700	490000	26.4575
661	436921	25.7099	701	491401	26.4764
662	438244	25.7294	702	492804	26.4953
663	439569	25.7488	703	494209	26.5141
664	440896	25.7682	704	495616	26.5330
665	442225	25.7876	705	497025	26.5518
666	443556	25.8070	706	498436	26.5707
667	444889	25.8263	707	499849	26.5895
668	446224	25.8457	708	501264	26.6083
669	447561	25.8650	709	502681	26.6271
670	448900	25.8844	710	504100	26.6458
671	450241	25.9037	711	505521	26.6646
672	451584	25.9230	712	506944	26.6833
673	452929	25.9422	713	508369	26.7021
674	454276	25.9615	714	509796	26.7208
675	455625	25.9808	715	511225	26.7395
676	456976	26.0000	716	512656	26.7582
677	458329	26.0192	717	514089	26.7769
678	459684	26.0384	718	515524	26.7955
679	461041	26.0576	719	516961	26.8142
680	462400	26.0768	720	518400	26.8328

Table K (continued)

N	N^2	$\sqrt{N}$	N	N^2	$\sqrt{N}$
721	519841	26.8514	761	579121	27.5862
722	521284	26.8701	762	580644	27.6043
723	522729	26.8887	763	582169	27.6225
724	524176	26.9072	764	583696	27.6405
725	525625	26.9258	765	585225	27.6586
726	527076	26.9444	766	586756	27.6767
727	528529	26.9629	767	588289	27.6948
728	529984	26.9815	768	589824	27.7128
729	531441	27.0000	769	591361	27.7308
730	532900	27.0185	770	592900	27.7489
731	534361	27.0370	771	594441	27.7669
732	535824	27.0555	772	595984	27.7849
733	537289	27.0740	773	597529	27.8029
734	538756	27.0924	774	599076	27.8209
735	540225	27.1109	775	600625	27.8388
736	541696	27.1293	776	602176	27.8568
737	543169	27.1477	777	603729	27.8747
738	544644	27.1662	778	605284	27.8927
739	546121	27.1846	779	606841	27.9106
740	547600	27.2029	780	608400	27.9285
741	549081	27.2213	781	609961	27.9464
742	550564	27.2397	782	611524	27.9643
743	552049	27.2580	783	613089	27.9821
744	553536	27.2764	784	614656	28.0000
745	555025	27.2947	785	616225	28.0179
746	556516	27.3130	786	617796	28.0357
747	558009	27.3313	787	619369	28.0535
748	559504	27.3496	788	620944	28.0713
749	561001	27.3679	789	622521	28.0891
750	562500	27.3861	790	624100	28.1069
751	564001	27.4044	791	625681	28.1247
752	565504	27.4226	792	627264	28.1425
753	567009	27.4408	793	628849	28.1603
754	568516	27.4591	794	630436	28.1780
755	570025	27.4773	795	632025	28.1957
756	571536	27.4955	796	633616	28.2135
757	573049	27.5136	797	635209	28.2312
758	574564	27.5318	798	636804	28.2489
759	576081	27.5500	799	638401	28.2666
760	577600	27.5681	800	640000	28.2843

Table **K** (continued)

N	N²	$\sqrt{N}$	N	N²	$\sqrt{N}$
801	641601	28.3019	841	707281	29.0000
802	643204	28.3196	842	708964	29.0172
803	644809	28.3373	843	710649	29.0345
804	646416	28.3549	844	712336	29.0517
805	648025	28.3725	845	714025	29.0689
806	649636	28.3901	846	715716	29.0861
807	651249	28.4077	847	717409	29.1033
808	652864	28.4253	848	719104	29.1204
809	654481	28.4429	849	720801	29.1376
810	656100	28.4605	850	722500	29.1548
811	657721	28.4781	851	724201	29.1719
812	659344	28.4956	852	725904	29.1890
813	660969	28.5132	853	727609	29.2062
814	662596	28.5307	854	729316	29.2233
815	664225	28.5482	855	731025	29.2404
816	665856	28.5657	856	732736	29.2575
817	667489	28.5832	857	734449	29.2746
818	669124	28.6007	858	736164	29.2916
819	670761	28.6182	859	737881	29.3087
820	672400	28.6356	860	739600	29.3258
821	674041	28.6531	861	741321	29.3428
822	675684	28.6705	862	743044	29.3598
823	677329	28.6880	863	744769	29.3769
824	678976	28.7054	864	746496	29.3939
825	680625	28.7228	865	748225	29.4109
826	682276	28.7402	866	749956	29.4279
827	683929	28.7576	867	751689	29.4449
828	685584	28.7750	868	753424	29.4618
829	687241	28.7924	869	755161	29.4788
830	688900	28.8097	870	756900	29.4958
831	690561	28.8271	871	758641	29.5127
832	692224	28.8444	872	760384	29.5296
833	693889	28.8617	873	762129	29.5466
834	695556	28.8791	874	763876	29.5635
835	697225	28.8964	875	765625	29.5804
836	698896	28.9137	876	767376	29.5973
837	700569	28.9310	877	769129	29.6142
838	702244	28.9482	878	770884	29.6311
839	703921	28.9655	879	772641	29.6479
840	705600	28.9828	880	774400	29.6648

Table **K** (continued)

N	N²	$\sqrt{N}$	N	N²	$\sqrt{N}$
881	776161	29.6816	921	848241	30.3480
882	777924	29.6985	922	850084	30.3645
883	779689	29.7153	923	851929	30.3809
884	781456	29.7321	924	853776	30.3974
885	783225	29.7489	925	855625	30.4138
886	784996	29.7658	926	857476	30.4302
887	786769	29.7825	927	859329	30.4467
888	788544	29.7993	928	861184	30.4631
889	790321	29.8161	929	863041	30.4795
890	792100	29.8329	930	864900	30.4959
891	793881	29.8496	931	866761	30.5123
892	795664	29.8664	932	868624	30.5287
893	797449	29.8831	933	870489	30.5450
894	799236	29.8998	934	872356	30.5614
895	801025	29.9166	935	874225	30.5778
896	802816	29.9333	936	876096	30.5941
897	804609	29.9500	937	877969	30.6105
898	806404	29.9666	938	879844	30.6268
899	808201	29.9833	939	881721	30.6431
900	810000	30.0000	940	883600	30.6594
901	811801	30.0167	941	885481	30.6757
902	813604	30.0333	942	887364	30.6920
903	815409	30.0500	943	889249	30.7083
904	817216	30.0666	944	891136	30.7246
905	819025	30.0832	945	893025	30.7409
906	820836	30.0998	946	894916	30.7571
907	822649	30.1164	947	896809	30.7734
908	824464	30.1330	948	898704	30.7896
909	826281	30.1496	949	900601	30.8058
910	828100	30.1662	950	902500	30.8221
911	829921	30.1828	951	904401	30.8383
912	831744	30.1993	952	906304	30.8545
913	833569	30.2159	953	908209	30.8707
914	835396	30.2324	954	910116	30.8869
915	837225	30.2490	955	912025	30.9031
916	839056	30.2655	956	913936	30.9192
917	840889	30.2820	957	915849	30.9354
918	842724	30.2985	958	917764	30.9516
919	844561	30.3150	959	919681	30.9677
920	846400	30.3315	960	921600	30.9839

Table **K** (continued)

N	N²	√N		N	N²	√N
961	923521	31.0000		981	962361	31.3209
962	925444	31.0161		982	964324	31.3369
963	927369	31.0322		983	966289	31.3528
964	929296	31.0483		984	968256	31.3688
965	931225	31.0644		985	970225	31.3847
966	933156	31.0805		986	972196	31.4006
967	935089	31.0966		987	974169	31.4166
968	937024	31.1127		988	976144	31.4325
969	938961	31.1288		989	978121	31.4484
970	940900	31.1448		990	980100	31.4643
971	942841	31.1609		991	982081	31.4802
972	944784	31.1769		992	984064	31.4960
973	946729	31.1929		993	986049	31.5119
974	948676	31.2090		994	988036	31.5278
975	950625	31.2250		995	990025	31.5436
976	952576	31.2410		996	992016	31.5595
977	954529	31.2570		997	994009	31.5753
978	956484	31.2730		998	996004	31.5911
979	958441	31.2890		999	998001	31.6070
980	960400	31.3050		1000	1000000	31.6228

answers
to the
exercises

PROLOGUE

p. 13   (1) See pp. 2 and 13. (2) See pp. 2–4. (3) See pp. 3–4. (4) See pp. 4–5. (5) See pp. 6–8. (6) See pp. 9–10. (7a) Subject bias could occur because subjects knew which substances they were given and thus expected to have different experiences. Therefore, subject expectancies would be an extraneous variable whose effects would be confounded with those of the marijuana-cigarette independent variable. (7b) Perhaps just giving the workers more attention and the belief that someone in management cared about them—not the music, decorations, or diversions per se—caused productivity to increase. (This is the classic Hawthorne effect.) (7c) No evidence is presented that sweat smells stronger at one point in the menstrual cycle than another. The study apparently demonstrates that something in sweat can mediate the synchrony phenomenon, but one is not certain that it is strength of odor, qualitative changes in odor, or even whether the effect is actually communicated by smell. (For example, the substance rubbed on the upper lip could have been tasted by participants.) Finally, in natural circumstances it is possible that the synchrony could be mediated by means other than smell (or taste) that were not tested.

CHAPTER 1

p. 35   (1a) Ordinal; (1b) ratio; (1c) nominal. (2) The difference centers on the fact that the Celsius scale is interval, while the Kelvin is ratio. (3) Student 6 has a rank of 2 and student 5 a rank of 3. Their ranks as well as their scores (24 and 25, respectively) differ by one point. In contrast, student 4 also differs from student 5 by a rank of one, but has 16 more score points. Thus, when scores are transformed to ranks, the equal interval property of the measurement scale will be lost. (4a) Continuous; (4b) discrete; (4c) continuous; (4d) discrete. (5a) 7.5 to 8.5; (5b) 7.95 to 8.05; (5c) 7.995 to 8.005; (5d) 15.25 to 15.35; (5e) 9.45 to 9.55; (5f) 99.945 to 99.955; (5g) 2.755 to 2.765; (5h) 14.005 to 14.015; (5i) 199.9995 to 200.0005. (6a) 1.6; (6b) 3.2; (6c) 6.0; (6d) 9.4; (6e) 10.0; (6f) 6.3. (7a) 25; (7b) 13; (7c) 16; (7d) 20; (7e) 45; (7f) 88; (7g) 151; (7h) 625. (8a) 75; (8b) 264; (8c) 135; (8d) 15; (8e) 21. (9a) $k + 1$; (9b) 1; (9c) $1 + \dfrac{\Sigma Z}{Nk(\Sigma Z + 1)}$.

CHAPTER 2

p. 56   (1)

Class Interval	Real Limits	Interval Size	Midpoint	f	Rel f	Cum f	Cum Rel f
88–95	87.5–95.5	8	91.5	3	.06	50	1.00
80–87	79.5–87.5	8	83.5	3	.06	47	.94
72–79	71.5–79.5	8	75.5	5	.10	44	.88
64–71	63.5–71.5	8	67.5	4	.08	39	.78
56–63	55.5–63.5	8	59.5	6	.12	35	.70
48–55	47.5–55.5	8	51.5	4	.08	29	.58
40–47	39.5–47.5	8	43.5	7	.14	25	.50
32–39	31.5–39.5	8	35.5	4	.08	18	.36
24–31	23.5–31.5	8	27.5	8	.16	14	.28
16–23	15.5–23.5	8	19.5	2	.04	6	.12
8–15	7.5–15.5	8	11.5	4	.08	4	.08

$N = 50$

Eleven intervals of size 8 were selected, but 9 to 11 intervals might have been picked as long as they covered the range of scores and the lower stated limit of the first interval was evenly divisible by the interval size. See Table 2–7, p. 47. **(2)** See examples of graphs in text, paying close attention to the points mentioned in the text. **(3)**

Class Interval	Real Limits	Interval Size	Midpoint	f	Rel f	Cum f	Cum Rel f
2.8–3.1	2.75–3.15	.4	2.95	2	.07	28	1.00
2.4–2.7	2.35–2.75	.4	2.55	3	.11	26	.93
2.0–2.3	1.95–2.35	.4	2.15	9	.32	23	.82
1.6–1.9	1.55–1.95	.4	1.75	4	.14	14	.50
1.2–1.5	1.15–1.55	.4	1.35	6	.21	10	.36
.8–1.1	.75–1.15	.4	.95	2	.07	4	.14
.4– .7	.35– .75	.4	.55	2	.07	2	.07
				$N = 28$			

Because of rounding, the *Rel f* column of problem 3 does not sum to 1.00.

CHAPTER 3

p. 81 **(1)** means = 5, 6, 6, 5; medians = 5.00, 6.50, 7.00, $3.5 + 2\frac{1}{2}/3 = 4.33$; modes = 7, 2, 3 and 8, 4. **(2)** The sum of the squared deviations about the mean is 100; about the median is 111. **(3a)** Median because distribution is skewed to right; **(3b)** mode because the distribution is likely to be bimodal as a result of having both boys and girls in it; **(3c)** mean, median, and mode are likely to be approximately the same. **(4a)** 6; **(4b)** 4.5; **(4c)** 3.0; **(4d)** $5.5 + 2\frac{1}{2}/4 = 6.125$; **(4e)** $2.5 + \frac{1}{2}/2 = 2.75$; **(4f)** 3.3.

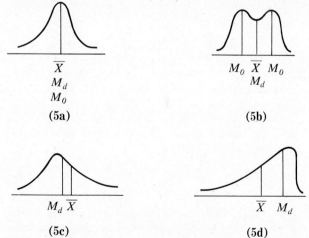

(5a)

(5b)

(5c)

(5d)

(6) See p. 69. **(7)** The range is based upon only the two most extreme scores, not all the scores. **(8a)** $\bar{X} = 7.00$, $s^2 = .67$, $s = .82$; **(8b)** $\bar{X} = 7.00$, $s^2 = 8.67$, $s = 2.94$; **(8c)** $\bar{X} = 7.00$, $s^2 = 11.43$, $s = 3.38$. **(9)** See p. 75 for a discussion of this issue. **(10)** The denominator $N - 1$ is used so that s^2 is an unbiased estimator of σ^2. **(11)** See p. 78.

ANSWERS TO THE EXERCISES

CHAPTER 4

p. 106 **(1a)** 56.5; **(1b)** 48.5; **(1c)** 75; **(1d)** 85.50; **(1e)** 54.75; **(1f)** 71.5; **(1g)** 59.0; **(1h)** 72.5. **(2a)** $P_{.2625}$; **(2b)** $P_{.5125}$; **(2c)** $P_{.7375}$; **(2d)** $P_{.65}$; **(2e)** $P_{.175}$; **(2f)** $P_{.3125}$; **(2g)** $P_{.85}$; **(2h)** $P_{.075}$. **(3a)** 90, 25, 5; **(3b)** 68, 25, 5; **(3c)** 800, 2500, 50; **(3d)** 40, 6.25, 2.5; **(3e)** -40, .391, .625. **(4a)** 3, .44, .67; **(4b)** 42, 64, 8; **(4c)** 75, 400, 20. **(5)** Percentiles reflect only ordinal position (i.e., they specify the proportion of the group falling within and below a given score) and do not indicate how far the other scores are from a given percentile value. Standard scores take into account the variability of the distribution. **(6)** See p. 94. **(7)** $\bar{X} = 6.00$; $s = 2.16$; $z_1 = -.93, -.46, 0, 1.39$; $\bar{z} = 0$; $s_z = 1.00$; yes, because a distribution of standard scores has a mean of 0 and a standard deviation of 1. **(8)** Relative frequency, or the proportion of cases falling between two specified values on the abscissa. **(9)** .50; .50; .1587; .8413. **(10a)** .6826; **(10b)** .1359; **(10c)** .0668; **(10d)** .0500; **(10e)** .0100. **(11)** ± 3 standard deviations; 37; 49. **(12)** .6915; .1587; .4207. **(13)** 48.64; 34.18. **(14)** .4495; .4374; .1935. **(15)** Section (d), because a score of 90 in this distribution yields the highest z, 2.5.

CHAPTER 5

p. 130 **(1)** Y on X: $a = .40$, $b = .67$, $\tilde{Y} = .67X + .40$; W on X: $a = 9.45$, $b = -.86$, $\tilde{W} = -.86X + 9.45$; W on Y: $a = 8.29$, $b = -.86$, $\tilde{W} = -.86Y + 8.29$. **(2a)** 4.39; **(2b)** 6.87; **(2c)** 5.71; **(2d)** $X = 12$ beyond range of original X values. **(3)** $s_{y \cdot x} = 1.69$, $s_{w \cdot x} = 1.24$; $s_{y \cdot w} = 1.75$. **(4)** See p. 120. **(5)** Money earned = .06(sales) + 400. **(6)** Both are measures of variability, but one reflects variability about a mean and the other variability about a regression line. **(7)** See p. 124. **(8a)** Impossible, line intersects X-axis at -5; **(8b)** possible; **(8c)** impossible, $s_{y \cdot x}$ must be less than or equal to s_y.

CHAPTER 6

p. 157 **(1)** For A and B: (I) = 702, (II) = 806, (III) = -303, $r = -.40$; For A and C: (I) = 702, (II) = 576, (III) = -153, $r = -.24$; For B and C: (I) = 806, (II) = 576, (III) = 0, $r = .00$; the r does not change. **(2)** (I) = 124, (II) = 118, (III) = 66, $r = .55$; after adding the point (12, 8), (I) = 783, (II) = 320, (III) = 420, $r = .84$; a score extremely deviant from $(\bar{X}, \bar{Y})$ will alter the r considerably; adding the score $(-12, 8)$ to the original data will produce a correlation of $-.62$. **(3)** See p. 150. **(4)** $\Sigma(\tilde{Y}_i - \bar{Y})^2$, the total sum of squares of points about their mean, can be partitioned into the following components:

$$\Sigma(Y_i - \bar{Y})^2 = \Sigma(\tilde{Y}_i - \bar{Y})^2 + \Sigma(Y_i - \tilde{Y}_i)^2$$

The $\Sigma(Y_i - \tilde{Y}_i)^2$ represents the squared deviations of the points about the regression line, that is, the error remaining after predicting Y from X. The $\Sigma(\tilde{Y}_i - \bar{Y})^2$ is the portion of the total that is not error, that is, the variability in Y_i associated with X. **(5)** The two regression lines result from the fact that minimizing the squared deviations between points and the line for Y_i does not minimize these deviations for X_i. However, the *degree* of linear relationship as reflected in r is the same regardless of the direction of prediction. **(6)** Extreme points influence the regression constants and r more than points near $(\overline{X}, \overline{Y})$ because their deviations from $(\overline{X}, \overline{Y})$, being large, contribute disproportionately to the numerators of b and r when they are squared. **(7)** Since $r = b_{yx}(s_x/s_y)$, when scores are standardized, $s_x = s_y = 1$, making $s_x/s_y = 1$. The result is $r = b_{z_y z_x}$. **(8)** Since $r = 1 - s_{y \cdot x}^2/s_y^2$, the correlation becomes larger as $s_{y \cdot x}^2/s_y^2$ becomes smaller. **(9a)** Possible; **(9b)** impossible, r and b must be the same sign; **(9c)** impossible, $s_{y \cdot x}$ must be $\leq s_y$; **(9d)** impossible, since $r = b(s_x/s_y)$, r should be .80 rather than .15 given the other information; **(9e)** impossible, if $r = 1.00$, $s_{y \cdot x}$ must be 0; **(9f)** possible (in fact, always true).

CHAPTER 7

p. 185 **(1)** Only (b) is not mutually exclusive. **(2)** (a), (b) when drawing is with replacement, and (c) are independent. **(3)** Events A and B are independent if and only if the conditional probability of B given A is the same as the probability of B [i.e., if $P(B|A) = P(B)$]. **(4a)** 1/3, 2/3, 2/3; **(4b)** 5/6, 5/6; **(4c)** 1/36; **(4d)** 3/4, 3/5, 3/10, 3/10, 3/5. **(5a)** 120; **(5b)** 20; **(5c)** 210. **(6)** 360. **(7a)** 1; **(7b)** 15; **(7c)** 35. **(8)** 70; 1680. **(9)** 1/220; 1/55. **(10)** 1/84; 1/504; 1/6. **(11)** $21\left(\frac{1}{3}\right)^5\left(\frac{2}{3}\right)^2 = .037$; .0439. **(12)** $(1/6)(2/6) + (4/6)(3/6) = 7/18$.

CHAPTER 8

p. 205 **(1a)** An 8 A.M. class appeals to only some students. One might wonder if the sample of students in an 8 A.M. class would be typical of college students and whether it would be composed of a different type of student than a 2 P.M. class would be. Also, students in the 8 A.M. class may be less alert than those in the 2 P.M. class. **(1b)** Are those students who volunteer for a jury experiment, as opposed to some other type, typical of all college students? Are college students typical of young adults? **(1c)** Are people who live in a major city and have telephones typical of people in other cities? people who live in smaller towns? or people who do not have telephones? Also, only people with phones will be contacted. **(2)** The standard deviations of the individual samples divided by $\sqrt{4} = 2$ and the standard deviation of the ten sample means (*not* divided by $\sqrt{4}$) each estimate the population standard error of the mean, $\sigma_{\bar{x}}$. Since the individual samples differ in the particular subjects that are included, these estimates will vary by sampling error. As the sample size increases, the estimates will be more accurate and their standard error will be smaller. The standard error reflects the sampling error in a statistic—that is, the extent to which the value of the statistic will change from sample to sample simply because of differences in sample composition. **(3a)** 4; **(3b)** 3; **(3c)** 3. **(4)** If the population distribution of raw scores is normal or if the N is large, the sampling distribution of the mean will be normal in form or approach normality. **(5)** Two variables are independent if they are unrelated in such a way that the value of one does not influence (or relate to) the value of the other. $\overline{X}$ and s_x^2 are independent if the population distribution of X's is normal (actually, symmetrical). **(6)** The theoretical relative frequency of an event in an idealized experiment is the probability of that event. **(7a)** .1151; **(7b)** .0228; **(7c)** .8413, .1587; **(7d)** .4796. **(8a)** .0548; **(8b)** .0082; **(8c)** .9987, .0013; **(8d)** .4895.

CHAPTER 9

p. 230   **(1)** Assumptions are held true throughout the hypothesis-testing procedure, whereas the hypotheses are a set of mutually exclusive alternatives, one of which is being tested and will or will not be rejected by the procedures. **(2)** *Chance* implies that the result is simply sampling error and that no differences exist in the population. **(3)** This ratio yields the values that define the critical region and therefore is critical to the decision between rejecting and not rejecting H_0. **(4)** H_0, not H_1, is being tested; one can find neither evidence to reject H_0 nor evidence for its validity. **(5)** A type I error occurs when H_0 is erroneously rejected; a type II error occurs when H_0 is erroneously not rejected. **(6)** α and β are inversely related. **(7)** Power is the probability that the test will correctly reject H_0. **(8)** If there is considerable evidence or a decisive theory that indicates that the result of the experiment will be in a specified direction (e.g., mean A will be larger than mean B), then a directional test can be performed. Otherwise, use a nondirectional test. **(9)** α is the probability of a type I error. As α becomes smaller, the probability of a type II error increases. See also p. 217. **(10)** Use z if σ_x is available; use t if σ_x is estimated with s_x. **(11a)** *Assumptions.* The members of the sample are randomly and independently selected and the population involved is normal with $\mu = 81$ and $\sigma = 10$. *Hypotheses.* H_0: $\overline{X}$ is computed on a sample from a population with $\mu = 81$. H_1: $\overline{X}$ is computed on a sample from a population with $\mu \neq 81$. *Formula.*

$$z = \frac{\overline{X} - \mu}{\sigma_{\overline{x}}} = \frac{\overline{X} - \mu}{\sigma_x / \sqrt{N}}$$

Significance level. Assume $\alpha = .05$. *Critical values.* -1.96 and $+1.96$ for a two-tailed test. *Decision rules.* If z_{obs} is between -1.96 and $+1.96$, do not reject H_0; if z_{obs} is less than or equal to -1.96 or greater than or equal to $+1.96$, reject H_0. *Computation.* $z_{obs} = 1.50$. *Decision.* Do not reject H_0, since the observed mean deviates from the population mean by an amount that is within the range of sampling error. **(11b)** One would then use the formula for t with $df = 24$ rather than the z distribution. The critical values of t would be ± 2.064 for a two-tailed test at $\alpha = .05$, and the decision rules would be: If t_{obs} is between -2.064 and $+2.064$, do not reject H_0. If t_{obs} is less than or equal to -2.064 or greater than or equal to $+2.064$, reject H_0. Because s_x is also 10, the observed t would be calculated by

$$t_{obs} = \frac{\overline{X} - \mu}{s_x / \sqrt{N}} = \frac{84 - 81}{10 / \sqrt{25}} = 1.50$$

This would result in the same decision (do not reject H_0) as in (a). **(11c)** *Assumptions.* The members of the sample are randomly and independently selected and the population of non-pierced males is normal with a mean of 62 inches. *Hypotheses.* H_0: $\overline{X}$ is computed on a sample from a population with $\mu \leq 62$. H_1: $\overline{X}$ is computed on a sample from a population with $\mu > 62$. (Notice that this is a directional alternative. Why?) *Formula.*

$$t = \frac{\overline{X} - \mu}{s_{\overline{x}}} = \frac{\overline{X} - \mu}{s_x / \sqrt{N}}$$

Significance level. Assume $\alpha = .05$. *Critical value.* From Table B with $df = N - 1 = 35$, one-tailed, $\alpha = .05$: $+1.691$ (approximately; obtained by interpolation). *Decision rules.* If t_{obs} is less than 1.691, do not reject H_0. If t_{obs} is greater than or equal to 1.691, reject H_0. *Computation.*

$$t_{obs} = \frac{\overline{X} - \mu}{s_x/\sqrt{N}} = \frac{64.5 - 62}{7/\sqrt{36}} = 2.14$$

Decision. Reject H_0, since the observed mean of 64.5 is greater than would be expected to occur from errors of sampling from a population with $\mu = 62$. This result does not constitute proof that the piercing and molding causes increased height, since the design of the study (observational rather than experimental) made it impossible to rule out the effects on height of such extraneous variables as diet and genetic stock. **(11d)** *Assumptions.* The sample of applicants is randomly and independently selected and the mean of the current student population is 113. *Hypotheses.* H_0: $\overline{X}$ is computed on a sample from a population with $\mu = 113$. H_1: $\overline{X}$ is computed on a sample from a population with $\mu \neq 113$. *Formula.*

$$t = \frac{\overline{X} - \mu}{s_{\overline{x}}} = \frac{\overline{X} - \mu}{s_x/\sqrt{N}}$$

Significance level. Assume $\alpha = .05$. *Critical values.* Given a nondirectional test, $df = N - 1 = 120$, $\alpha = .05$: $t_{crit} = \pm 1.98$. *Decision rules.* If t_{obs} is between -1.98 and $+1.98$, do not reject H_0. If t_{obs} is less than or equal to -1.98 or greater than or equal to $+1.98$, reject H_0. *Computation.*

$$t_{obs} = \frac{\overline{X} - \mu}{s_x/\sqrt{N}} = \frac{117 - 113}{16/\sqrt{121}} = 2.75$$

Decision. Reject H_0, since the observed mean is too deviant from $\mu = 113$ to be simply a function of sampling error. **(12a)** 17.52 to 42.48, 13.014 to 46.986; **(12b)** $s_{\overline{x}} = 12/\sqrt{16} = 3$, 39.607 to 52.393, 37.159 to 54.841.

CHAPTER 10

p. 255 **(1)** The assumption of normality is made so that the standard normal or the t distribution may be used to determine the required probability. This assumption can be made if the population distribution(s) of raw scores are normal or if the sample size is sufficiently large. **(2)** When the same subjects produce both groups of scores, the individual differences that characterize those subjects influence the scores in both groups, causing them to be correlated to some extent. This correlation affects the accuracy of the estimate of the standard error of the difference between means based upon independent groups, so another procedure must be used. **(3)** Test for the difference between independent means. *Hypotheses.* H_0: $\mu_1 \leq \mu_2$. H_1: $\mu_1 > \mu_2$. *Assumptions.* The subjects are randomly and independently sampled, the groups are independent, variances are homogeneous, and X is normally distributed. *Decision* *rules*. Given the .05 level and $df = N_1 + N_2 - 2 = 17$, directional test. If $t_{obs} < 1.740$, do not reject H_0. If $t_{obs} \geq 1.740$, reject H_0. *Computation*.

$$t_{obs} = (\bar{X}_1 - \bar{X}_2) \bigg/ \sqrt{\left[\frac{(N_1 - 1)s_1^2 + (N_2 - 1)s_2^2}{N_1 + N_2 - 2}\right] \cdot \left[\frac{1}{N_1} + \frac{1}{N_2}\right]}$$

$$t_{obs} = (7.10 - 3.22) \bigg/ \sqrt{\left[\frac{(10 - 1)3.66 + (9 - 1)2.44}{10 + 9 - 2}\right] \cdot \left[\frac{1}{10} + \frac{1}{9}\right]} = 4.81$$

Decision. Reject H_0; the observed difference in means is too great to be a simple result of sampling error. The dissonance theory is supported. **(4)** Test of the difference between two correlated means. *Hypotheses*. H_0: $\mu_1 \leq \mu_2$. H_1: $\mu_1 > \mu_2$. *Assumptions*. The data are in the form of pairs of scores that were randomly and independently sampled and the population of the D_i is normally distributed. *Decision rules*. Given .05 level, $df = N - 1 = 8$, directional test. If $t_{obs} < 1.860$ do not reject H_0. If $t_{obs} \geq 1.860$ reject H_0. *Computation*.

$$t_{obs} = \frac{\sum D}{\sqrt{\left[N \sum D^2 - (\sum D)^2\right] / (N - 1)}} = \frac{5}{\sqrt{[9(33) - (5)^2]/8}} = .86$$

Decision. Do not reject H_0; the observed difference between means is too small and could be sampling error. No evidence exists that mothers are better at detecting hunger cries. **(5)** Test of difference between independent means. *Hypotheses*. H_0: $\mu_1 = \mu_2$. H_1: $\mu_1 \neq \mu_2$. *Assumptions*. The subjects are randomly and independently sampled, the groups are independent, variances are homogeneous, and X is normally distributed. *Decision rules*. Given .05 level, $df = N_1 + N_2 - 2 = 22$, nondirectional test. If $-2.074 < t_{obs} < 2.074$, do not reject H_0. If $t_{obs} \leq -2.074$ or $t_{obs} \geq 2.074$, reject H_0. *Computation*.

$$t_{obs} = \frac{(4.25 - 1.33)}{\sqrt{\left[\frac{(12 - 1)8.93 + (12 - 1)5.88}{12 + 12 - 2}\right] \cdot \left[\frac{1}{12} + \frac{1}{12}\right]}}$$

$$t_{obs} = 2.61 \text{ to } 2.63 \text{ (depending on rounding)}$$

Decision. Reject H_0; the difference in observed means is too great to be simply a function of sampling error. There is a difference between the two therapeutic approaches. **(6)** Verify that the following set of numbers satisfies the conditions of this problem.

A	B
2	0
4	3
6	4
8	7
10	8
12	9
14	13
16	15

(7) *Hypotheses.* H_0: $\rho = 0$. H_1: $\rho \neq 0$. *Assumptions.* The subjects are randomly and independently sampled and the population distributions of the two variables are normal. *Decision rules.* Given .05, $df = N - 2 = 7$, nondirectional test. If $-.6664 < r_{obs} < .6664$, do not reject H_0. If $r_{obs} \leq -.6664$ or $r_{obs} \geq .6664$, reject H_0. *Computation.* $r_{obs} = .52$. *Decision.* Do not reject H_0; the observed correlation is too small and may be a function of sampling error. (8) *Hypotheses.* H_0: $\rho_1 \leq \rho_2$. H_1: $\rho_1 > \rho_2$. *Assumptions.* The subjects are randomly and independently sampled, the groups are independent, the population distributions of X and Y for each correlation are normal, and N_1 and N_2 are both greater than 20. *Decision rules.* Given .05, directional test. If $z_{obs} < 1.645$, do not reject H_0. If $z_{obs} \geq 1.645$, reject H_0. *Computation.*

$$ z_{obs} = \frac{z_{r_1} - z_{r_2}}{\sqrt{1/(N_1 - 3) + 1/(N_2 - 3)}} = \frac{1.333 - .590}{\sqrt{1/(33 - 3) + 1/(28 - 3)}} $$

$$ z_{obs} = 2.74 \text{ or } 2.75 \text{ (depending on rounding)} $$

Decision. Reject H_0; the difference between the correlations is too large to be a simple function of sampling error. The correlation between the IQs of identical twins is higher than between the IQs of fraternal twins. (9a) For the new curriculum: pretest, $\overline{X} = 9.89$, $s^2 = 6.86$; posttest, $\overline{X} = 12.78$, $s^2 = 4.69$; for the old curriculum: pretest, $\overline{X} = 10.00$, $s^2 = 6.22$; posttest, $\overline{X} = 10.4$, $s^2 = 6.71$. For the new curriculum: $df = 8$, $t_{crit} = \pm 2.306$, $t_{obs} = -2.87$, reject H_0 (note that t_{obs} is negative only because the pretest was considered group 1 in the statistical computations). For the old curriculum: $df = 9$, $t_{crit} = \pm 2.262$, $t_{obs} = 1.31$, do not reject H_0. (9b) $df = 17$, $t_{crit} = \pm 2.110$, $t_{obs} = 2.16$, reject H_0. (9c) For the new curriculum: $df = 7$, $r_{crit} = .5822$, $r_{obs} = .22$, do not reject H_0. For the old curriculum: $df = 8$, $r_{crit} = .5494$, $r_{obs} = .93$, reject H_0. (9d) $z_{crit} = \pm 1.96$, $z_{obs} = -2.58$, reject H_0. The difference in pretest and posttest r's indicates that while the old curriculum influenced each pupil equally over the entire range of initial ability as reflected on the pretest, the new curriculum did not.

CHAPTER 11

p. 284 (1a) *Hypotheses.* H_0: $\mu_1 = \mu_2 = \mu_3 = \mu$. H_1: not H_0. *Assumptions.* Random and independent sampling, independent groups, homogeneity of group population variances, normality of population distribution of scores. *Decision rules.* Given .05 significance level and $df = 2, 14$, if $F_{obs} < 3.74$, do not reject H_0. If $F_{obs} \geq 3.74$, reject H_0.
Computation.

$T_1 = 25$	$T_2 = 20$	$T_3 = 42$	$T_{total} = 87$
$n_1 = 5$	$n_2 = 5$	$n_3 = 7$	$N = 17$
$\overline{X}_1 = 5.0$	$\overline{X}_2 = 4.0$	$\overline{X}_3 = 6.0$	
$\sum X_{i1}^2 = 163$	$\sum X_{i2}^2 = 110$	$\sum X_{i3}^2 = 284$	$\sum\sum X_{ij}^2 = 557$
$\dfrac{T_1^2}{n_1} = 125$	$\dfrac{T_2^2}{n_2} = 80$	$\dfrac{T_3^2}{n_3} = 252$	$\sum\left(\dfrac{T_j^2}{n_j}\right) = 457$

(I) = 445.2353	(II) = 557.0000	(III) = 457.0000

$SS_{between}$ = 11.7647 $df_{between}$ = 2 $MS_{between}$ = 5.8824
SS_{within} = 100.0000 df_{within} = 14 MS_{within} = 7.1429
SS_{total} = 111.7647 df_{total} = 16

Source	df	SS	MS	F
Between Groups	2	11.7647	5.8824	.82
Within Groups	14	100.0000	7.1429	
Total	16	111.7647		

Decision. Do not reject H_0; the observed means are similar enough so that their small differences could be attributable to sampling error. **(1b)** Adding different constants to the groups is analogous to introducing different treatment effects. Reanalyzing, MS_{within} remains the same (MS_{within} = 7.1429), while $MS_{between}$ reflects the introduction of mean differences ($MS_{between}$ = 79.1176). F_{obs} = 11.08, which is greater than the critical value of 3.74 and leads to rejecting H_0. **(1c)** Adding 20 to the last score in each group increases within-group variability (MS_{within} = 103.0612) but influences between-group variability very little ($MS_{between}$ = 1.5126). F_{obs} = .01; do not reject H_0. **(2)** See p. 264. **(3)** $X_{ij} = \bar{X} + (\bar{X}_j - \bar{X}) + (X_{ij} - \bar{X}_j) = 12$.
(4)

Source	df	SS	MS	F
Between Groups	2	38.9333	19.4667	7.79**
Within Groups	12	30.0000	2.5000	
Total	14	68.9333		

CHAPTER 12

p. 318 **(1)** See pp. 289–92. **(2)** See pp. 308–09. **(3a)** *Hypotheses.* Factor A. H_0: $\alpha_1 = \alpha_2 = 0$. H_1: not H_0. Factor B. H_0: $\beta_1 = \beta_2 = 0$. H_1: not H_0. AB Interaction. H_0: $\alpha\beta_{11} = \alpha\beta_{12} = \alpha\beta_{21} = \alpha\beta_{22} = 0$. H_1: not H_0. *Assumptions.* The groups are independent and randomly sampled with $n > 1$ and all n's equal from populations having normal distributions and homogeneous variances. The factors are fixed. *Decision rules.* For all tests df = 1, 12, and α = .05. If $F_{obs} < 4.75$, do not reject H_0. If $F_{obs} \geq 4.75$, reject H_0. *Computation.* (I) = 451.5625, (II) = 511.0000, (III) = 453.1250, (IV) = 451.6250, (V) = 456.2500.

Source	df	SS	MS	F
A	1	1.5625	1.5625	.34
B	1	.0625	.0625	.01
AB	1	3.0625	3.0625	.67
Within	12	54.7500	4.5625	
Total	15	59.4375		

Decision. Do not reject H_0 for any test. **(3b)** *Computation.* (I) = 976.5625, (II) = 1161.0000, (III) = 1103.1250, (IV) = 976.6250, (V) = 1106.2500.

Source	df	SS	MS	F
A	1	126.5625	126.5625	27.74**
B	1	.0625	.0625	.01
AB	1	3.0625	3.0625	.67
Within	12	54.7500	4.5625	
Total	15	184.4375		

By adding a treatment effect to A, only the results for A and the total are changed. *Decision.* Reject H_0 for factor A. **(3c)** *Computation.* (I) = 976.5625, (II) = 1271.0000, (III) = 1053.1250, (IV) = 1071.6250, (V) = 1216.2500.

Source	df	SS	MS	F
A	1	76.5625	76.5625	16.78**
B	1	95.0625	95.0625	20.84**
AB	1	68.0625	68.0625	14.92**
Within	12	54.7500	4.5625	
Total	15	294.4375		

Adding an interaction effect changes all results except within. *Decision.* Reject H_0 for all tests. **(4)** *Computation.* (I) = 378.4500, (II) = 467.0000, (III) = 382.5000, (IV) = 378.9000, (V) = 383.0000.

Source	df	SS	MS	F
User	1	4.05	4.05	.77
Drug	1	.45	.45	.09
User × Drug	1	.05	.05	.01
Within	16	84.00	5.25	
Total	19	88.55		

(5) *Computation.* (I) = 651.0417, (II) = 759.0000, (III) = 691.0833, (IV) = 669.4167, (V) = 724.5000.

Source	df	SS	MS	F
User	1	40.0416	40.0416	23.21**
Drug	1	18.3750	18.3750	10.65**
User × Drug	1	15.0417	15.0417	8.72**
Within	20	34.5000	1.7250	
Total	23	107.9583		

CHAPTER 13

p. 350 **(1)** A nonparametric test might be used if nominal or ordinal data are involved or if some of the assumptions of the parametric test (e.g., homogeneity of variance, normality) cannot be met. **(2)** Power-efficiency is the probability that the statistical test will correctly reject the null hypothesis. For a given N, parametric tests are more powerful than analogous nonparametric tests. **(3)** Chi square 2×3 table. $\chi_{obs}^2 = 16.32$ (obtained if only two decimal places are used in all calculations; answers will vary with rounding practices), which exceeds the critical value of 5.99 ($df = 2$). Reject H_0. **(4)** Chi square 2×2 table. $\chi_{obs}^2 = 5.05$, $df = 1$, $\chi_{crit}^2 = 3.84$. Reject H_0. **(5)** Mann-Whitney U test. $U_{obs} = 79.5$, $U_{crit} = 16$ and 74 for N's = 9 and 10, .05 level, nondirectional test. Reject H_0. **(6)** For group A, $\bar{X} = 19.57$, $M_d = 13$; for group B, $\bar{X} = 19.29$, $M_d = 19$. t test for the difference between independent groups. $t_{obs} = .04$. Do not reject H_0. Mann-Whitney U test. $U_{obs} = 42$. $U_{crit} = 8, 41$. Reject H_0. The means of the groups are very similar but the medians are somewhat different. Because of the distributions (e.g., $X = 62$) the median is more appropriate. The $X = 62$ elevates the mean of group A and makes its variance quite large. The t test is not significant. When the data are ranked, the 62 does not exert such an influence and the U test is significant. The median and U test both use predominantly the ordinal characteristics of the scale. **(7)** Kruskal-Wallis test for three independent samples. $H_{obs} = 11.77$, $H_{crit} = 5.99$, $df = 2$. Reject H_0. **(8)** Wilcoxon test. $W_{obs} = 7.5$, $W_{crit} = 13$ for $N = 11$, .05 level, directional. Reject H_0. **(9)** Spearman rank-order correlation. $r_S = .71$. Critical value of $r_S = .504$ for $N = 12$, at .05 level, directional. Reject H_0. This result suggests that regardless of starting score, children improved more or less the same amount. **(10a)** Wilcoxon test. $W_{obs} = 1$, $W_{crit} = 3$, $N = 8$, .05 level, nondirectional. Reject H_0. **(10b)** Spearman rank-order correlation. $r_S = -.76$. $r_{crit} = \pm .715$, $N = 8$, .05 level, nondirectional. Reject H_0. **(10c)** For the third week: $H_{obs} = .06$, $H_{crit} = 5.99$, $df = 2$. Do not reject H_0. For the fifth week: $H_{obs} = 10.31$, $H_{crit} = 5.99$, $df = 2$. Reject H_0.

APPENDIX I

p. 367 **(1a)** 3/5; **(1b)** 7/6; **(1c)** 11/16; **(1d)** 41/35; **(1e)** 3/7; **(1f)** 5/8; **(1g)** 13/45; **(1h)** $-6/77$. **(2a)** 1/8; **(2b)** 4/15; **(2c)** 1/7; **(2d)** 3/4; **(2e)** 6/7; **(2f)** 6. **(3a)** 120; **(3b)** 12; **(3c)** 30. **(4a)** 8; **(4b)** 36; **(4c)** 13; **(4d)** 56; **(4e)** 2^5; **(4f)** 72; **(4g)** 2; **(4h)** $3^{-1} = 1/3$; **(4i)** 8; **(4j)** 9/25; **(4k)** 160; **(4l)** 1/576. **(5a)** $x^2 + 2xy + y^2$; **(5b)** $p^2 - 2pq + q^2$; **(5c)** $X^2 - 2X\bar{X} + \bar{X}^2$; **(5d)** $(ab)^2 - 2ab^2c + (bc)^2$; **(5e)** 0; **(5f)** $a - c/b$; **(5g)** $(a + d)/(a - b)$; **(5h)** $2X$; **(5i)** $Y\sqrt{3}$; **(5j)** $X\sqrt{3X}$. **(6a)** Sum = 8, squared sum = 64, sum of squares = 30; **(6b)** sum = 14, squared sum = 196, sum of squares = 74; **(6c)** sum = 11, squared sum = 121, sum of squares = 115.

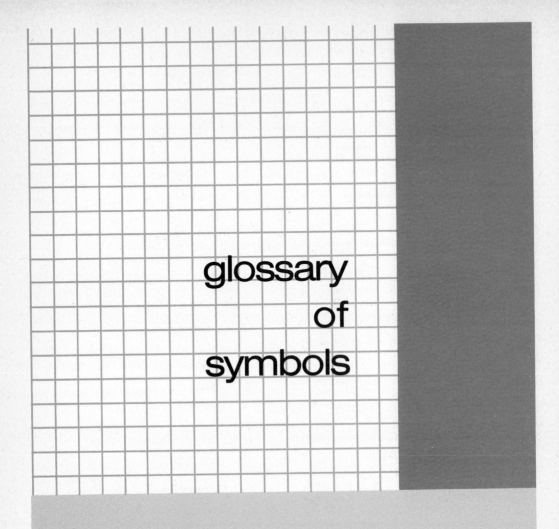

glossary
of
symbols

The following glossary is arranged in approximately alphabetical order, with English and Greek letters intermixed. Non-letter symbols have been placed at the end.

Symbol	Meaning
a	Regression constant representing the y-intercept. It is also written with paired subscripts (a_{yx}, a_{xy}), which denote whether the regression is Y on X or X on Y.
α	Greek alpha, the significance level in hypothesis testing, which is also the probability of a type I error.
b	Regression constant indicating the slope of the regression line. It is also written with paired subscripts (b_{yx}, b_{xy}) which indicate whether the regression is Y on X or X on Y, and with multiple subscripts $(b_{z_y z_x})$ indicating the value is for standardized variables z_y and z_x.
β	Unsubscripted Greek beta represents the probability of a type II error.
c	Often used to denote any nonzero constant.
$°C$	Degrees Celsius.
$_nC_r$	Number of combinations (order irrelevant) of n things taken r at a time.
χ^2	Greek chi squared, a test statistic used in several nonparametric tests. It is often subscripted to indicate whether the value is an observed (χ^2_{obs}) or a critical one (χ^2_{crit}).
Cum f	Cumulative frequency.
Cum Rel f	Cumulative relative frequency.
D_i	Difference between pairs of scores (e.g., in the t test for the difference between means for correlated groups).
d_i	Difference between ranks of paired scores (e.g., in Spearman's rank-order correlation).
df	Degrees of freedom.
E_{jk}	Expected frequency for the jkth cell.

Symbol	Meaning
e	Base of Napierian logarithms, $e = 2.7183 \ldots$.
F	Test statistic, usually the ratio of two independent variances. It is sometimes subscripted to indicate whether the value is an observed (F_{obs}) or a critical one (F_{crit}).
f	Frequency.
$°F$	Degrees Fahrenheit.
H	Test statistic for the Kruskal-Wallis test.
H_0	Null hypothesis.
H_1	Alternative hypothesis.
i	Size of a class interval. Also, a subscript indicating the ith subject or score.
k	Often used to denote any nonzero constant.
L	Lower limit of the interval containing the median or a specified score value.
$\log_e$	Logarithm to the base e, also called the natural or Napierian logarithm. Sometimes written ln in other contexts.
$\log_{10}$	Logarithm to the base 10. Most frequently written without subscript.
μ	Greek mu, the population mean. It is also written with subscripts (for example, $\mu_{\bar{x}}$, indicating the population mean of the distribution of sample means).
M_d	Median.
M_o	Mode.
MS	Mean square or variance estimate in the analysis of variance. It is often written with subscripts indicating the mean square for a particular source of variability (e.g., $MS_{between}$, MS_{AB}).

Symbol	Meaning	Symbol	Meaning
N	Total number of subjects, observations, or paired observations (depending on the research design).		dard deviation applies (for example, s_x, the standard deviation of the X's; $s_{\bar{x}}$, the standard error of the mean; $s_{\bar{x}_1-\bar{x}_2}$, the standard error of the difference between two means).
n	Number of subjects or observations within a specific subgroup of a larger sample. Sometimes written n_j to indicate the number of subjects in the jth group.	s^2	The sample variance. It is also written with subscripts (see s).
n_b	Number of scores falling below the lower limit of the interval containing the desired value.	σ	Lower-case Greek sigma, the population standard deviation. It is also written with subscripts (see s).
n_w	Number of scores within the interval containing the desired value.	σ^2	Lower-case Greek sigma squared, the population variance. It is also written with subscripts (see s).
O_{jk}	Observed frequency for the jkth cell.	$s_{y \cdot x}$	The sample standard error of estimate in regression.
$P(A)$	Probability of event A.		
$P(B\|A)$	Conditional probability of the event B given that event A has already occurred.	$\sigma_{y \cdot x}$	The population standard error of estimate, read "sigma sub y dot x."
p	Probability that the observed data could be obtained if the null hypothesis is true.	SS	Sum of squares. It is often written with subscripts indicating the particular source of variability (e.g., SS_{between}, SS_{AB}).
P_n	The nth percentile point.	$\sum_{i=1}^{N} X_i$	Capital Greek sigma means to sum the X_i for $i = 1$ to N. It is also written without limits on the summation sign and subscripts, ΣX.
$_nP_r$	Number of permutations (order considered) of n things taken r at a time.		
π	Greek pi, constant equal to 3.1415	T	Total, sometimes written with subscripts indicating which scores are summed.
r	The sample Pearson product-moment correlation coefficient.	t	Student's t test statistic. It is also written with subscripts indicating the value is an observed (t_{obs}) or a critical one (t_{crit}), or indicating the value of t at $p = .05$ ($t_{.05}$), etc.
r^2	Estimated proportion of variance in Y attributable to X (the square of the correlation coefficient).		
ρ	Greek rho, the population product-moment correlation coefficient.	U	Test statistic for the Mann-Whitney U test. It is also written with subscripts indicating the value is an observed (U_{obs}) or a critical one (U_{crit}).
r_s	The sample Spearman rank-order correlation coefficient.		
ρ_s	The population Spearman rank-order correlation coefficient.	X_i	The ith score of variable X. Other letters (e.g., Y_i, W_i) are also used to denote variables.
$Rel\,f$	Relative frequency.		
S	Universal set or sample space.	$\bar{X}$	The sample mean. Any variable with a bar over it signifies the mean of that variable (e.g., $\bar{Y}$). It is also written with subscripts indicating the levels and factors involved (e.g. $\bar{X}_j$, $\bar{X}_{.k}$, $\bar{X}_{..}$).
s	The sample standard deviation. It is also written with subscripts that indicate the variable or statistic for which the stan-		

Symbol	Meaning

$\tilde{Y}$ — Value of Y predicted on the basis of the regression line. One may also find $\tilde{X}$ when the regression is for X on Y.

W — Test statistic for the Wilcoxon test. It is also written with subscripts indicating if the value is an observed (W_{obs}) or a critical one (W_{crit}).

z — Standard score or standard normal deviate. It is also written with subscripts indicating if the value is an observed (z_{obs}) or a critical one (z_{crit}).

z_r — Transformed value of the correlation coefficient, r.

A' — Prime after a captial letter; A' indicates the set complement to A (i.e., all elements not in A).

$\subseteq$ — The sign $\subseteq$ in $A \subseteq B$ indicates that A is a subset of B.

$\cup$ — The sign $\cup$ in $A \cup B$ indicates union, the set of elements that are in A, in B, or in both A and B.

$\cap$ — The sign $\cap$ in $A \cap B$ indicates intersection, the set of elements that are in A and in B.

$=$ — Is equal to.

$\neq$ — Is not equal to.

$<$ — Is less than.

$\leq$ — Is less than or equal to.

$>$ — Is greater than.

$\geq$ — Is greater than or equal to.

$\pm$ — Plus or minus.

$\sqrt{}$ — Square root of.

$|c|$ — The absolute value of c.

∞ — Infinity.

$\varnothing$ — Null, or empty, set.

*, **, *** — The observed value of the test statistic is significant at the .05(*), the .01(**), or the .001(***) level.

index

(continued from front endpaper)

Difference between two correlated groups	$t_{obs} = \dfrac{\Sigma D_i}{\sqrt{\dfrac{N\Sigma D_i^2 - (\Sigma D_i)^2}{N-1}}}$ $\qquad df = N - 1$	239

Significance of the correlation coefficient

See Table C, Appendix II $\qquad$ 243

Difference between two independent correlation coefficients

$$z_{obs} = \frac{z_{r_1} - z_{r_2}}{\sqrt{\dfrac{1}{N_1 - 3} + \dfrac{1}{N_2 - 3}}} \qquad 246$$

For z_{r_1} and z_{r_2}, see Table D, Appendix II

Simple analysis of variance $\qquad$ 259

Given: (I) $\dfrac{T^2}{N}$ $\quad$ (II) $\displaystyle\sum_{j=1}^{p}\left(\sum_{i=1}^{n_j} X_{ij}^2\right)$ $\quad$ (III) $\displaystyle\sum_{j=1}^{p}\left(\frac{T_j^2}{n_j}\right)$

Source	df	SS	MS	F
Between Groups	$p - 1$	(III) − (I)	SS_b / df_b	MS_b / MS_w
Within Groups	$N - p$	(II) − (III)	SS_w / df_w	
Total	$N - 1$	(II) − (I)		

Two-factor analysis of variance $\qquad$ 287

Given:

(I) $\dfrac{T^2}{N}$ $\quad$ (II) $\displaystyle\sum_{i=1}^{n}\sum_{j=1}^{p}\sum_{k=1}^{q} X_{ijk}^2$ $\quad$ (III) $\dfrac{\displaystyle\sum_{j=1}^{p} T_{j.}^2}{nq}$ $\quad$ (IV) $\dfrac{\displaystyle\sum_{k=1}^{q} T_{.k}^2}{np}$ $\quad$ (V) $\dfrac{\displaystyle\sum_{j=1}^{p}\sum_{k=1}^{q} T_{jk}^2}{n}$

Source	df	SS	MS	F
A	$p - 1$	(III) − (I)	$\dfrac{SS_A}{df_A}$	$\dfrac{MS_A}{MS_{within}}$
B	$q - 1$	(IV) − (I)	$\dfrac{SS_B}{df_B}$	$\dfrac{MS_B}{MS_{within}}$
AB	$(p-1)(q-1)$	(V) + (I) − (III) − (IV)	$\dfrac{SS_{AB}}{df_{AB}}$	$\dfrac{MS_{AB}}{MS_{within}}$
Within	$N - pq$	(II) − (V)	$\dfrac{SS_{within}}{df_{within}}$	
Total	$N - 1$	(II) − (I)		